Microbiomics and Sustainable Crop Production

Microbiomics and Sustainable Crop Production

Mohammad Yaseen Mir
Centre of Research for Development
University of Kashmir
Srinagar, India

Saima Hamid
Department of Environmental Science
University of Kashmir
Srinagar, India

WILEY

This edition first published 2023

Registered Offices
John Wiley & Sons, Inc., 111 River Street, Hoboken, NJ 07030, USA
John Wiley & Sons Ltd, The Atrium, Southern Gate, Chichester, West Sussex, PO19 8SQ, UK

For details of our global editorial offices, customer services, and more information about Wiley products visit us at www.wiley.com.

Wiley also publishes its books in a variety of electronic formats and by print-on-demand. Some content that appears in standard print versions of this book may not be available in other formats.

Library of Congress Cataloging-in-Publication Data is applied for
Hardback ISBN 9781119799313

Cover Design: Wiley
Cover Image: © nd3000/Shutterstock.com

Set in 9.5/12.5pt STIXTwoText by Straive, Pondicherry, India

Contents

Preface

Microbiomes formed by ages of evolution in soils play an important role in sustainability of crop production by enriching soil and alleviating biotic and abiotic stresses. This diversity is an essential part of the agroecosystems, which are being pushed to edges by pumping agrochemicals and constant soil disturbances. Consequently, efficiency of cropping system has been decreasing, aggravated further by the increased incidence of abiotic stresses due to changes in climatic patterns. Thus, the sustainability of agriculture is at stake. Understanding the microbiota inhabiting phyllosphere, endosphere, rhizosphere, and non-rhizosphere and its utilization could be a sustainable crop production strategy. With the advent of omics technologies and gene editing, along with increasing efforts towards transdisciplinary research across various agriculturally relevant microbiomes, major breakthroughs and microbial-based innovations for agriculture are greatly anticipated. We are currently experiencing a reinvigoration of microbial biotechnologies that utilize microbial benefits to improve agroecosystem functioning, for example through enhanced soil health, crop vigour, and pest protection. This book will be immensely helpful to the students of plant biotechnology, agricultural sciences, agricultural engineering, and plant sciences at both undergraduate and postgraduate levels in universities and colleges. Besides, the book will be quite supportive to researchers who work in the field of plant biotechnology, plant sciences, and agricultural sciences.

Key features are:

1) Engineering of the plant microbiome
2) Advanced fundamental research on microbe–microbe and plant–microbe interactions (bridging the lab–field gap)
3) Rhizosphere engineering and sustainable agriculture
4) Multi-omics strategies to address stress alleviation
5) Endosphere and endophyte communities
6) Core microbiomes: for sustainable agroecosystems

About the Authors

Dr. Mohammad Yaseen Mir
Senior Researcher
Centre of Research for Development
University of Kashmir Srinagar-190006
Teacher
Department of School Education, J&K Government, India

Dr. Mohammad Yaseen Mir was working as a senior researcher in a research project funded by Ministry of Environment, Forest & Climate Change (MoEF & CC), Government of India along with G.B. Pant National Institute of Himalayan Environment & Sustainable Development under National Mission on Himalayan Studies (NMHS) and has recently joined Department of Education J&K as a teacher. He did his MPhil and PhD Programmes from the University of Kashmir. He was awarded Doctorate Merit Scholarship by the University of Kashmir to pursue his PhD programme. He was awarded certificate of appreciation by the

University of Kashmir for participation in national seminar on 'Environment Pollution: Join the Race to Make the World a Better Place' on the Eve of Observance of World Environment Day 2016. He earned a PhD in Botany with a specialization in plant biotechnology on the topic 'Role of elicitors for in vitro induction of secondary metabolites in suspension cultures of *Artemisia amygdalina* D' from the University of Kashmir, India. He is currently editor and reviewer of many reputed journals. Dr. Mir is also subject course expert for environmental science approved CEC-Moocs by MHRD GoI, New Delhi. He has published many research articles, review papers, book chapters in reputed, referred international and national journals like Springer/Elsevier/Hindwaii/Frontiers. He has also authored and edited many books like *Sustainable Agriculture: Biotechniques in Plant Biology* with Springer, *Nano-Technological Intervention in Agricultural Productivity* with John Wiley, and *Core Microbiome: Improving Crop Quality and Productivity* with John Wiley.

Dr. Saima Hamid
Senior Research Fellow
Department of Environmental Science
University of Kashmir Srinagar-190006, India

Dr. Saima Hamid holds Doctorate as well as Master's Degree in Environmental Science from the University of Kashmir, India. She has qualified UGC-NET exams multiple times and other state level exams. Her research area includes Plant molecular biology, stress adaptations, bioactive compound isolation and characterization, climate change and microbial biotechnology. She has published more than thirty research and review articles in highly reputed journals like Elsevier, Frontier's in Plant Sciences, Springer, MDPI, Wiley and other peer reviewed journals. She is the Co-founder of CERD Foundation, member of various international organisations which focuses on Plant adaptations and climate change.

She is an international forum speaker from six years and columnist of various international magazines and local newsletters in different countries like Island Chief of Maldives, Sri Lanka, Bangladesh and many others.

Recent Publication

Din, S., Hamid, S., Yaseen, A. et al. (2022). Isolation and Characterization of Flavonoid Naringenin and Evaluation of Cytotoxic and Biological Efficacy of Water Lilly (Nymphaea mexicana Zucc.). *Plants*. 11, 3588. https://doi.org/10.3390/plants11243588

Hamid, S. and Mir, M.Y, (2021). Global Agri-Food Sector: Challenges and Opportunities in COVID-19 Pandemic. *Front. Sociol.* 6:647337. doi: 10.3389/fsoc.2021.647337

Hamid, S., Mir, M.Y., Rohela, G.K. (2020). Novel coronavirus disease (COVID-19): a pandemic (epidemiology, pathogenesis and potential therapeutics). New Microbes New Infect. Apr 14;35:100679.

Sahu, P.K., Jayalakshmi, K., Tilgam, J. et al. (2022). ROS generated from biotic stress: Effects on plants and alleviation by endophytic microbes. *Front. Plant Sci.* 13:1042936. doi: 10.3389/fpls.2022.1042936

Sahu, P.K., Tilgam, J., Gupta, A. et al. (2022). Surface sterilization for isolation of endophytes: Ensuring what (not) to grow. *Basics of microbiology*, 62 (6), 647–668.

1

Agricultural Microbiomes

Functional and Mechanistic Aspects

CONTENTS

Microbiomics and Sustainable Crop Production, First Edition. Mohammad Yaseen Mir and Saima Hamid.

1.1 Introduction

In recent years, the microbial environment has gotten a lot of attention because lower sequencing costs have allowed for more in-depth study of the structure and dynamics of host-associated microbiota. It is widely acknowledged that microbes have immense ability to improve host well-being in both humans and plants. Targeted application of beneficial microbial cocktails can be a sustainable way to mitigate biotic and abiotic stress conditions and maintain yield stability in the potential vision of precision agriculture. Many beneficial microbes, on the other hand, have similar pathogenic relatives, and it is unknown how the plant immune system distinguishes between pathogenic and beneficial microbes in order to combat infection by the former and promote colonization by the latter. It is possible that even the earliest eukaryotes were overwhelmed by a variety of prokaryotes, and that eukaryotic immune systems developed to distinguish between beneficial and pathogenic bacteria. As a result, a deep and complex interaction between microbes and hosts is predicted, affecting every aspect of eukaryote biology. Traditional as well as systems biological 'omics' and computational modelling methods would be needed to understand microbe–host interactions.

1.2 Model Microbiome–Plant Systems

Plant microbiome researchers must design new model systems as well as draw on existing systems to integrate microorganism populations as an emerging group. To create culture collections, large-scale microbial isolation activities and genome sequencing programmes will be required, as will concerted group efforts to build a set of uniform protocols and growth platforms. A small flowering angiosperm in the mustard family, *Arabidopsis thaliana*, is an example of a popular, albeit non-agricultural, model for plant microbiome study. The *Arabidopsis* scheme, however, has shortcomings, including a lack of symbiotic relationships with nodulating nitrogen-fixers and mycorrhizal fungi, as well as genomic and phenotypic variations from essential monocot crops. As a result, there would be a need for multiple model systems. To figure out which processes can transfer to crops at different evolutionary distances, a collection of model plant species is needed [1]. The legumes *Medicago* [2], *Populus* [3], rice [4], *Sorghum* [5], *Miscanthus* [6], maize [7], and tomato [8] have all made progress towards the development of model host–microbiome systems. Many of these model organisms have completely sequenced genomes and burgeoning scientific populations, allowing them to be used in more microbiome studies. The complete maturation of model systems necessitates coordinated initiatives to create public services, such as repositories and curated databases for sequenced culture collections of related microbiota. A multidisciplinary group of academic and commercial plant microbiome scientists, with funding from funding agencies, is needed to develop successful model systems for elucidating plant–microbiome interactions with full interoperability – a key move towards building a knowledge base for long-term high-yielding agriculture innovation.

1.2.1 Plant Perception of Microbes

Pathogens and endophytes must first resolve systemic obstacles such as cell walls [9], waxy epidermal cuticles [10], and constitutive antimicrobial compounds like phytoanticipins. The evolutionary proximity of beneficials and pathogens can be explained in part by this common criterion. Plant surface receptors called pattern-recognition receptors (PRRs) detect the presence of microbes close to the cell membrane [11]. As per findings of Boller and Felix [12] and Macho and Zipfel [13], in the defence mechanism of pathogen- or microbe-triggered immunity (PTI/MTI), the intracellular signalling helps in culmination of conserved pathogen- or microbe-associated molecular patterns (PAMPs/MAMPs); for example bacterial flagellin or elongation factor thermo unstable (EF-Tu) as MTI mechanism embrace production of reactive oxygen species (ROS), nitrogen oxide species (NO_x), nutrient allocation, metabolite release against microbial protection, and release signalling molecules for defence purposes and other functions, including transcriptional modifications. Beneficials and pathogens, on the other hand, have common, if not identical, molecular patterns, making separation by unique PRRs impossible. FLS2, which recognizes flg22, the most conserved motif in bacterial flagellin, is one of the most important models for studying PRR function [14]. BAK1 is a coreceptor that FLS2 has to activate downstream signalling. BAK1 is also a coreceptor for BRI1, a leucine-rich repeat receptor kinase (LRR-RK) that detects plant brassinosteroids (BR) and serves as an integrator between defence and growth signalling [15]. Certain portions of the protein are recognized by other receptors. Tomato can sense flgll-28 in an FLS2-independent manner by FLS3 [16], and the rice pathogen *Acidovorax avenae* has a separate flagellin motif, CD2-1, whose receptor is still unknown [17]. Interestingly, some *A. avenae* strains are resistant to flagellin glycosylation detection [18]. In comparison to viruses, certain beneficials have epitopes that resist detection by either or both receptors [19]. Apart from MAMP-masking or avoidance processes, certain beneficials are possibly identified by their flagellin and inhibit full-fledged immune responses by yet unexplained mechanisms. Most genes caused by experience of distilled flg22 in *Arabidopsis* were downregulated in response to colonization by the commensal *Rhizobium* sp. 129E [20]. According to their findings, this commensal is capable of interfering with MAMP-induced transcriptional responses through alternative pathways. Since this *Rhizobium* strain lacks the type III secretion system (T3SS) or Nod factor biosynthesis genes, signalling from other heteromeric PRR complexes is likely to play a role.

1.2.2 Molecular Plant

Interactions between symbionts and plants point to pathways that underpin the distinction between friend and foe. When arbuscular mycorrhizal fungi (AMF) and rhizobia come into contact for the first time, they cause temporary defence-like responses that are easily suppressed [21]. Signalling by the Myc and Nod factors is thought to be essential in this repression [22]. Both symbiotic signals are characterized by their ability to induce nuclear calcium oscillations through a signalling cascade involving many conserved symbiotic proteins [23]. Lysine motif (LysM) receptors, like kinases (RLK), detect Nod factors in hosts, and it is thought that related receptors exist for Myc factors [24]. Any of these receptors tend to be involved in pathogen identification. OsCERK1 is a LysM-RLK that is needed for the establishment of mycorrhizal root symbiosis as well as resistance to rice blast fungus [25], indicating that it functions as a 'molecular transition' between symbiotic and defence responses. According to Gourion et al. [22] the exact functioning is not clear, although LysM-RLK is found to be of some specific mechanism with frequent occurrence. NFP is a Nod factor receptor found in *Medicago truncatula* that regulates vision and defence against the fungus *Colletotrichum trifolii*, as well as the oomycetes *Aphanomyces euteiches* and *Phytophthora palmivora* [26, 27]. Combinatorial physical interactions between receptors and coreceptors are essential for signal specificity and integration, according to detailed studies of exemplary PRRs and LysM-RLK. In nature, plant roots come into contact with a slew of MAMPs as well as a smorgasbord of signalling molecules. As a result, it is conceivable, if not likely, that a personalized response to complex microbial assemblages is installed through a network of interacting receptors' combinatorial and quantitative perceptions of the various signalling molecules. As a result, PRR signalling would include the use of interconnected global systems approaches. Smakowska-Luzan et al. [28] performed a proteome-scale interactome analysis, which is a significant step towards a complete understanding of this important plant perception mechanism. Also, 225 LRR-RKs (CSILRR) formed a physical cell surface interaction network in *A. thaliana*, which they mapped using biochemical pull-down experiments. CSILRR showed that all LRR-RKs are highly interconnected, clustering into many modules of unknown biological significance.

Importantly, the authors demonstrated that not only direct connections but also indirect network effects modulate downstream signalling performance, and that the entire network contributes to the plant immune system's well-balanced responses. Understanding plant immunity would need a better understanding of the LRR-RK network's automated information processing.

1.2.3 Bacterial Signalling: Quorum Sensing and Symbiosis Factors

Plants use metabolites, volatiles, symbiosis cues, and quorum sensing (QS) molecules to detect bacterial contact in addition to sensing conserved microbial patterns [29]. *N*-Acyl homoserine lactones (AHL) are important components of bacterial contact that plants can detect. This was shown with the beneficial *Acidovorax radicis* N35, where the AHL-producing wild form was able to dampen the barley defence response, while flavonoid defence was unregulated after inoculation with the non-AHL-producing mutant [30]. Other examples demonstrate the growth-promoting and priming effects of AHLs on host plants such as *Medicago*, tomato, *Arabidopsis*, and barley [31].

According to Cha et al. [32] and von Bodman et al. [33], since pathogenic bacteria often manufacture AHL, these signalling substances are unlikely to provide enough information for the plant to modulate its defensive responses on their own. It is possible that the QS molecule compositions and concentrations suggest an unbalanced microbial composition. Although the biochemical effects of AHLs have been studied in depth, the processes and mechanisms by which plants perceive these bacterial molecules are yet to be discovered [34]. Even in plants that do not form symbiosis, lipochitooligosaccharides, i.e. Myc and Nod symbiosis factors, may promote root production, seed germination, and plant growth. As a result, the detection and signalling mechanism for symbiosis factors is partly independent of the host's symbiosis competence. More research is required to understand how the plant interprets the wide range of rhizosphere signals emitted by microorganisms, and how different molecules can interact synergistically or antagonistically to affect plant growth and stress resistance [35].

1.2.4 Hormone Signalling in Microbe–Host Interactions

Signalling by phytohormones is crucial to almost all plant processes. Salicylic acid (SA), jasmonic acid (JA), and ethylene (ET) are the primary mediators of defence responses. JA and ET mediate induced systemic resistance (ISR) and protection against necrotrophs and insects, while SA mediates structural activity relationships (SAR) and defence against biotrophic and hemibiotrophic pathogen invasion [36]. Auxin, gibberellins (GA), BR, or cytokinins (CK) are hormones that primarily regulate developmental processes or abiotic stress responses (abscisic acid [ABA]). Hormone signalling is heavily integrated, and multiple hormones control every mechanism of concern, beyond these seemingly clean classifications [37]. Phytohormones are therefore essential for bidirectional communication between plants and microbes. Strigolactones are exuded from roots in response to phosphate or nitrogen deficiency in order to attract AM fungi, and their biosynthesis is downregulated after the fungus has colonized [38]. GA, SA, and ET, on the other hand, inhibit both AM and root nodule symbiosis, while auxin and ABA promote AM production in a concentration-dependent manner. For nodule forming, CK and localized auxin signalling are necessary. Depending on the circumstances and plant organisms, the position of JA in symbiosis establishment is uncertain and can be positive, negative, or neutral. Beneficial and pathogenic bacteria also influence the hormone signalling pathway. Coronatine (COR) is a toxin developed by the pathogenic *Pseudomonas syringae* pv. tomato DC3000 (Pst) that is similar to plant JA-isoleucine (JA-Ile) but is more active [39]. The required SA-mediated defences against the hemibiotrophic Pst are suppressed as a result of the activation of JA-dependent defence mechanisms [40]. Pathogens control plant signalling in general to inhibit defensive responses and redirect nutrients to infested tissues for long-term pathogenic colonization [41]. According to Singh et al. [42] and Li et al. [43], presence of beneficial strains including *Pseudomonas fluorescens Pf4*, *Pseudomonas aeruginosa Pag*, or *Bacillus velezensis* LJ02 enhances endogenous SA levels while some strains are proved to show opposed results in *A. thaliana* as well as decreasing JA-Ile levels as reported by Srivastava et al. [44]; however, *Paraburkholderia phytofirmans* PsJN decreases expression of JA-biosynthesis and wound-induced JA accumulation [45]. As a result, phytohormones derived from microbes have a wide range of effects, based on the plant–microbe mix. While various studies show contradictory findings, the SA signalling mechanism appears to be important for shaping the root microbiome. SA mutants had mild effects on microbiome structure, according to one report [46]. On the other hand, according to Lebeis et al. [47] the rhizosphere microbiota of *A. thaliana* mutants lacking SA synthesis or vision was altered, whereas the corresponding JA and ET mutants had no such impact. Many plant-growth promoting rhizobacteria (PGPRs) influence plant development, especially root growth, by producing auxins, GA, or CK, in addition to influencing defences, which is common to pathogens and beneficials. To unravel the underlying complexity, systems biology tools such as metabolomics, global network analysis, hormone profiling, and specialized quantitative modelling of molecular processes in plants and soil will be necessary. Auxin signalling in the plant root is being extensively investigated, and sophisticated models are available [48]. Detailed mechanistic information as well as fluorescent auxin reporters that offer time-resolved data on auxin distribution permitted the construction of such quantitative models [49]. Quantitative time-resolved models are built on the foundation of both. Although preliminary evidence on the effects of auxin concentrations on receptor pairs is available, quantitative data on the chemicals and receptors that convert a given auxin concentration into specific transcriptional responses are often lacking [50]. A model of the SA signalling system will be useful in studying microbe–host interactions. NPR1, NPR3, and NPR4 are newly discovered SA receptors that work together to modulate responses to various SA concentrations [51]. BOP1 and BOP2, on the other hand, appear to have no role in SA signalling but have been linked to developmental programmes in legumes such as blooming and nodule formation [52]. Simultaneously, the biochemical control of NPR1 and presumably its paralogs is complicated, including several cellular compartments, redox potential, phosphorylation, and degradation. Although critical aspects for model creation such as thermogravimetric analysis (TGA) transcription factors and signalling network components are known [53], our knowledge of this important immune signalling system is still limited. For quantitative modelling of SA signalling, the creation of fluorescent SA sensors as well as quantitative protein level and binding data are crucial. All hormone signalling pathways are linked, and only a few biological reactions are mediated by a single hormone. The integrative research of Tsuda et al. [54] constitutes great efforts in unravelling the interaction of SA, JA, and ET during immunity in *Arabidopsis*. They separated the hormone signalling network into four sectors (SA, JA, ET, and PAD4) and tested immunity in all potential mutants from each sector following stimulation with a panel of MAMPs and effectors. Their findings revealed substantial hormone network component interactions, including additive, synergistic, and compensatory interactions. Later research by the same group led to the conclusion that the PTI signalling network is strongly buffered against interference, such as pathogen effectors [55].

1.2.5 Interactome Network Analysis

Molecular interaction network techniques can be useful in the absence of quantitative dynamic models for identifying modules, routes, components, and system-level patterns of molecular host-microbe interactions [56]. A reference protein network is necessary to situate host–microbe interaction data in the context of host biology. With the publishing of the first experimental map of physical protein–protein interactions among several thousand *Arabidopsis* proteins, plant interactome analysis began:

Arabidopsis Interactome-1 (AI-1) [57] offered a first integrated organizational view of plant molecular connectivity. Since then, numerous specialized and complementary maps have been developed to aid in the investigation of certain processes. Using the split-ubiquitin technique, a map of roughly 12 000 protein–protein interactions was created for membrane proteins [58]. A G-protein interactome showed a novel role for G-proteins in the regulation of cell-wall modification, a critical defensive mechanism [59]. Interolog mapping was used to create a protein–protein interaction network for the fungus *Phomopsis longicolla*, which causes *Phomopsis* seed damage in soybean [60]. To control plant defence and physiology, pathogens and beneficial microorganisms can release hundreds of (virulence) effector proteins into the cytosol and apoplast of the host plant [12]. To fully appreciate host–microbe interactions, their functions must be comprehended in a holistic and time-resolved manner. Small-scale investigations of plant-targeted pathogen effectors revealed that virulence effectors affect host protein activities to interfere with immune responses and cause illness, a phenomenon known as effector-triggered susceptibility (ETS) [61]. Effector-triggered immunity (ETI) is initiated when a pathogen's effectors are recognized by a host resistance protein (R protein) [61]. A large-scale interactome study (PPIN-1) mapped the interactions of virulence effectors from the bacterial pathogen Pst and the oomycete pathogen *Hyaloperonospora arabidopsidis* with proteins in the AI-1 host network to get a systems-level perspective on effector functions [62]; a follow-up study later added interactions of effectors from the biotrophic ascomycete *Golovinomyces orontii* [63]. The findings demonstrated that pathogen effectors partially converge on common host proteins, many of which are highly linked hubs in the host network. The host proteins showed genetic validation rates ranging from 100% for the most strongly targeted proteins to 40% for the less highly targeted proteins, depending on the degree of convergence. In addition to convergence, numerous effectors targeted proteins throughout the host network, most likely as a result of the immunological signalling network's high buffering [55]. Positive and balanced selection was seen in the immediate network neighbourhood of the highly targeted proteins, according to population genetic analysis. As a result, pathogen-induced selection pressure appears to be absorbed by the network that surrounds the effector targets [63]. These data support the idea that host–microbe interactions are mediated by a complex network that can only be partially comprehended by looking at individual routes. Pathogens appear to alter host networks rather than deconstructing network integrity, according to research on the *Yersinia pestis* interactome [64]. Pathogens are not the only ones that produce effector proteins. Plant immune responses and symbiotic relations can be modulated by mycorrhizal fungi, endophytic fungi, and nitrogen-fixing rhizobia effector proteins [65]. Many PGPRs, such as *Pseudomonas simiae* WCS417, and many proteobacterial strains are anticipated to contain functioning T3SS and effectors in complex microbiome data sets [66]. It is recognized that virulence effectors are necessary for productive and beneficial interactions between the beneficial fungus *Serendipita indica* and rhizobial bacteria. *Bradyrhizobium elkanii* T3SS-delivered effectors even allowed soybean nodulation without the need for

Nod factor [67]. Many proteobacteria contain type IV and type VI secretion systems, which may transfer bacterial protein into hosts and other microorganisms in addition to T3SS. *P. simiae* WCS417 possesses two T6SS loci [66] and may send effectors to both its plant host and other competing microorganisms to alter the microbiota. Proteomic techniques can help researchers better understand the variety of bacteria effector repertoire [68]. In a study comparing the genomes of a beneficial soil fungus, *Colletotrichum tofieldiae*, and a closely related pathogenic counterpart, *Colletotrichum incanum*, researchers discovered that while their secretomes were similar, the beneficial fungus had 50% fewer effector genes and lower activation of pathogenicity-related genes in plants [69]. As a result, microbial secretomes, as well as the quantity and kind of secreted effectors, may serve as a key point of distinction between beneficials and pathogens. Non-pathogenic interactions are most likely influenced by the beneficial effectors' complement. Understanding the global dynamics of effectors targeting different areas of the host network, and how this dynamic connects to ETS and ETI, as well as the systems-level and dynamic variations between pathogen and helpful effectors' secretion, will be a major issue for systems biology. RNA, which is transported to the host through extracellular vesicles, has emerged in recent years as a key communication molecule between hosts and microorganisms (EVs). EVs were initially discovered in mammalian cells and are now found in bacteria, archaea, and eukaryotes. In *Arabidopsis*, small RNA from the fungus *Botrytis cinerea* was shown to target host defence genes [70]. Plant EVs and multivesicular bodies accumulate around plasmodesmata during fungal infections to facilitate callose deposition at infection sites via host-induced gene silencing (HIGS) using dsRNA. Another layer of communication is formed by EVs and their RNA payload, whose relevance is just becoming apparent [71].

1.2.6 Transcriptional Regulatory Networks

Transcriptional profiling is frequently utilized, and the findings of significant research are referenced throughout this article. While comparative transcriptomics is prevalent, causal regulatory networks and co-expression correlation networks are less so. Co-expression networks are based on the idea that time series transcript profiles might reveal causal links between transcripts. Weighted Gene Correlation Network Analysis (WGCNA) is a widely used approach for grouping genes into co-expression modules using hierarchical clustering [72]. Signalling network connections, metabolic pathways, and phenotypic features are all used to compare these modules. Saelens et al. [73] examined 42 alternative clustering, decomposition, biclustering, and iterative network inference approaches in addition to WGCNA. In *Arabidopsis*, these methods were used. Kim et al. [74] used *A. thaliana* and other plants including maize and wheat to study their interactions with microorganisms.

The discovered modules give a preliminary look at genes with similar functions and can aid in the understanding of processes related to infection or commensalism.

1.2.7 Metabolic Exchanges and Nutrient Competition in the Soil

Metabolic exchanges are one of the core concepts of microbiome–host interactions. Plant roots deliver up to 40% of the complex carbons generated by photosynthesis to the rhizosphere, nourishing the microbiome [75]. Fungi and bacteria, on the other hand, aid in the solubilization and absorption of critical elements like phosphorus, nitrogen, and iron by the plant [76]. Plant reprogramming by pathogens via effectors and hormone signalling aims to relocalize nutrients as the metabolism of an individual organism has been studied using genome-scale metabolic modelling, and community-level response modelling is in progress but difficult [77]. Metabolic modelling of prokaryotes is increasingly commonplace [78]; metabolic models for *Arabidopsis*, barley, maize, sorghum, sugarcane, and canola have been developed on the plant side. By comparing metabolic capacities of beneficials and pathogens, the metabolic capacities of beneficials and pathogens may be assessed [79]. *P. syringae* has evolved to be metabolically adapted for a plant-pathogenic lifestyle, according to Mithani et al. [80]. The pathogenic *P. syringae* is metabolically extremely similar to its benign cousin *P. fluorescens* Pf-5, according to a comparison of metabolic networks for nine *Pseudomonas* strains, suggesting that metabolism may not be a crucial differentiating trait. A genome-scale metabolic model for the oomycete *Phytophthora infestans* was recently created, which predicts biochemical processes in various cellular compartments and in the context of the pathogen's stage [81]. To provide a more exact picture of the metabolic changes caused in plant and microbe during colonization, these models will need to be constrained by metabolite levels.

1.2.8 Integrated Multi-omics Modelling

Although reciprocal benefit between plants and their microbiome is evident, and a 'cry for assistance' might recruit bacteria to aid the host, it is unknown how the plant combines microbial identification with nutrient-related signals at this time. Phosphorus is frequently abundant in soil, while plant-absorbable orthophosphate is rare [82]. Castrillo et al. [83] explored the relationship between diet and defence in a stunning multi-omics systems biology exercise. They demonstrated that the plant phosphate starvation response (PSR) plays a crucial role in modifying the root microbiome by combining 16S rRNA sequencing, genome-wide expression analysis, analysis and modelling of SynComs, and functional experiments. When phosphate uptake-deficient and phosphate hyperaccumulating *Arabidopsis* mutants were compared to wild type, they found that they constructed distinct root-associated microbiomes. The transcription factors PHR1, and probably PHL1, are integrators of PSR and immune responses, as phr1 and phr1; phl1 mutant plants were more resistant to the oomycete and bacterial pathogens. The relationship between PSR and plant immunity appears to be influenced not only by the surrounding microbiota but also by pathogens, bringing new issues concerning the distinctions between helpful and pathogenic bacteria [84].

1.2.9 From Systems Biology to Crop Protection

Understanding crop–microbe connections is becoming easier because of conceptual and molecular advancements in microbe–host biology. McGrann et al. [85] utilized a draught genome assembly to estimate a secretome of roughly 1000 proteins in the developing foliar fungal barley pathogen *Ramularia collocygni*, which causes *Ramularia* leaf spot. They postulated that *R. collocygni* first acts as an endophyte without generating disease symptoms before transitioning to a necrotrophic phase, based on the reduced amount of plant cell-wall-degrading enzymes and the presence of genes associated with chitin recognition avoidance. Systems biological analysis will rely heavily on understanding such dynamics as well as the underlying molecular processes and signals. The host specialization of four *Rhynchosporium* species on grasses was studied in another research [85]. *Rhynchosporia* are hemibiotrophic fungi that invade the intercellular matrix of host plants gradually and without causing symptoms. Six unique effector proteins from *Rhynchosporium commune* were revealed to be important for maintaining the biotrophic growth stage in favour of the necrotrophic destructive stage, giving therapy leads. Beneficial microorganisms' impacts on enhanced biomass and greater tolerance to biotic and abiotic stressors in monocot crops were explored in a groundbreaking study that included multi-omics techniques. Using phenotyping, transcriptomic, molecular, and metabolomic techniques, Fiorilli et al. [86] investigated the three-way interactions between *Xanthomonas translucens*, the protective symbiotic AM fungus, and the host. They proposed a two-step process for conferring *Xanthomonas* resistance to AM-treated wheat: first, the activation of a broad-spectrum defence (BSD) response in the roots and leaves of AM-treated plants, and, second, a switch to pathogen-specific defence (PSD) upon bacterial infection, which ultimately leads to pathogen resistance.

1.3 Stability, Resilience, and Assembly of Agricultural Microbiomes

Knowing the identification and functional features of microorganisms present in crop plant 'core' microbiomes is important, but understanding the processes by which microorganisms assemble into such communities is crucial for any attempt to control or regulate the agricultural microbiome. The tremendous diversity of plant microbial communities creates a formidable barrier to properly comprehending the ecology of plant-associated microbiomes, and as a result, the relationships between microbiome assembly and plant performance remain little known [87]. As a result, we urge that researchers focus their efforts on developing synthetic microbial communities that can colonize plant organs and remain long enough in natural conditions to provide advantages to the host. What characteristics of a synthetic microbial population make it more likely to colonize plant organs is a fundamental topic. Microbial genes that are consistently represented in core metagenomes – and that are enriched in plant microbiomes when compared to soil – are promising candidates for important activities that boost colonization capacity. Community characteristics such as phylogenetic diversity and species richness may also influence colonization success, and these might be investigated in the lab utilizing gnotobiotic model systems and culture collections [88]. What qualities enable a synthetic community to survive invasion and displacement by plentiful microorganisms in the surrounding environment once it has colonized a plant and what characteristics provide resistance to abiotic stresses, or at the very least the capacity to recover from them, both are questionable. These traits might be the same as or different from the ones that allowed the colonization to begin in the first place. All of these issues might be investigated in the lab using gnotobiotic systems before being tested in greenhouse and outdoor studies. Years of agronomic work refining single-strain distribution via seed coatings, clay particles, and peat have shown that, in addition to genetic features of the microbiome, the technique of microbiome inoculation may play a large role in its effectiveness. Seeding numerous described microbial species into resilient particles for distribution by air, water, soil, or new delivery mechanisms such as insect vectors might be used to develop successful microbiome inoculation tactics for agriculture. However, attempts to increase colonization ability should also guarantee that synthetic communities do not overrun local ecosystems and have a detrimental impact on soil health, nearby plants, or future crops [89]. Finally, what external variables influence a positive synthetic microbiome's success? Synthetic communities may need to be adaptable to changes in host phenotypic features that affect assembly, or they may need to be customized to colonize certain crop species. Similarly, the environment will influence a community's ability to colonize or its resilience. Any attempt to regulate agricultural microbiomes would require an understanding of both abiotic (e.g. temperature, light, acidity, nutrient, and water availability) and biotic (e.g. competition, predation, parasitism, and mutualism) factors within the microbiome elements impacting its assembly [89]. It will be especially challenging to test the impact of biotic factors on microorganisms that cannot be cultivated independently. Strong interactions, for example, can 'link' organisms to other microorganisms (e.g. mycoparasitism).

Network analysis is particularly beneficial for discovering related microorganisms and 'hub' microorganisms, similar to keystone species that interact with a large number of other bacteria and hence have a significant influence on the community's structure and function [90]. Testing the effects of biotic variables on microorganisms that cannot be grown on their own will be particularly difficult. For example, strong interactions can 'link' organisms to other microbes (e.g. mycoparasitism). Network analysis is particularly useful for identifying related microorganisms and 'hub' microorganisms, which are comparable to keystone species in that they interact with a large number of other bacteria and so have a considerable impact on the structure and function of the community [90].

1.4 Core Plant Microbiome and Metagenome

Targeted investigations of agricultural microbiomes in the field, in addition to model systems and culture collections, provide critical fundamental information that can lead to innovation in a variety of ways. First, identifying the 'core' microbiome – the set of microbial species identified in the majority of samples of a given set of plants [91–95] – will aid in identifying plant-associated microorganisms that should be prioritized for future research, inclusion in culture collections, and manipulation studies. Although the plant microbiota is varied, not all of these microorganisms play functionally essential roles in the biology of their hosts. Researchers can filter out transitory connections and focus on stable taxa that have a better chance of impacting host phenotype by defining the core microbiome. In contrast to very deep sequencing of a few plant microbiomes, culture-independent surveys (such as sequencing internal transcribed spacer (ITS) and 16S rRNA amplicons) of large numbers of microbiomes of the same plant species from a variety of environments would improve progress towards this goal and could be followed-up by selective culturing of candidate core microbiota. Second, finding a functional core microbiome in addition to a taxonomically defined core microbiome based on phylogenetic differences is critical since component bacteria are likely to be adaptable. Using metagenomic and metatranscriptomic techniques to uncover shared projected functions that are likely relevant for the collection of plants investigated, the functional core may be found [96]. Extrapolating functional information from phylogenetic marker genes like ITS offers only a limited amount of information while metatranscriptomics, metaproteomics, and metabolomics, on the other hand, show the functional community phenotype, whereas focused metagenomics of functional genes and shotgun metagenomics provide a deeper understanding of community functional potential [97]. We can detect fundamental community functionality and the level of taxonomic functional redundancy using a combination of multi-omic techniques across large sample sizes of plants. Third, comparing core microbiomes or metagenomes between important plant groups and genotypes of the same plant species may reveal host-driven variations in microbiome assembly. The soil is the most important source of microorganisms that make up the plant microbiome [93], while there have been rare reports of seed-borne vertical transmission [98], and the atmosphere also contributes to above-ground plant microbiota [99]. Plants of various species or genotypes generate a mainly shared core microbiota, derived from the same environmental 'inoculum' at taxonomic levels of family and above. To what extent does the host impact microbiome assembly, and is within-species host genetic variation sufficient for breeding superior microbiome associations? Broad-sense heritability is often found to be between 5 and 7% [7, 100], implying that the potential of microbiome-related variables to react to conventional artificial selection may be restricted. Microbiomes might, however, be altered to boost heredity [101], and focused breeding efforts might be successful if we understand the host molecular basis of microbiome construction. Fourth, comparing the core microbiota of genetically diverse plant groupings may uncover plant genes and functional features that drive microbiome formation. Although plant functional features such as cuticle composition [102], root length and exudates [103, 104], and plant defences (immunity) [105] have been implicated, the methods by which hosts winnow the ambient community to establish their microbiota are not entirely known. Individual microorganisms (e.g. rhizobia, mycorrhizae) may have indirect impacts on later-arriving microorganisms, whereas cuticle and root characteristics should directly impact colonization by a wide variety of microbial species [106]. Furthermore, crop species or cultivars that consistently assemble various microbiomes may be relying on their microbiota to meet a variety of demands, especially if the plants are suited to distinct environmental difficulties [107].

1.5 Interactions Among the Microbes, Environment, and Management

All organisms are inhabited by microorganisms including archaea, bacteria, fungi, and viruses; this microbiota plays a key role in host health and development [104, 108]. The microbiome associated with plants is considered its second genome. It is determinant for plant health, growth, fitness, and consequently productivity [109]. Each environment associated with the plant – rhizosphere, endosphere, and phyllosphere – presents a specific microbial community with specific functions [4].

These culture-independent methods show that plant microbiome can reach densities greater than the number of plant cells and also greater expressed genes than the host cells. Metagenomics analysis using next-generation sequencing technologies shows that only 5% of bacteria have been cultured by current methods, revealing how many microorganisms and their functions remain unknown [110]. The first step in plant–microbe interaction is microbial recognition of plant exudates in the soil. There is a hypothesis that plants are able to recruit microorganism by plant exudates, which are composed of amino acids, carbohydrates, and organic acids that can vary according to the plant and its biotic or abiotic conditions [111]. Different plants select specific microbial communities as reported by Berg et al. [104] when comparing rhizosphere colonization of two medicinal plants: chamomile (*Matricaria chamomilla*) and nightshade (*Solanum distichum*); despite being cultivated under similar conditions, they presented different structural (analysing 16S rRNA genes) and functional (analysing nitrogen-fixing nifH genes) microbial communities. Moreover, plant exudate of the same plant varies according to plant developmental stages selecting specific microbial communities [112]. Researchers already identified some plant exudate compounds responsible for specific interactions such as flavonoids in Legume-Rhizobia [48] and Strigolactone as a signal molecule for AMF [113].

Reinhold-Hurek et al. [114] proposed a model for microorganism colonization. In bulk soil, the microbial community presents a great diversity and is influenced only by soil type and environmental factors. Getting closer to plant roots (rhizosphere), where there are root exudates, there are fewer species and a more specialized community. And only a few species are able to enter plant root and establish in the plant. Furthermore, after entering the plant, microbial community varies among the different organs: top leaves, fruits, bottom leaves, flowers, stems, and roots [115]. Mutualistic microorganisms can protect plants from pathogen either by inducing plant resistance or by antibiosis. The ISR in plants leads to high tolerance to pathogens. There are soils that even if there is the pathogen the disease does not occur; the mechanisms of these disease suppressives are still being investigated. In this way, Mendes et al. [116] analysed the microbiome of a soil suppressive to the fungal pathogen *Rhizoctonia solani* that causes damping off in several agricultural crops. Using a 16S rDNA oligonucleotide microarray (PhyloChip), they were able to identify more than 33 000 bacterial and archaeal taxa in the sugar beet seedlings rhizosphere grown in suppressive soil and in conductive soil. These analyses revealed the bacterial groups present only in the suppressive soil. The authors reported that γ-proteobacteria, especially Pseudomonadaceae, were all more abundant in suppressive soil than in conductive soil, focusing thereby on this bacterial group. Using random transposon mutagenesis technic in *Pseudomonas* sp. they were able to identify genes responsible for the biosynthesis of an antifungal: nine amino acid chlorinated lipopeptide produced by *Pseudomonas* sp. and that controls the pathogen. From the same PhyloChip diversity analysis, Cordovez et al. [117] identified other antifungal, this time produced by rhizosphere-associated streptomycetes. These *Streptomyces* isolates were able to produce chemically diverse volatile organic compounds (VOCs) with an antifungal activity as well as plant growth-promoting properties. Showing that different bacteria groups can have similar roles in the same environment, another example was reported by Ardanov et al. [118] who showed that the inoculation of *Methylobacterium* strains also protected plants against pathogen attack and affected endophyte communities. Therefore, using this concept, researchers started inoculating plants with a pool of microorganism with complementary traits, for example with different mechanisms of control; however, it is a challenge to find the right players to be inoculated [119].

In order to define which microorganisms should be inoculated, several approaches were used. The first approach seeks to define a core microbiome of a healthy host, or understand the function of microbiomes by sequencing approach, that can be followed by experiments on gnotobiotic host manipulating the microbiome with a selection factor (for example, antibiotics, salinity, and UV light) or transferring microbiomes between hosts [101]. In this way, researchers are starting to study 'microbiome engineering', modulating microbial community. This modulation can occur either by performing plant breeding programmes selecting a beneficial interaction between plant lines and rhizosphere microbiome or by redirecting rhizosphere microbiome by stimulating or introducing beneficial microorganisms [101, 119]. The microbiome engineering can occur by altering ecological processes such as modulation in community diversity and structure changing microbe–interaction networks and by altering the evolutionary processes that include extinction of microbial species in the microbiome, horizontal gene transfer, and mutations that can restructure microbial genomes [120].

Summarizing, plant phenotype is the sum of plant response to the environment and to the present microbiome (including endophytes and pathogens); this microbiome also responds to the environment and interacts with each other [121]. Mendes and Raaijmakers [108] suggest a similarity between gut and plant rhizosphere microbiomes. They are both open systems, with a gradient of oxygen, water, and pH resulting in a large number and diversity of microorganism due to the different existing conditions. There are differences between gut and plant rhizosphere microbiome composition, therefore there are some similarities related to nutrient acquisition, immune system modulation, and protection against infections. Berg et al. [122] point seven similarities between host-associated microbiome ecology, among them: different abiotic conditions shape the structure of microbial communities; host and its microbiome co-evolute; core microbiome can be transmitted vertically; during life cycle, the microbiome structure varies; host-associated microbiomes are composed of bacteria, archaea, and eukaryotic microorganisms; functional diversity is key in a microbiome; and microbial diversity is lost by human interventions.

1.5.1 Secondary Metabolism

Microorganisms produce a large variety of compounds known as secondary metabolites that do not play an essential role in growth, development, and reproduction of the producing organism. Nevertheless, these metabolites are often bioactive compounds and can perform important functions in defence, competition, signalling, and ecological interactions [123, 124]. To establish a microbial interaction network, microorganisms usually respond by metabolic exchange, which leads to complex regulatory responses involving the biosynthesis of secondary metabolites. These interactions can be parasitic, antagonistic, or competitive and the metabolites involved and their functions have been specially studied recently as a result of the advent of tools such as metabolomics and imaging mass spectrometry (IMS) technology [125, 126]. Siderophores are related to competitive and cooperative microbial interactions and can also play other roles, such as signalling and antibiotic activity [127, 128]. Hopanoids play an important role in bacterial interaction, conferring tolerance and improving the adaptation of bacteria in different environments [129, 130]. In fungi, the compounds differentially regulated in an interaction are often bioactive secondary metabolites, such as diketopiperazines, trichothecenes, atranones, and polyketides [131, 132]. Nevertheless, there is still a lot to understand about the mechanisms involved and the role of many secondary metabolites and genes differentially expressed during the interaction. In this section, we present examples of studies on secondary metabolites involved in different types of microbial interactions.

1.5.2 Endophyte–Phytopathogen–Plant Interaction

The metabolites and mechanisms involved in the interactions between endophyte, phytopathogen, and host plant are still very unclear and are predicted to involve many secondary metabolites. Endophytic fungi are known to produce a large variety of bioactive secondary metabolites [133, 134] that are probably related to the endophyte complex interactions with the host and the phytopathogens and can perform important ecological functions, for example, in the plant development (as growth promoters) and in defence, acting against phytopathogens [135, 136]. This interaction has been studied in co-cultures of the phytopathogen *Moniliophthora roreri* and the endophyte *Trichoderma harzianum* that cohabit in cacao plants [127]. *T. harzianum* is extensively used as a biocontrol agent and has known ability to antagonize *M. roreri*. Tata et al. [127] identified four secondary metabolites (T39 butenolide, harzianolide, sorbicillinol, and an unknown substance) whose production was dependent on the phytopathogen presence and was spatially localized in the interaction zone [127]. T39 butenolide and harzianolide have been reported to have antifungal activity. Sorbicillinol is an intermediate in the biosynthesis of bisorbicillinoids, a family of secondary metabolites that present diverse activities [137].

Trichoderma atroviride, commonly used as a biocontrol agent, produces acetic acid-related indole compounds that may stimulate plant growth. Colonization of *Arabidopsis* roots by *T. atroviride* promotes growth and enhances systemic disease resistance conferring resistance against hemibiotrophic and necrotrophic phytopathogens [138]. Other co-cultured studies were performed with bacteria. Araújo et al. [139] isolated a great number of *Methylobacterium* strains from asymptomatic citrus plants (with *Xylella fastidiosa* but without disease); then Lacava et al. [140] showed that *Methylobacterium mesophilicum* SR1.6/6 and *Curtobacterium* sp. ER1.6/6 isolated from health and asymptomatic plants inhibited the growth of the phytopathogen *X. fastidiosa*, the causal agent of citrus variegated chlorosis. Moreover, transcriptional profile of *X. fastidiosa* was evaluated during in vitro co-cultivation with a citrus endophytic strain of *M. mesophilicum*. It was shown that genes related to growth, such as genes involved in DNA replication and protein synthesis, were downregulated, while genes related to energy production, stress, transport, and motility, such as fumarate hydratase, dihydrolipoamide dehydrogenase (Krebs cycle), pilY transporter, clpP peptidase, acriflavin resistance, and toluene tolerance genes, were upregulated [141].

Another approach to study endophyte–phytopathogen plant interaction is based on the genome sequencing and transposon mutagenesis of an endophyte strain of *Burkholderia seminalis*, which suppresses orchid leaf necrosis by *Burkholderia gladioli*, which revealed eight loci related to biological control. A web cluster related to the synthesis of extracellular polysaccharides of the bacterial capsule was identified [142]. Extracellular polysaccharides are known to be key factors in bacterial–host interactions [143, 144]. In addition, gene clusters putatively related to indole-acetic acid and ET biosynthesis were identified in the sequenced genome of the endophyte strain, suggesting that this strain might interact with the plant by altering hormone metabolism [142].

1.5.3 Hopanoid

Hopanoids compose the cell membrane of some bacteria, [145] presenting the same function of eukaryotes cholesterol. They are responsible for stabilization of the membrane and regulate its fluidity and permeability [146]. Experiments that knockout biosynthesis genes such as hnpF (squalene hopene cyclase: shc) gene show that the absence of hopanoids does not influence bacterial growth [146, 147] but affects tolerance to several stress conditions, such as extremely acidic environments [148] or toxic compounds such as dichloromethane (DCM) [149]; it also affects the resistance to antibiotics [64] and antimicrobial lipopeptide [63], playing a role in multidrug transport [83] and bacterial motility [84]. Hopanoids act in increasing bacteria tolerance to adverse environments, conferring resistance to stress conditions including extreme pH and temperature and exposure to detergents and antibiotics [129, 130]. In this way, hopanoids may be involved in bacteria–plant interaction, being responsible for adaptation of bacteria in aerobic microenvironment and low pH culture medium [131] as well as involved in nitrogen metabolism in *Frankia* sp. [150]. For example, a type of hopanoids produced by the nitrogen-fixing bacteria *Bradyrhizobium diazoefficiens* is essential for its symbiosis with the host *Aeschynomene afraspera*, a tropical legume. In this case, the synthesis of C35 hopanoids is related to evasion of plant defence, utilization of host photosynthates, and nitrogen fixation [151].

1.5.4 Parasitic Interaction

The study of the mycoparasitic interaction between *Stachybotrys elegans* and *R. solani* revealed many secondary metabolites differentially expressed in the interaction [132]. During the interaction, *S. elegans* produces cell-wall-degrading enzymes and expresses genes associated with parasitism [152, 153] while *R. solani* responds with an elevated level of the pyridoxal reductase-encoding gene [152]. A metabolomic study showed the profile of the induced secondary metabolites during the interaction. It showed a significant effect of the mycoparasite on *R. solani* metabolism: the biosynthesis of many antimicrobial compounds were downregulated, possibly as a result of the interaction, and only a few diketopiperazines were induced [132]. Diketopiperazines are known to have antimicrobial properties, among others biological activities [154]. The mycoparasite *S. elegans* produced several mycotoxins, mainly trichothecenes and atranones. It was hypothesized that the trichothecenes were triggered by *R. solani* and were responsible for the alteration in its metabolism, growth, and development [132]. Trichothecenes are a major class of mycotoxins and have been reported to inhibit eukaryotic protein biosynthesis and generate oxidative stress.

1.5.5 Microbial Community's Interaction

Actinomycetes are noteworthy as producers of many natural products (NPs) with a wide range of bioactivities [126]. A study on *Streptomyces coelicolor* interacting with other actinomycetes showed that most of the compounds produced in each interaction was unique, revealing a differential response in each case. Many unknown molecules and an extended family of acyl-desferrioxamine siderophores never described before in *S. coelicolor* were identified. They identified 227 compounds differentially produced in interactions; half of these were known metabolites: prodiginines, actinorhodins, coelichelins, and acyl-desferrioxamines. Thus, actinomycetes interspecies interaction seems to be very specific and complex [155]. It has been shown that fungal–bacterial interactions can lead to the production of specific fungal secondary metabolites and not only diffusible compounds act in this communication but also there is a contribution from physical interaction [156]. Schroeckh et al. [157] demonstrated that an intimate physical interaction between *Aspergillus nidulans* and the actinomycete *Streptomyces rapamycinicus* leads to the activation of fungal secondary metabolite genes related to the production of aromatic polyketides, which were otherwise silent. A PKS gene required for the biosynthesis of the archetypal polyketide orsellinic acid, lecanoric acid (typical lichen metabolite), and the compounds F-9775A and F-9775B (cathepsin K inhibitors) was identified [156]. It was later reported that alterations in fungal histone acetylation via the Saga/Ada complex are triggered by the actinomycete leading to the induction of the otherwise silent PKS cluster. This result shows that bacteria can trigger alterations of histone acetylation in fungi [156].

1.5.6 Siderophore

The production and acquisition of siderophores by microorganisms is a crucial mechanism to obtain iron. Many microorganisms secrete siderophores in the environment that when loaded are recognized by cell surface receptors and then transported into the microbial cell [158]. Thus, they are related to competitive and cooperative microbial interactions. In addition, many siderophores can also present other functions, for example, they can function as sequesters of a variety of metals and even heavy metal toxins, as signalling molecules, as agents in regulating oxidative stress, and as antibiotics, which were reviewed by Johnstone and Nolan. In some *Pseudomonas* species, a group of siderophores called pyoverdines is essential for infection and biofilm formation, probably helping to regulate bacterial growth [159]. Pyoverdines have been reported to act as signalling molecules triggering a cascade that results in the production of several virulence factors, such as exotoxin A, PrpL endoprotease, and pyoverdine itself [160].

In the marine environment, exogenous siderophores affect the synthesis of induced siderophores and other iron acquisition mechanisms by other microbial species, working as signalling compounds that influence the growth of marine bacteria under iron-limited conditions. Many strains of marine bacteria were reported to produce siderophores and iron-regulated outer membrane proteins only in the presence of exogenous siderophores produced by other species, such as *N,N*-bis(2,3-dihydroxybenzoyl)-*O*-serylserine from a *Vibrio* sp., even under very low iron concentrations [161].

1.5.7 Symbiotic Interaction

A remarkably complex inter-kingdom interaction is the symbiotic relationship between *Burkholderia*, a genus of bacteria, and *Rhizopus*, a genus of phytopathogen fungi that causes rice seedling blight. The endosymbiotic bacteria *Burkholderia* spp. is responsible for the production of the phytotoxin rhizoxin, the causal agent of rice seedling blight [162]. It was reported that in the absence of the endosymbiont, *Rhizopus* is not capable of producing spores, indicating that the fungus is dependent on factors produced by the symbiont to complete its life cycle [163]. This complex symbiont–pathogen–plant interaction is still poorly understood regarding the metabolites and mechanisms involved in the communication and interaction. A study on exopolysaccharide (EPS), which usually plays key roles in interactions, produced by *Burkholderia rhizoxinica* described a previously unknown structure of EPS. However, the loss of EPS production did not affect the endosymbiotic interaction with *Rhizopus microsporus*, as shown by a targeted knockout mutant experiment [164]. *B. gladioli* produces enacyloxins (polyketides with potent antibiotic activity) in co-culture with *R. microsporus*. The fungus induces the growth of *B. gladioli* resulting in an increased production of bongkrekic acid, which inhibited the growth of the fungus [164].

1.6 Microbiome Innovation in Agriculture: Insect Pest Management

Insects are associated with diverse microbial communities and in many cases, these associations are crucial for insect survival and development. Symbiotic microbes in the gut, hemolymph, as well as in specialized cells carry an arsenal of enzymes that provide specialized services to the insect hosts [165]. Supplies of essential nutrients (particularly amino acids and B vitamins) by endosymbionts have been well documented in a number of crop pests, particularly plant sap-sucking Hemipteran insects such as aphids, whiteflies, and psyllids [166–169], and in human disease vectors and urban pests such as tsetse flies in the genus *Glossina* and the common bed bug (*Cimex lectularius*). Some symbionts can degrade complex polysaccharides or recycle nitrogen for insects, such as the termites [170, 171] and cockroaches [172, 173]. The production of antimicrobials by symbionts aids the immune system to fight against invading pathogens, as was shown in the beewolf digger wasps [174] and cotton leafworm [175]. Besides nutritional and immune services, symbionts can shape the ecological interactions between insects and their natural enemies. For instance, the secondary symbiont of aphids *Hamiltonella defensa* increased the chance of host survival from parasitoid wasp attacks by disrupting wasp embryogenesis, mediated by its bacteriophage-encoded toxins [176–180]. *H. defensa* was also shown to attenuate volatile release in aphid-infested plants, thus reducing parasitic wasp recruitment [181]. Similarly, symbiont manipulation of plant physiology that facilitates insect colonization was observed in whiteflies and the Colorado potato beetle [182, 183]. Modification of body colour by facultative symbionts may determine aphid susceptibility to predation or parasitism [184]. In particular, *Rickettsiella* infection in the pea aphid *Acyrthosiphon pisum* increased the synthesis of blue-green polycyclic quinone pigments, turning the host from red to green. This symbiotic-dependent colour variation is believed to affect the aphid's relative risks between predation and parasitism, as their predators such as the ladybird beetles preferentially prey on the red morphs, while parasitoids preferentially attack the green morphs [185–187]. Termites [171, 172] and cockroaches have symbionts that can digest complex polysaccharides or recycle nitrogen for them [173]. As proven in the beewolf digger wasps [174] and cotton leafworms [174], symbionts produce antimicrobials that help the immune system combat invading infections [175]. In different insect pests, symbionts were also shown to impact pesticide resistance. Resistance to organophosphorous pesticides is related to the direct detoxification by their symbionts in the beans bug (*Riptortus pedestric*) and the eastern fruit fly (*Bactrocera dorsalis*) [176, 177]. *Bacillus thuringiensis* (Bt) insecticide action was demonstrated to depend on the presence of symbiotic mid-gout bacteria in gypsy moth larvae *Lymantria dispar* [188]. Field microbiome studies showed that Bt resistance in bollworm cotton (*Helicoverpa armigera*) was related with different microbiome compositions [178–180]. The diamond back moth (*Plutella xylostella*) was diagnosed with varying levels of sensitivity to Chlorpyrifos treated with antibiotics and then decolonized with various gut-associated bacteria [189, 190]. In mosquitoes that demonstrated decreased lethality of pesticides, various intestinal commensal bacteria were found to have been contributing to pesticide resistance [191]. Collective data suggest that bacteria have a greater role as previously assumed in the formation of insect behaviour [192]. Long-term scatter, egg position, mattress, hosted search, and kin recognition are insect behaviours that are found to be altered by microorganisms [193]. Studies further imply that microbiomes might alter host behaviour by the formation of host neuro-endocrine circuit-acted metabolites [194–196], a phenomenon termed the 'gut-brain axis'. There has been a lot of study on the gut-brain axis, with the majority of it being on mammalian systems. This field, however, is still in its infancy. A recent study found that the microbiome of the *Drosophila melanogaster* alters the host's olfactory-guided foraging preferences towards meals with various microbial content [197]. Farine et al. [198] and Qiao et al. [199] showed similar microbiome-priming effects on fly behaviour in following investigations. Scientists have achieved two breakthroughs in insect microbiome research thanks to advances in high-throughput sequencing and functional genomics: (i) investigate previously unidentified microbiomes in a wider range of insects, resulting in a better knowledge of the host and environmental variables that influence insect microbiome diversity and composition – some examples include microbial communities associated with Drosophilid and Tephritid fruit flies [200–202], ants [203, 204], bees [205], mosquitoes [206, 207], ticks [208], beetles [209], and midges [210], among others;

and (ii) assign particular microbial taxa or consortia with microbiome roles. While insect microbiomes differ in terms of diversity and stability, there is widespread agreement that microbial effect on insect invasive characteristics exists. For example, the invasiveness of the sweet potato whitefly (*Bemisia tabaci*) was aided by the introduction of a *Rickettsia* sp. into the pest population (from 1% infected in 2000 to 97% in 2006), which resulted in faster development, a higher survival rate to adulthood, and increased host fecundity [197, 211–214]; it was found that the microbiome of the strong lab model *D. melanogaster* accelerates larval development, impacts host foraging choice, and reproduction [197, 213]. *Drosophila suzukii*, often known as spotted wing drosophila (SWD), is a tiny fruit pest that relies on the microbiome to flourish [215]. Microbial symbiosis was initially discovered in the olive fruit fly *Bactrocera oleae*, a very damaging agricultural pest belonging to the Tephritidae family [216]. Unlike other fruit-feeding *Bactrocera* species, *B. oleae* has an obligatory bacterial symbiont (Candidatus *Erwinia dacicola*) that is maintained in the midgut caeca of the larvae. The symbiont is yet to be grown, but studies have demonstrated that it aids fly growth and reproduction by delivering necessary amino acids and metabolizing urea from a variety of sources, including bird droppings, making nitrogen accessible to adult flies [217, 218]. It also aids the development of larvae in unripe olives by inhibiting the production of oleuropein, a plant defence molecule [217, 219]. The symbiont was shown to be missing in domesticated *B. oleae* grown on antibiotic-laced artificial medium, illustrating the influence of upbringing on symbiont selection [220]. *Pantoea* sp. and *Burkholderia* sp. are two more bacterial species found in the intestines of *B. oleae*, although their nutritional significance is unknown [202]. The medfly (*Ceratitis capitata*) and apple maggot fly both have microbiome-dependent larval development (*Rhagoletis pomonella*). Diazotrophs that express the nitrogen reductase gene (nifH) in the stomach of medflies are involved in microbial food supply [221]. *Citrobacter*, *Klebsiella*, *Pectobacteria*, *Enterobacter*, and *Pantoea* are among the bacteria found in medflies [200]. When provided as probiotics, the community has been found to boost fly growth, reproduction, and lifespan, as well as boost male copulatory success [200, 218]. A modest but persistent population associated with the medfly stomach contains *Pseudomonas* spp., in addition to the dominating Enterobacteriaceae. Malacrin et al. [222] used a metabarcoding strategy to find changes in the microbial community at different instar stages of the medfly [222]. *Burkholderia* was discovered to be prevalent in early instars and adults, and it has been postulated that it may play a role in nitrogen fixation in *Tetraponera* ants [223]. Similarly, bacteria from the genera *Sphingomonas* and *Pseudomonas*, as well as an unidentified bacterium from the family Methylobacteraceae, were found to be more abundant in late instars of the medfly, whereas bacteria from the families *Leuconostoc*, *Weissella*, *Acetobacter*, *Gluconobacter*, and an unidentified bacterium from the family Xanthomonadaceae, were more abundant at pupal stage. In addition, medflies fed on different host plants had a diverse microbial community. Malacrin et al. [222] observed that medfly larvae fed *Ficus carica* fruits had *Acinetobacter* and *Gluconobacter*, whereas *Acetobacter* and *Leuconostoc* were more prevalent when given *Prunus persica* (peaches). It has been proposed that *Acinetobacter* and *Gluconostoc* are involved in phenolic glycoside detoxification [224]. Similarly, enterobacteria such as *Pantoea*, *Klebsiella*, and *Enterobacter* are found in the guts of apple maggot flies [225]. During oviposition, microorganisms are deposited into the fruit, supplying necessary nutrients and proteins for larval growth [225, 226]. Symbiotic bacteria enhance larval development in several Tephritids of the subfamilies Dacinae and Trypetinae by metabolizing carbohydrates, raising organic nitrogen levels, and producing vitamins [217, 225]. However, their functions in adult flies are unclear [218].

1.6.1 Manipulation of Insect-Associated Microbiomes for Pest Management

The contribution of microbiomes to a wide range of insect invasiveness-related features implies a wealth of resources that might be targeted for pest control. The use of biochemicals to eradicate or disrupt insect symbiosis is a straightforward method [227]. Antibiotics such as tetracycline and penicillin, for example, have been demonstrated to render tsetse flies infertile by inhibiting the obligatory mutualist *Wigglesworthia*, obstruct the development of young ticks, and reduce adult tick reproduction by lowering their symbiont load [228]. Antimicrobial peptides (AMPs) have also been investigated for manipulating insect symbionts, albeit they are more typically utilized to combat human or plant infections transmitted by insects [229]. Insect innate immunity relies heavily on AMPs (which include a variety of amphiphilic and cationic oligopeptides). They provide immunity against bacteria, fungi, and viruses, among other microorganisms. Cecropin was the first AMP recovered from *Hylaophora cecropia* pupae [230, 231], and since then over 150 AMPs have been purified from insects [232]; for example, the α-helical peptides (e.g. moricin and cecropin), cysteine-rich peptides (e.g. defensin and drosomycin), proline-rich peptides (e.g. apidaecin, drosocin, and lebocin), and glycine-rich proteins (e.g. apidaecin, drosocin [e.g. attacin and gloverin). Some AMPs (such as defensins, cecropins, attacins, and proline-rich peptides) are found in all insect orders, whereas others (such as moricin and gloverin) are exclusively found in particular insect orders (e.g. Lepidoptera). AMPs' antibacterial activity is due to their positively charged surface, which allows them to bind negatively charged microbial surfaces via charge–charge interactions, disrupting bacterial cell wall integrity [233]. Resistance to bacterial, fungal, and certain eukaryotic parasites has been conferred by AMPs in plants and insects. Insect defensins (gallerimycin from *Galleria mellonella*) and cecropin (sarcotoxin-IA from *Sarcophaga peregrine*), for example, have been found to provide harmful fungal resistance in transgenic tobacco [234]. In rice and tomato plants, transgenic production of cecropins has also been demonstrated to give resistance to fungal and bacterial diseases [235, 236]. Cecropins have also been shown to be effective against protozoan parasites including *Plasmodium* and *Trypanosoma* [237, 238]. The use of transgenic cecropin expression in *Anopheles gambiae*, a human parasite vector, has been found to lower the quantity of *Plasmodium berghei* oocysts by 60% [239]. Furthermore, transgenic co-expression of defensin-A and cecropin-A in *Aedes aegypti* has been found to prevent *Plasmodium* parasite transmission cooperatively [240]. Non-target effects, which can lead to disturbance of the natural microbiome of non-target insects, are a significant drawback of utilizing antibiotics or AMPs. Low bioavailability, instability, and antimicrobial resistance are some of the other drawbacks of employing AMPs [241]. Nonetheless, approaches such as fusion with antigen-specific antibody fragments [242], protein engineering, and synthetic biology approaches can be used to create AMPs with improved specificities (e.g. substitution of amino acids, chemical modifications) [243]. Nanotechnology-assisted distribution of AMPs to many biological systems is also being investigated [244]. Insect trait manipulation can also be accomplished by introducing a foreign microbe or replacing a symbiont with another bacterium. In stink bugs and aphids, experimental substitution of certain cultivated and uncultured insect symbionts has been demonstrated. According to Hosokawa et al. [245], exchanging the Ishikawaella symbionts between the stink insect *Megacopta punctatissima*, a common pest of soybean and other legumes, and a closely related non-pest species, *Megacopta cribraria*, resulted in poor *M. punctatissima* egg hatching on the plants. The major symbiont Buchnera was replaced with a new genotype by microinjection in the pea aphid in one experiment, and the pest's heat tolerance was altered [246].

1.6.2 Incompatible Insect Technique (IIT)

Several techniques have been developed to produce gnotobiotic insects in the lab, including *Drosophila* [213], mosquitoes [247], and honey bees [248]. Cleaning insect eggs to remove maternally deposited bacteria on the surface or treating larval or adult insects with antibiotics, followed by feeding on food seeded with cultured microbes or microbe-laden materials were all common parts of the technique (e.g. faecal transplantation). While the gnotobiotic approach has aided in the discovery of microbiome impact on insect traits such as development, physiology, behaviour, and insecticide resistance [190, 213, 247–249], its application in pest management is still largely theoretical. The use of *Wolbachia* to control mosquitoes and mosquito-borne disease pathogens is one notable exception. *Wolbachia* is a widespread, vertically transmitted endosymbiont found in arthropods, infecting nearly 60% of all insects [250]. The bacteria are most recognized for inducing cytoplasmic incompatibility (CI), feminization, parthenogenesis, and male death in its hosts [251]. *Wolbachia*-infected females have a reproductive advantage over uninfected females because their sex ratios are skewed towards females, making it easier for them to spread across a community. *Wolbachia*-induced CI is used in the incompatible insect technique (IIT) to reduce mosquitoes and other insect pests [259[251, 252]. CI causes embryonic death [253] and can be caused unidirectionally in crossings between *Wolbachia*-infected males and uninfected females, or bidirectionally in crosses between infected animals carrying different *Wolbachia* strains [254]. As demonstrated in *A. aegypti*, *Wolbachia*-induced sterility has little effect on male mating competitiveness or survival [255]. *Wolbachia*-infected males are returned into the wild on a regular basis at IIT [256, 257]. Several insect pests, including *Rhagoletis cerasi*, *C. capitata*, the tsetse fly, and disease vectors, such as *Culex pipiens*, *Aedes albopictus*, and *Culex quinquefasciatus*, have been studied extensively using IIT [256, 257]. *Wolbachia* can be transinfected into a new host that is not natively infected with CI-inducing *Wolbachia* strains [258–260]. *Wolbachia* strain wSuz, as naturally, infects *D. suzukii* but does not cause CI [261–263]. *D. suzukii* has been successfully introduced to two CI-inducing *Wolbachia* strains (wHa and wTei) from different *Drosophila* species, paving the way for IIT for this pest [261, 264]. By distorting sex ratios toward females, Wolbachia-infected females have a reproductive advantage over uninfected females, facilitating their propagation in a population. The incompatible insect technique (IIT) employs Wolbachia-induced CI as a strategy to control mosquitoes and other insect pests [251, 252]. In addition, because host genotype influences *Wolbachia* density and phenotypic manifestation of infection in hosts, including CI, the genotype of IIT insects should be examined [265, 266]. Recent research has made substantial progress in identifying the molecular mechanisms that cause *Wolbachia*-mediated CI. When expressed dually in uninfected males, the wMel genes cifA and cifB (encoded by WO prophage) functionally repeated CI, according to LePage et al. [267]. Both genes are unable to cause CI on their own. Further research revealed that transgenic production of the cifA gene in *Drosophila* recovers CI and eliminates the embryonic lethality induced by wMel *Wolbachia* [268]. Beckmann et al. [269] found that the *Wolbachia* deubiquitylating enzymes (DUB) cidA and cidB combine to cause CI in transgenic *Drosophila*. Aside with CI, *Wolbachia*'s pathogen-blocking capabilities are another key feature for mosquito control. When transinfected with *Wolbachia* generated from *Drosophila* or other mosquitoes including *A. albopictus* and *C. quinquefasciatus*, *A. aegypti*, the vector for several clinically important arboviruses, showed dramatically decreased competence for dengue, chikungunya, yellow fever, and Zika viruses [270–273], as well as *Plasmodium* and filarial nematodes [274, 275].

The specific mechanism of *Wolbachia*-mediated pathogen inhibition is still being researched. Priming of the immune system, alterations in cholesterol and lipid droplet formation and trafficking [276], and (viral) RNA degradation are among the possibilities that have been offered [277]. In the year 2011, in the wild in Cairns, Australia, *A. aegypti* harbouring the wMel strain were released, marking the first experiment of microbiome modification of a wild insect population with the objective of lowering vector competence [278]. The *Wolbachia* infection has firmly established in the mosquito population, according to a follow-up experiment two years later [278]. More crucially, the discharge effectively halted dengue transmission in Cairns and its environs in northern Queensland, Australia. Meanwhile, *Wolbachia* frequencies in the original Cairns populations are at 95%, with a 96% decline in dengue incidence as of late 2019. *Wolbachia* has now been found in northern Queensland, as well as in Yogyakarta, Indonesia, and Kuala Lumpur, Malaysia, as a result of further releases [279–281].

1.6.3 Paratransgenesis

Paratransgenesis, a similar method that has gained popularity in recent years, involves genetically modifying microorganisms to express desired effects in insects [282–284]. Paratransgenesis avoids the problems of fitness cost associated with introducing a transgene into insects and transgenic instability in insect genomes by not changing the insects (i.e. transgenesis). This method is best for bacteria that can be cultivated, altered, and reintroduced into insect hosts with ease. Although paratransgenesis was first postulated in the early 1990s, the majority of study has focused on human disease vectors and a few Hemipteran crop pests. Beard et al. [285, 286] demonstrated that the triatomine bug's gut symbiont *Rhodococcus rhodnii* may be genetically engineered to express effector molecules (cecropin A and similar pore-forming molecules) against the protozoan *Trypanosoma cruzi*, which causes Chagas disease. Inoculating eggshells or food with excrement seeded with the designed symbiont allows the symbiont to be introduced to insect offspring. Durvasula et al. [287] used an anti-trypanosome single-chain antibody to alter the symbiont and found a considerable decrease in parasite burden. Following the positive findings of laboratory investigations, field trials were conducted to assess the transmission efficiency of modified *R. rhodnii* to the triatomine bug using CRUZIGARD, a simulated triatomine-faecal substance made of an inert guar gum matrix painted with India ink [288]. To manage *Rhodnius prolixus*, a study recently combined paratransgenesis with RNA interference (RNAi) technology. In *R. prolixus*, oral administration of an *Escherichia coli* strain HT115 or *R. rhodnii* engineered to express dsRNA targeting the antioxidant genes-heme-binding protein (RHBP) and catalase (CAT) genes caused systemic RNAi to silence these genes, resulting in poor nymph development and reduced female fecundity [289]. Using the engineered symbiont *Sodalis glossinidius*, which produced antigen-binding molecules targeting *Trypanosoma brucei*, the causative agent of sleeping sickness, similar paratransgenic techniques have been explored on tsetse flies. *Sodalis* may be transferred vertically by the milk glands and is present in the hemolymph, midgut, and milk gland [290, 291]. Using bacteria and fungi obtained from mosquito midguts and ovaries, different paratransgenic techniques have been investigated in mosquitoes to limit the transfer of malaria-causing *Plasmodium* parasites. The Gram-negative *Asaia bogorensis* was chosen for paratransgenesis against *P. berghei* because it has been demonstrated to survive in mosquito midguts and spread swiftly both vertically and horizontally within a population [292, 293]. The siderophore receptor gene was fused with anti-plasmodial effector genes to create genetically engineered *Asaia* strains. The scorpine AMP and a synthetic antiPbs21 scFv-Shiva1 immunotoxin made up of a single-chain antibody (scFv) against the *P. berghei* ookinete surface protein 21-Shiva1 fusion protein were among the genes identified. *Anopheles stephensi* mosquitoes fed with the altered *Asaia* and challenged with *P. berghei*-infected blood showed a substantial reduction in parasite growth [294]. *Pantoea agglomerans*, common mosquito symbiotic bacteria, was previously modified to produce anti-*Plasmodium* effector proteins utilizing an *E. coli*-derived Type I hemolysin secretion system. In the midgut of *Anopheles* mosquitoes, these modified *P. agglomerans* strains were shown to suppresses the growth of *P. falciparum* and *P. berghei* [295]. *Metarhizium anisopliae*, an entomopathogenic fungus, has also been engineered to release the antibiotic scorpine and anti-plasmodial peptide SM1, which inhibits *Plasmodium* parasite growth [196]. Novel *Serratia* sp. AS1 colonizes the ovaries and gut of *A. stephensi* as AS1 strain was both sexually and vertically transmitted, persisting for at least three generations. Mosquitoes infected with an engineered *Serratia* AS1 containing five different anti-plasmodium effector molecules (Shiva1, a cecropin-like synthetic AMP; MP2, midgut peptide 2; EPIP, enolase–plasminogen interaction peptide (lysine-rich enolase peptide); scorpine, scorpion *Pandinus imperator* venom AMP; and mPLA2, inactive bee venom phospholipase A2) displayed a reduction in the oocyte load by 93% [296]. The Glassy-winged sharpshooter (GWSS), *Homalodisca coagulata*, demonstrates the potential of paratransgenesis in crop protection against insect pests or insect-vectored illnesses. GWSS is a vector for *X. fastidiosa*, a bacterial pathogen that causes Pierce's disease in grapes by manufacturing EPSs, which assist the infection to colonize the xylem of its host plant and restrict the flow of the xylem [297–299]. *Alcaligenes xylosoxidans* var. denitrificans (Axd) was identified as a candidate for genetic modification among the several bacterial species identified from GWSS. It has always been detected in the xylem of host plants, in the same spot where the pathogen lives. GWSS was successfully given with genetically engineered Axd harbouring a DsRed fluorescent protein gene from injected stems. It was discovered to invade the foregut of insects, implying that a paratransgenic method to eradicate *X. fastidiosa* from GWSS is possible [298]. However, as this genus of bacteria has been recognized as a nosocomial human pathogen linked in causing lung infection in cystic fibrosis patients, the usage of altered Axd in plants has significant downsides [300]. To alleviate the safety concern, an endophytic bacteria from grapes, *P. agglomerans* E325 (an EPA-approved agent for managing fire blight in pears and apples), was genetically engineered to express anti-*Xylella* effector proteins melittin and a scorpine-like AMP, and colonized the foregut of GWSS using an artificial feeding system (AFS) [301]. In addition, in simulated field settings, targeted administration of recombinant *P. agglomerans* E325 to the gut of GWSS utilizing a microencapsulation approach was established to treat Pierce's illness. The microencapsulation approach might be effective in the field since it might restrict the transmission of alien genetic material [301]. Leonard et al. [302] devised a paratransgenic technique in honeybees and showed in the lab that it increased bee survival against viral infection and *Varroa* mites. The researchers created symbiotic gut bacteria called *Snodgrassella alvi* that produces dsRNA that targets bee, virus, and mite genes. The selection of microorganisms, the genetic design, and the implementation of the treated insects all have a role in the success of gnotobiotic or paratransgenic techniques for insect pest management. Another barrier is the association's stability, which should be particular to target insects or innocuous to non-target hosts. Although persistent association guarantees that microbial-mediated effects on host insects are long-lasting, certain microorganisms may be 'lost' from the insects owing to environmental selection pressure or antagonistic interactions with other germs. Despite these cautions, it is expected that gnotobiotic and paratransgenic insect research and development will continue to expand.

1.6.4 Exploiting the Chemical Inventories of Microbiomes to Develop New Biopesticides

The evolution of pesticide resistance is a major barrier for crop security, since many insect pests have no other treatment options. The difficulties of turning a lead compound into a product that can pass tight environmental and safety laws have made the development of new synthetic pesticides more expensive and complicated. According to a recent research, developing a synthetic pesticide in the United States currently costs over $300 million and takes about 12 years [200]. The demand for novel insecticidal chemicals is necessitated by the need to address resistance issues and improve sustainable agriculture. The effort to produce innovative insecticides with little environmental effect has sparked renewed interest in biopesticides in recent years (i.e. pesticides based on living organisms or their NPs, including their genes and metabolites). The global market for biopesticides is now estimated at

$3 billion and is predicted to expand by 15% over the next four years, exceeding the growth of synthetics by a factor of 10 [303]. Biopesticides are divided into three types: biochemical pesticides, which are naturally occurring chemicals that can change pest behaviour or physiology; microbial pesticides, which use pathogenic or toxic microbes as the active ingredient; and plant-incorporated protectants (PIPs), which are transgenic plants that produce pesticidal compounds [304].

1.6.5 Microbial Insecticides and Plant-Incorporated Protectants

Historically, microbial insecticides have taken the form of living microbes or spores applied directly to the field. Among the most common microbial insecticides are the entomopathogenic fungi, or EPFs, which are used in half of all classical biological control programmes [305]. Several attributes of EPFs make them ideal candidates for biological control agents: many can be mass cultured in vitro, fungal spores have a long shelf life compared to other biological control agents, and they are often capable of persisting in the host population without repeated introductions [306]. *Metarhizium anisopliae*, *Beauveria bassiana*, and *Entomophaga maimaiga* are among the most commercially important soil-dwelling Hypocreales. Entomopathogenic nematodes (EPNs) from the Steinernematidae and Heterorhabditidae families, as well as their bacterial symbionts *Xenorhabdus* and *Photorhabdus*, are essential biological control agents for insects. EPNs infect a wide spectrum of hosts, kill quickly, and actively seek for new victims to infect. They do, however, have a limited shelf life and are susceptible to environmental factors such as low moisture levels and ultraviolet (UV) radiation [303]. Entomopathogens have several advantages over conventional insecticides as biological control agents: they are often host-specific, reducing non-target species impact, and are usually harmless to humans; they can be used in organic farming and Integrated Pest Management Programmes; and they are less susceptible to pest resistance issues than conventional products. They are, however, difficult and costly to mass produce, as many of them require a host to complete their life cycle; they are particularly sensitive to environmental variables such as sunlight and humidity; and they have a short shelf life. *B. thuringiensis* (Bt), a Gram-positive, spore-forming species often found in soils, is the source of the majority of bacterial insecticidal compounds (Table 1.1). In 1901, Bt was identified as the cause of sotto sickness in silkworms (*Bombyx mori*) [306]. However, until the 1950s, when it was found that the Cry proteins of certain strains are poisonous to insects, the mechanism of Bt-induced insect death was unknown [313]. Until 1995, when transgenic crops expressing Cry proteins were commercially accessible, Bt remained a niche product utilized exclusively in forestry and organic farming. Bt application was fundamentally revolutionized with the development of Bt transgenic crops: the toxin became constitutively produced, covered all tissues, and

Table 1.1 List of some of entomopathogens.

Pathogen	Target pests	Mechanism	Virulence factors	Commercial application	References
Bacillus thuringiensis various strains	Lepidoptera, Coleoptera, and Diptera	Colonization and perforation of midgut epithelium	Cry proteins; Vegetative Insecticidal Proteins (VIP)	Vegetative Insecticidal Proteins (VIP); Spore suspension (liquid and granular); Cry protein expression in GM crops	[307]
Serratia entomophila	*Costelytra giveni*	Colonization of foregut and cessation of feeding	sepA, sepB, and sepC on pADAP	Soil-applied granules	[308]
Bacillus popilliae; *Bacillus lentimorbus*	*Popillia japonica*	Spores germinate in the larval gut; vegetative cells penetrate into the haemocoel leading to death			[309]
Beauveria bassiana	Aphids, fungus gnats, mealy bugs, mites, thrips, and whiteflies	Conidia land on host and germinate into cuticle; mortality caused by fungal toxicity and colonization of vital tissues		Spore suspension; promotion of existing natural populations	[307]
Verticillium lecanii	*Trialeurodes vaporariorum*, Hemiptera: Aphididae	Conidia land on host and germinate into cuticle; mortality caused by fungal toxicity and colonization of vital tissues		Spore suspension	[310]

(*Continued*)

Table 1.1 (Continued)

Pathogen	Target pests	Mechanism	Virulence factors	Commercial application	References
Nosema locustae	European corn borer caterpillars, grasshoppers, and mormon crickets	Infects fat body tissues, disrupting host metabolism and energy storage		Spores applied to bait	[311]
Ns NPV (*Neodiprion sertifer* nuclear polyhedrosis virus)	Pine sawfly	Capsid dissolves in the alkaline midgut to release the virus particle causing cell lysis	Polyhedral capsid	Suspension concentration of virus	[312]

non-target effects were eliminated because only insects that devour the plants were harmed [314]. Bt maize, cotton, soy, and eggplant types are extensively farmed now, and new Bt varieties are being created all the time [315]. Bt maize and cotton types are grown on more than 80% of maize and cotton land in the United States [316]. Moreover, approximately 300 different variants of the cry gene have been found so far, with specificity for lepidopteran, dipteran, coleopteran, or nematode hosts [306]. However, the evolution of Bt resistance has posed a danger to the efficacy of Bt crops in recent years, highlighting the relevance of insect resistance management (IRM) and the need to find novel microbial insecticides. Spinosad is a commercial pesticide developed from microbial NPs that has proven to be effective. It is the outcome of Dow Agrosciences' natural product discovery initiative, which began in the mid-1980s. The larval mosquitoes and Lepidoptera were exposed to fermentation broths made from soil samples as part of the programme. Insecticidal activity was found in a soil sample taken from a Virgin Islands rum distillery, which was shown to be caused by secondary metabolites generated by an unknown actinomycete eventually identified as *Saccharopolyspora spinosa* [317]. The spinosyns, which are metabolites of nicotinic acetylcholine and gamma-aminobutyric acid (GABA) receptors, have been discovered to be broad-spectrum insecticides with minimal vertebrate toxicity and unique method of action. Spinosad, a formulation that combines spinosyn A and spinosyn D in a 5 : 1 ratio, was authorized for over 150 conventional and organic crops in the United States by 2004 [318]. This small number of microbial insecticides is unlikely to reflect the full variety of microbial insecticidal compounds. Insecticidal screenings of microbial NPs are uncommon in the literature; the bulk of NP discovery efforts are focused on potential medicines. Furthermore, because of the rapid pace of rediscovery of known compounds over the last 30 years, detecting new NPs via traditional, culture-dependent screens has become more difficult [319]. Given that the fraction of successfully cultivated microbial species is typically considered to be less than 1%, many bioactive microbial NPs are likely to remain undiscovered [307]. In the realm of NP discovery, the emergence of culture-independent, high-throughput metagenomics approaches has ushered in a revival. Extracting DNA from environmental samples, cloning onto plasmids, and expressing in recombinant expression systems might be used to 'mine' microbial toxins. This strategy, however, implies that the expression system has the molecular engineering required to heterologously express and ply the desired genetic products, including regulatory elements (for example, promotors and ribosome sites), chemicals, and biosynthetic precursors. For example, a Gabor et al. [320] investigation recovered only <40% of enzyme activities in the *E. coli* vector by shotgun cloning of environmentally microbial DNA. This problem can be avoided by doing functional tests in several expression systems, often by using a shuttle vector that is particularly developed [319].

1.6.6 Microbial Semiochemicals

To adapt their comportments in reaction to the environment, insects rely on chemical communication. These include foraging, matching, risk prevention, identification of children, and social interactions. Semiochemicals are defined as substances that transmit signal between organisms that cause changes in the behaviour of recipients. The use of these semiochemicals to modify insect behaviour is a sustainable technique for plague management. Pheromones of the insects represent a big market share of attractants for plant and urban pests, and are classified as semiochemical products that promote inter-specific contact. Both the monitoring and management of a range of agricultural and forestry and urban pests, such as the *Anthonomus grandis* cotton boll, the *L. dispar* gypsum moth, and Musca home-fly, are utilized for pheromone boiled traps [320–323]. Field applications of synthetic pheromones also showed that multiple lepidopteran pesticides, including *Cydia pomonella* codling moth, *Pectinophora gossypiella* rose bollworm, and diamond bolt moth, can be effectively controlled through an impediment to their mates, a process called a disruption of the mating [324]. Insect pheromones are not many commercially utilized semi-organic products, but organic plant or microbial volatile compounds (mVOCs) [325]. VOCs are a vast family of low-molecular weight chemicals that are characterized by low vapour and high volatility. VOCs may be made by various metabolisms, such as fermentation, catabolism, and reduction of sulphur and terpenoid biosynthesis [325, 326]. A lot of VOCs are efficient semiochemicals that only require little concentrations to cause insect reactions, which have very sensitive (particularly olfactory) chemosensory systems. VOCs enable broad communication through air transport and act as significant methods for the sense of the environment. The discovery and deployment of mVOCs has recently become more and more relevant as a tool for pest management. This tendency has been supported by several causes. First, an online library of over 2000 microbial volatiles from about 1000 microbial species is reported as mVOC 2.0 [326, 327]; <10% of known mVOCs are studied or assigned with functions. Since an estimated 1018 microbial species are expected to exist on earth, many VOCs still remain to be identified. Second, the demand for NPs is growing in the field of agriculture pesticide management, because they are generally better perceived by the public in terms of safety and environmental sustainability than synthetic chemicals.

Over three billion years, microbes have existed and their metabolism has been very varied and dynamic through evolutionary processes. The metabolism of microbiomes often goes much beyond their eukaryotic hosts' capacity. Genetic engineering techniques such as recombinant and synthetic approaches can also be used to improve microbial metabolism. This means that microbial production of desirable metabolites (including mVOCs) may easily be scaled up to suit commercial requirements. In a variety of ways, advances in 'omics' tools and computational biology for studying microbial metabolism can aid in the discovery of mVOCs. High-throughput sequencing, in particular, is shedding light on the metabolic potential of microbiomes found in plants, soils, and other natural resources that are not farmed. Assembling complicated metagenomes and metatranscriptomes as well as predicting biosynthetic gene clusters from large data sets of DNA/RNA sequences are all becoming more computationally feasible [328, 329].

Biologically active NPs can be obtained from uncultured microbes by cloning environmental DNA into plasmid vectors and expressing in recombinant systems such as *E. coli*, yeast, or baculovirus [329]. In recent years, this technique has been used to screen for new antibiotics and medicinal medications, and it may theoretically be extended to the discovery of microbial-based agricultural compounds, such as novel semiochemicals and biopesticides [326]. Microbes can impact insect behaviour through modifying plant VOC emission patterns, in addition to directly affecting insects. In response to stimulation by JA, a key plant hormone involved in anti-pest defence, growth, and development, lima bean plants colonized with Rhizobia produced a distinct VOC profile from non-colonized plants [330]. Increased indole production was related to the unique VOC profile of colonized plants, which corresponded to considerably stronger *Epilachna varivestis* (Mexican bean beetle) repellency [330]. Several symbiotic fungi, such as endophytic fungi and arbuscular mycorrhizas, have also been demonstrated to influence plant volatile emissions and, as a result, plant susceptibility to certain insect pests [331, 332]. These findings suggest that symbiotic microorganisms may have indirect positive impacts on plant defence that may be turned into plant probiotics. Some pathogenic microorganisms, on the other hand, are known to alter plant volatile emissions in order to attract pests and vectors. Huanglongbing (HLB), commonly known as citrus greening disease, is caused by the bacteria *Candidatus Liberibacter asiaticus*, which infects citrus trees. Citrus trees infected with *Candidatus L. asiaticus* produced more methyl salicylate, which was thought to attract the Asian citrus psyllid (*Diaphorina citri*), its carrier insect [333]. A variety of plant viruses, including the Cucumber mosaic virus and the Barley yellow dwarf luteovirus, have been discovered to alter plant volatile emissions in order to attract their vector insects (aphids) and therefore promote transmission [334]. Laboratory research and field experiments have supported the use of microbial-based attractants and repellents in pest management. The invasive *D. suzukii*, for example, is a major pest of tiny fruits, and it differs from other non-pest *Drosophila* species in that it affects intact, mature, and maturing fruits [335]. *Drosophila* species are attracted to fermentation products [335], and adding fermenting sugar–yeast combination to a popular non-microbial trap bait (apple cider vinegar + 10% ethanol) boosted *D. suzukii* capture rate by up to 15-fold in both laboratory and field circumstances [336]. In addition, it has been observed that yeast wine–vinegar–sugar mixes are more effective attractants than wine–vinegar alone. Ishii et al. [337] genetically engineered the acetic acid bacteria (AAB) *Komagataeibacter europaeus* to produce more acetoin (a major element in vinegar) and found that the mutant strain captured substantially more *Drosophila* than other strains put on sticky traps. Tephritid fruit flies are attracted to microbial semiochemicals in the same way [338–340]. For example, alcohols, ketones, pyrazines, phenols, and acids were found in *Klebsiella pneumoniae* fermented broth that attracted the Mexican fruit fly (*Anastrepha ludens*) [333]. Female Caribbean fruit flies (*Anastrepha suspensa*) were shown to be attracted to the VOCs 3-methyl-1-butanol and ammonia released by *Enterobacter agglomerans*, a bacterium isolated from larval-infected fruits and fly adults [340]. Insect repellents might be made using some mVOCs, particularly those generated by fungus. Mould emits geosmin, which *D. melanogaster* detects and avoids [338]. It was eventually discovered that it causes unpleasant behaviours in SWD, and that it may be used to prevent oviposition on crops [341]. The development of particular fungus species (*Phoma* spp., *Fusarium* spp., or *Rhizopus* spp.) on chicken faeces inhibited oviposition by the housefly, *Musca domestica*, and the impact was linked to dimethyl trisulphide and 2-phenylethanol emission [342]. VOCs might come from plant-associated symbionts as well. *Muscodor vitigenus*, for example, generates naphthalene and repels insects as an endophyte of the tropical liana *Paullinia paullinoides* [343]. Microbial catalysis of repellent chemicals has been established in fungi such as *Penicillium*, *Aspergillus*, and *Fusarium*, which can convert dihydrojasmone and *cis*-jasmone to bioactive molecules like [+]-[*R*]-4-hydroxydihydrojasmone and [−]-4-hydroxyjasmone [344]. In a food-choice behavioural experiment, these chemicals were found to repel aphids. [±]-β-aryl-γ-et hylidene-γ-lactones are the compounds that have previously been determined to have phagodeterrent, antifungal, antibacterial, and anticancer activities, which were enantioselectively hydrolysed by a strain of *Aspergillus ochraceus* to form [−]-[*S*]-γ-ethylidene-γ-lactones and [+]-[*R*]-γ-ketoacids [345]. The smaller mealworm *Alphitobius diaperinus* was used to test the phagodeterrent/repellent capabilities of these derived compounds. Many entomopathogenic bacteria, including *B. thuringiensis* (Bt), *Pseudomonas entomophila*, and nematode-associated *Xenorhabdus* and *Photorhabdus* species, have been proven to repel insects [346]. A recent study found that a cocktail of bioactive secondary metabolites (mainly fabclavines) derived from *Xenorhabdus budapestensis* was more effective than DEET (chemical name, N,N-diethyl-meta-toluamide) in repelling *A. aegypti* and required a lower dose to produce the same repellent effect [347]. Leroy et al. [347] have documented several examples of microbial semiochemicals and their interactions with insects. Scientists will be able to make reliable inferences about microbial regulatory networks in mVOC generation by identifying distinct pathways from large-scale omics data. Dissecting synergistic multi-species interactions (i.e. consortia of microorganisms) in metabolite synthesis is another developing subject [348]. These findings suggest that microbial-based insect attractants or repellents might lead to more diversified, effective, and versatile agricultural products. Incorporating microbial-based semiochemicals into push–pull tactics is one of the applications. A target pest is repelled or discouraged away ('pull') from a protected source (a valuable crop or farm animal) using unpleasant stimuli in a standard push–pull system. In the meantime, the pest is drawn away from the protected supply by appealing stimuli ('pull'). Different crop plants (intercrops or trap crops), visual cues, pheromones, synthetic volatiles, phagostimulants/repellents, and anti-feedants are all typical stimuli [349]. To increase the efficacy of future push–pull techniques, microbial products can be utilized instead of or in addition to these stimuli.

1.6.7 Combining Microbial-Based Biopesticides with Nanotechnologies

Nanotechnology advancements are supporting the successful and long-term usage of biopesticides in field applications. When entomopathogenic bacteria, such as Bt and *Photorhabdus luminescens*, were administered to insect pests in the form of nanoparticles, their efficacy rose dramatically [350]. It has been found that larvae exposed to Bt nanoparticles died quicker and more often than larvae exposed to un-homogenized Bt powder, owing to greater Cry toxin solubility in the alkaline midguts. Furthermore, the development of nanoformulation delivery technologies such as nanoemulsion, nanocapsule, and nanosuspension has improved the longevity and stability of biopesticides in a variety of environments, including UV radiation and humidity [351]. Because they are frequently constructed of biodegradable natural polymers, these delivery techniques are environmentally beneficial. Smart nanopesticides are an emerging subject stemming from nanoformulation research, in which active chemicals are contained in stimulus-responsive carriers and their release is regulated by stimuli like light, temperature, humidity, or pH [352, 353]. The stimuli-responsive administration strategy allows for more accurate spatial administration while reducing pesticide dosage and frequency, resulting in less environmental effect. For VOCs, nanoformulations offer the potential to fine-tune the compounds' thermal stability, resulting in optimal release and semiochemical lifetime. There have been several comprehensive reviews written on various types of nanopesticides and stimuli-responsive delivery systems [339]. The application of nanotechnology in agrochemicals is still in its early stages. Commercial items in this category, on the other hand, are starting to appear on the market. Seltima R, a fungicide introduced by BASF in 2016 to protect rice crops, is one example. Seltima encapsulates pyraclostrobin, a fungicide agent that is very poisonous to aquatic organisms, using humidity-responsive polymers. This encapsulation is water resistant, enabling for regulated delivery of the fungicide to the rice leaf surface while reducing pollution of the aquatic habitats nearby. There will be unanticipated barriers in the processing, storage, and distribution of these products for agricultural purposes, based on the anticipation that new microorganisms and microbiome-derived compounds will be found at an exponential rate. The combination of microbial products with nanotechnology offers a number of options for improving efficacy and stability while also regulating environmental dispersal. Microbiome mining and nanotechnology, when coupled, have a lot of promise for developing breakthrough bioinsecticide applications in the near future.

1.6.8 Microbial Interventions to Improve Fitness of Mass-Reared Insects for Autocidal Programmes

Engel and Moran [193] have described probiotics ways for promoting beneficial insect populations (e.g. pollinators, natural enemies of pests). The IIT (discussed in Section 1.6.1) and the sterile insect technique (SIT) are two microbiological techniques for promoting insect development for autocidal programmes. SIT is an environmentally acceptable insect pest management technology that involves mass-rearing the target pests and generating sterile male insects using non-lethal ionizing radiation exposure [354]. The pest population is then suppressed or eradicated by releasing large numbers of totally infertile males to mate with wild females [354]. Inherited sterility (IS), a derivation of the SIT, was created for insect pests, mostly lepidopterans, which require a high-radiation dosage to establish total sterility. In IS, partly sterile males are created by exposing them to sub-sterilizing amounts of radiation and then mating them with wild females, resulting in low egg viability and mostly male progeny [355]. This method eliminates the negative effects of high-dose radiation on the insects. For some lepidopteran (moth) species, studies have demonstrated that males with hereditary sterility reduce wild populations more successfully than entirely sterile males. The primary screwworm (*Cochliomyia hominivorax*), the medfly (*C. capitata*), the Mexican fruit fly (*A. ludens*), several Lepidoptera (moths), and tsetse flies have all been successfully controlled by SIT over the years [356]. However, because huge numbers of treated insects must be released to compete with wild males, and often repeated releases are required, the expense of pest management through autocidal programmes is a substantial drawback. SIT necessitates a ratio of up to 100 sterile insects per wild insect. Although there are fewer data on IIT, a study on *Wolbachia*-infected mosquitoes found that releasing sterile to wild males at a 10 : 1 ratio was enough to suppress the local mosquito population [357]. Irradiation and handling in mass raising facilities further jeopardize insect survival and performance, adding to the high expense of autocidal operations. The alteration of mass-reared and irradiated males' gut microbiomes has been associated with lower performance, such as competence in attracting and mating with wild females. Laboratory-reared agricultural pests and disease vectors have fewer gut microbial taxa than their wild counterparts [358]. Insects are thought to obtain their gut bacteria from their natural habitats and food sources; however, these bacteria are usually absent from the artificial environment and diets, resulting in a loss in species richness or variety [359]. Furthermore, the insect gut microbiome composition is subjected to a different selection pressure due to artificial rearing. Antimicrobials are routinely added as preservatives to insect diets in the laboratory, for example. As demonstrated by the *B. oleae* example, these antimicrobials are capable of wiping out the majority of bacteria normally associated with insects. If the microbiome is seen as an ecosystem, decreases in species richness or variety may result in poor host outcomes, particularly if the bacteria offer specialized services to the host. While there is a substantial amount of literature on the effects of artificial rearing on the microbiomes of various insects, nothing is known about how irradiation affects the microbiome of insects. Because numerous prominent pests belonging to this category have been the focus of SIT applications, the majority of the study has been undertaken on them. In medflies, recently eclosed irradiated males showed decreased levels of some prominent gut bacteria (particularly *Klebsiella* sp.) but a greater level of *Pseudomonas* sp., which was not found in wild flies [360]. Following a few days after irradiation, however, some of these bacteria appeared to rebound in numbers. Irradiation therapy increased the number of *Lactococcus* and *Orbus* but decreased *Lactobacillus*, members of the Orbacecae family, and *Morganella* in the oriental fruit fly [361]. More study is needed to figure out how these temporary alterations in the gut microbiota affect fly recovery after irradiation, as well as their long-term implications on fly fitness. Scientists have been looking at the impact of nutritional supplementation with bacteria on the performance of irradiated tephritid fruit flies, based on the notion that reversing the microbiome alterations associated with artificial rearing or irradiation therapy might increase the insects' fitness. The variations in these trials show that the effects of probiosis therapy for insects such as host age, genotype, microbial strain, nutrition, and raising conditions may be affected by diverse aspects. It is also worth mentioning that previous research has only looked at single bacteria or bacterial mix supplements. Other methods for manipulating the microbiomes of artificially grown or irradiated insects are still being investigated. Microbiome transplanting, either by growing the treated insects on a diet seeded with excrement from newly caught wild insects or by co-housing treated insects with wild insects, might be one way; however, it would be difficult to scale up for mass-rearing. Another alternative is to change the diet formula to encourage the growth of beneficial microorganisms in the insect gut ('prebiotics'), which would include adjusting the nutritional content and maybe integrating natural food ingredients into their meals. It is obvious that further study still has to be done to determine the fitness of sterile, artificially grown probiotics and to find criteria for treatment that impact the results of insects in response to the various ways indicated. Insects are good models for dissecting the intricacies of microbial interactions and processes from a fundamental research standpoint, because their microbiomes are often less diversified than in mammalian systems, where much probiotics research has been undertaken.

References

1 Chang, C., Bowman, J.L., and Meyerowitz, E.M. (2016). Field guide to plant model systems. *Cell* 167 (2): 325–339. https://doi.org/10.1016/j.cell.2016.08.031 PMID: 27716506.

2 Stanton-Geddes, J., Paape, T., Epstein, B. et al. (2013). Candidate genes and genetic architecture of symbiotic and agronomic traits revealed by whole-genome, sequence based association genetics in Medicago truncatula. *PLoS One* 8 (5): e65688. https://doi.org/10.1371/journal.pone.0065688 PMID: 23741505.

3 Hacquard, S. and Schadt, C.W. (2015). Towards a holistic understanding of the beneficial interactions across the populus microbiome. *New Phytol.* 205 (4): 1424–1430. https://doi.org/10.1111/nph.13133 PMID: 25422041.

4 Spence, C., Alff, E., Johnson, C. et al. (2014). Natural rice rhizospheric microbes suppress rice blast infections. *BMC Plant Biol.* 14: 130. https://doi.org/10.1186/1471-2229-14-130 PMID: 24884531.

5 Ramond, J.-B., Tshabuse, F., Bopda, C.W. et al. (2013). Evidence of variability in the structure and recruitment of rhizospheric and endophytic bacterial

communities associated with arable sweet sorghum (Sorghum bicolor (L) Moench). *Plant Soil* 372 (1–2): 265–278.
6 Li, D., Voigt, T.B., and Kent, A.D. (2016). Plant and soil effects on bacterial communities associated with Miscanthus X giganteus Rhizosphere and Rhizomes. *GCB Bioenergy* 8 (1): 183–193.
7 Peiffer, J.A., Spor, A., Koren, O. et al. (2013). Diversity and heritability of the maize rhizosphere microbiome under field conditions. *Proc. Natl. Acad. Sci.* 110 (16): 6548–6553.
8 Tian, B.-Y., Cao, Y., and Zhang, K.-Q. (2015). Metagenomic insights into communities, functions of endophytes, and their associates with infection by root-knot nematode, meloidogyne incognita, in tomato roots. *Sci. Rep.* 5: 17087.
9 Miedes, E., Vanholme, R., Boerjan, W., and Molina, A. (2014). The role of the secondary cell wall in plant resistance to pathogens. *Front. Plant Sci.* 5: 358.
10 Yeats, T.H. and Rose, J.K.C. (2013). The formation and function of plant cuticles. *Plant Physiol.* 163: 5–20.
11 VanEtten, H.D., Mansfield, J.W., Bailey, J.A., and Farmer, E.E. (1994). Two classes of plant antibiotics: phytoalexins versus "phytoanticipins". *Plant Cell* 6: 1191.
12 Boller, T. and Felix, G. (2009). A renaissance of elicitors: perception of microbe-associated molecular patterns and danger signals by pattern-recognition receptors. *Annu. Rev. Plant Biol.* 60: 379–406.
13 Macho, A.P. and Zipfel, C. (2014). Plant PRRs and the activation of innate immune signaling. *Mol. Cell* 54: 263–272.
14 Chinchilla, D., Bauer, Z., Regenass, M. et al. (2006). The Arabidopsis receptor kinase FLS2 binds flg22 and determines the specificity of flagellin perception. *Plant Cell* 18: 465–476.
15 Nam, K.H. and Li, J. (2002). BRI1/BAK1, a receptor kinase pair mediating brassinosteroid signaling. *Cell* 110: 203–212.
16 Fliegmann, J. and Felix, G. (2016). Immunity: flagellin seen from all sides. *Nat. Plants* 2: 16136.
17 Katsuragi, Y., Takai, R., Furukawa, T. et al. (2015). CD2-1, the C-terminal region of flagellin, modulates the induction of immune responses in rice. *Mol. Plant-Microbe Interact.* 28: 648–658.
18 Hirai, H., Takai, R., Iwano, M. et al. (2011). Glycosylation regulates specific induction of rice immune responses by Acidovorax avenae flagellin. *J. Biol. Chem.* 286: 25519–25530.
19 Gomez-Gomez, L., Felix, G., and Boller, T. (1999). A single locus determines sensitivity to bacterial flagellin in Arabidopsis thaliana. *Plant J.* 18: 277–284.
20 Garrido-Oter, R., Nakano, R.T., Dombrowski, N. et al. (2018). Modular traits of the Rhizobiales root microbiota and their evolutionary relationship with symbiotic rhizobia. *Cell Host Microbe* 24: 155–167.e5.

21 Libault, M., Farmer, A., Brechenmacher, L. et al. (2010). Complete transcriptome of the soybean root hair cell, a single-cell model, and its alteration in response to Bradyrhizobium japonicum infection. *Plant Physiol.* 152: 541–552.

22 Gourion, B., Berrabah, F., Ratet, P., and Stacey, G. (2015). Rhizobiumlegume symbioses: the crucial role of plant immunity. *Trends Plant Sci.* 20: 186–194.

23 Singh, S. and Parniske, M. (2012). Activation of calcium- and calmodulin-dependent protein kinase (CCaMK), the central regulator of plant root endosymbiosis. *Curr. Opin. Plant Biol.* 15: 444–453.

24 Buendia, L., Wang, T., Girardin, A., and Lefebvre, B. (2016). The LysM receptor-like kinase SlLYK10 regulates the arbuscular mycorrhizal symbiosis in tomato. *New Phytol.* 210: 184–195.

25 Zhang, X., Dong, W., Sun, J. et al. (2015). The receptor kinase CERK1 has dual functions in symbiosis and immunity signalling. *Plant J.* 81: 258–267.

26 Rey, T., Chatterjee, A., Buttay, M. et al. (2015). Medicago truncatula symbiosis mutants affected in the interaction with a biotrophic root pathogen. *New Phytol.* 206: 497–500.

27 Rey, T., Nars, A., Bonhomme, M. et al. (2013). NFP, a LysM protein controlling Nod factor perception, also intervenes in Medicago truncatula resistance to pathogens. *New Phytol.* 198: 875–886.

28 Smakowska-Luzan, E., Mott, G.A., Parys, K. et al. (2018). An extracellular network of Arabidopsis leucine-rich repeat receptor kinases. *Nature* 553: 342–346.

29 Jourdan, E., Henry, G., Duby, F. et al. (2009). Insights into the defenserelated events occurring in plant cells following perception of surfactin-type lipopeptide from Bacillus subtilis. *Mol. Plant-Microbe Interact.* 22: 456–468.

30 Han, S., Li, D., Trost, E. et al. (2016). Systemic responses of barley to the 3-hydroxy-decanoyl-homoserine lactone producing plant beneficial endophyte Acidovorax radicis N35. *Front. Plant Sci.* 7: 1868.

31 Mathesius, U., Mulders, S., Gao, M. et al. (2003). Extensive and specific responses of a eukaryote to bacterial quorum-sensing signals. *Proc. Natl. Acad. Sci. U. S. A.* 100: 1444–1449.

32 Cha, C., Gao, P., Chen, Y.C. et al. (1998). Production of acyl-homoserine lactone quorum-sensing signals by gram-negative plant-associated bacteria. *Mol. Plant-Microbe Interact.* 11: 1119–1129.

33 von Bodman, S.B., Bauer, W.D., and Coplin, D.L. (2003). Quorum sensing in plant-pathogenic bacteria. *Annu. Rev. Phytopathol.* 41: 455–482.

34 Schikora, A., Schenk, S.T., and Hartmann, A. (2016). Beneficial effects of bacteria-plant communication based on quorum sensing molecules of the N-acyl homoserine lactone group. *Plant Mol. Biol.* 90: 605–612.

35 Tanaka, K., Cho, S.H., Lee, H. et al. (2015). Effect of lipochitooligosaccharide on early growth of C4 grass seedlings. *J. Exp. Bot.* 66: 5727–5738.

36 Glazebrook, J. (2005). Contrasting mechanisms of defense against biotrophic and necrotrophic pathogens. *Annu. Rev. Phytopathol.* 43: 205–227.

37 Nguyen, D., Rieu, I., Mariani, C., and van Dam, N.M. (2016). How plants handle multiple stresses: hormonal interactions underlying responses to abiotic stress and insect herbivory. *Plant Mol. Biol.* 91: 727–740.

38 Yoneyama, K., Xie, X., Kim, H.I. et al. (2012). How do nitrogen and phosphorus deficiencies affect strigolactone production and exudation? *Planta* 235: 1197–1207.

39 Katsir, L., Schilmiller, A.L., Staswick, P.E. et al. (2008). COI1 is a critical component of a receptor for jasmonate and the bacterial virulence factor coronatine. *Proc. Natl. Acad. Sci. U. S. A.* 105: 7100–7105.

40 Wasternack, C. and Hause, B. (2013). Jasmonates: biosynthesis, perception, signal transduction and action in plant stress response, growth and development. An update to the 2007 review in Annals of Botany. *Ann. Bot.* 111: 1021–1058.

41 Ma, K.-W. and Ma, W. (2016). Phytohormone pathways as targets of pathogens to facilitate infection. *Plant Mol. Biol.* 91: 713–725.

42 Singh, U.P., Sarma, B.K., and Singh, D.P. (2003). Effect of plant growthpromoting Rhizobacteria and culture filtrate of Sclerotium rolfsii on phenolic and salicylic acid contents in chickpea (Cicer arietinum). *Curr. Microbiol.* 46: 131–140.

43 Li, Y., Gu, Y., Li, J. et al. (2015). Biocontrol agent Bacillus amyloliquefaciens LJ02 induces systemic resistance against cucurbits powdery mildew. *Front. Microbiol.* 6: 883.

44 Srivastava, S., Chaudhry, V., Mishra, A. et al. (2012). Gene expression profiling through microarray analysis in Arabidopsis thaliana colonized by Pseudomonas putida MTCC5279, a plant growth promoting rhizobacterium. *Plant Signal. Behav.* 7: 235–245.

45 Pinedo, I., Ledger, T., Greve, M., and Poupin, M.J. (2015). Burkholderia phytofirmans PsJN induces long-term metabolic and transcriptional changes involved in Arabidopsis thaliana salt tolerance. *Front. Plant Sci.* 6: 466.

46 Bodenhausen, N., Bortfeld-Miller, M., Ackermann, M., and Vorholt, J.A. (2014). A synthetic community approach reveals plant genotypes affecting the phyllosphere microbiota. *PLoS Genet.* 10: e1004283.

47 Lebeis, S.L., Paredes, S.H., Lundberg, D.S. et al. (2015). PLANT MICROBIOME. Salicylic acid modulates colonization of the root microbiome by specific bacterial taxa. *Science* 349: 860–864.

48 Clark, N.M., de Luis Balaguer, M.A., and Sozzani, R. (2014). Experimental data and computational modeling link auxin gradient and development in the Arabidopsis root. *Front. Plant Sci.* 5: 328.

49 Liao, C.Y., Smet, W., Brunoud, G. et al. (2015). Reporters for sensitive and quantitative measurement of auxin response. *Nat. Methods* 12: 207–210.

50 Fendrych, M., Leung, J., and Friml, J. (2016). TIR1/AFB-Aux/IAA auxin perception mediates rapid cell wall acidification and growth of Arabidopsis hypocotyls. *elife* 5: e19048.

51 Castello, M.J., Medina-Puche, L., Lamilla, J., and Tornero, P. (2018). NPR1 paralogs of Arabidopsis and their role in salicylic acid perception. *PLoS One* 13: e0209835.

52 Magne, K., George, J., Berbel Tornero, A. et al. (2018). Lotus japonicus NOOT-BOPCOCH- LIKE1 is essential for nodule, nectary, leaf and flower development. *Plant J.* 94: 880–894.

53 Seyfferth, C. and Tsuda, K. (2014). Salicylic acid signal transduction: the initiation of biosynthesis, perception and transcriptional reprogramming. *Front. Plant Sci.* 5: 697.

54 Tsuda, K., Sato, M., Stoddard, T. et al. (2009). Network properties of robust immunity in plants. *PLoS Genet.* 5: e1000772.

55 Hillmer, R.A., Tsuda, K., Rallapalli, G. et al. (2017). The highly buffered Arabidopsis immune signaling network conceals the functions of its components. *PLoS Genet.* 13: e1006639.

56 Marin-de la Rosa, N. and Falter-Braun, P. (2015). Primer on protein–protein interaction maps. *eLS.* https://doi.org/10.1002/9780470015902.a0006205.pub2.

57 Consortium, Arabidopsis (2011). Evidence for network evolution in an "Arabidopsis" interactome map. *Science* 333: 601–607.

58 Jones, A.M., Xuan, Y., Xu, M. et al. (2014). Border control–a membrane-linked interactome of Arabidopsis. *Science* 344: 711–716.

59 Klopffleisch, K., Phan, N., Augustin, K. et al. (2011). Arabidopsis G-protein interactome reveals connections to cell wall carbohydrates and morphogenesis. *Mol. Syst. Biol.* 7: 532.

60 Li, S., Musungu, B., Lightfoot, D., and Ji, P. (2018). The interactomic analysis reveals pathogenic protein networks in Phomopsis longicolla underlying seed decay of soybean. *Front. Genet.* 9: 104.

61 Dou, D. and Zhou, J.M. (2012). Phytopathogen effectors subverting host immunity: different foes, similar battleground. *Cell Host Microbe* 12: 484–495.

62 Mukhtar, M.S., Carvunis, A.R., Dreze, M. et al. (2011). Independently evolved virulence effectors converge onto hubs in a plant immune system network. *Science* 333: 596–601.

63 Wessling, R., Epple, P., Altmann, S. et al. (2014). Convergent targeting of a common host protein-network by pathogen effectors from three kingdoms of life. *Cell Host Microbe* 16: 364–375.

64 Crua Asensio, N., Munoz Giner, E., de Groot, N.S., and Torrent Burgas, M. (2017). Centrality in the host-pathogen interactome is associated with pathogen fitness during infection. *Nat. Commun.* 8: 14092.

65 Miwa, H. and Okazaki, S. (2017). How effectors promote beneficial interactions. *Curr. Opin. Plant Biol.* 38: 148–154.

66 Berendsen, R.L., Vismans, G., Yu, K. et al. (2018). Disease-induced assemblage of a plant beneficial bacterial consortium. *ISME J.* 12: 1496–1507.

67 Okazaki, S., Kaneko, T., Sato, S., and Saeki, K. (2013). Hijacking of leguminous nodulation signaling by the rhizobial type III secretion system. *Proc. Natl. Acad. Sci. U. S. A.* 110: 17131–17136.

68 Schumacher, J., Waite, C.J., Bennett, M.H. et al. (2014). Differential secretome analysis of Pseudomonas syringae pv tomato using gel-free MS proteomics. *Front. Plant Sci.* 5: 242.

69 Hacquard, S., Spaepen, S., Garrido-Oter, R., and Schulze-Lefert, P. (2017). Interplay between innate immunity and the plant microbiota. *Annu. Rev. Phytopathol.* 55: 565–589.

70 Weiberg, A., Wang, M., Lin, F.M. et al. (2013). Fungal small RNAs suppress plant immunity by hijacking host RNA interference pathways. *Science* 342: 118–123.

71 An, Q., Ehlers, K., Kogel, K.H. et al. (2006). Multivesicular compartments proliferate in susceptible and resistant MLA12-barley leaves in response to infection by the biotrophic powdery mildew fungus. *New Phytol.* 172: 563–576.

72 Langfelder, P. and Horvath, S. (2008). WGCNA: an R package for weighted correlation network analysis. *BMC Bioinf.* 9: 559.

73 Saelens, W., Cannoodt, R., and Saeys, Y. (2018). A comprehensive evaluation of module detection methods for gene expression data. *Nat. Commun.* 9: 1090.

74 Kim, M.S., Zhang, H., Yan, H. et al. (2018). Characterizing co-expression networks underpinning maize stalk rot virulence in Fusarium verticillioides through computational subnetwork module analyses. *Sci. Rep.* 8: 8310.

75 Whipps, J.M. (1990). Carbon economy. In: *The Rhizosphere* (ed. J.M. Lynch), 59–97. Chichester, UK: Wiley.

76 Jacoby, R., Peukert, M., Succurro, A. et al. (2017). The role of soil microorganisms in plant mineral nutrition current knowledge and future directions. *Front. Plant Sci.* 8: 1617.

77 Topfer, N., Kleessen, S., and Nikoloski, Z. (2015). Integration of metabolomics data into metabolic networks. *Front. Plant Sci.* 6: 49.

78 Heavner, B.D. and Price, N.D. (2015). Comparative analysis of yeast metabolic network models highlights progress, opportunities for metabolic reconstruction. *PLoS Comput. Biol.* 11: e1004530.

79 Botero, D., Alvarado, C., Bernal, A. et al. (2018). Network analyses in plant pathogens. *Front. Microbiol.* 9 (35).

80 Mithani, A., Hein, J., and Preston, G.M. (2011). Comparative analysis of metabolic networks provides insight into the evolution of plant pathogenic and nonpathogenic lifestyles in *Pseudomonas*. *Mol. Biol. Evol.* 28: 483–499.

81 Rodenburg, S.Y.A., Seidl, M.F., de Ridder, D., and Govers, F. (2018). Genome-wide characterization of *Phytophthora infestans* metabolism: a systems biology approach. *Mol. Plant Pathol.* 19: 1403–1413.

82 Raghothama, K.G. (1999). Phosphate acquisition. *Annu. Rev. Plant Physiol. Plant Mol. Biol.* 50: 665–693.

83 Castrillo, G., Teixeira, P.J., Paredes, S.H. et al. (2017). Root microbiota drive direct integration of phosphate stress and immunity. *Nature* 543: 513–518.

84 Lu, Y.T., Li, M.Y., Cheng, K.T. et al. (2014). Transgenic plants that express the phytoplasma effector SAP11 show altered phosphate starvation and defense responses. *Plant Physiol.* 164: 1456–1469.

85 McGrann, G.R., Andongabo, A., Sjokvist, E. et al. (2016). The genome of the emerging barley pathogen Ramularia collo-cygni. *BMC Genomics* 17: 584.

86 Fiorilli, V., Vannini, C., Ortolani, F. et al. (2018). Omics approaches revealed how arbuscular mycorrhizal symbiosis enhances yield and resistance to leaf pathogen in wheat. *Sci. Rep.* 8: 9625.

87 van der Heijden, M.G.A. and Hartmann, M. (2016). Networking in the plant microbiome. *PLoS Biol.* 14 (2): e1002378. https://doi.org/10.1371/journal.pbio.1002378 PMID: 26871440.

88 Bai, Y., Muller, D.B., Srinivas, G. et al. (2015). Functional overlap of the arabidopsis leaf and root microbiota. *Nature* 528: 364–369.

89 Schlaeppi, K. and Bulgarelli, D. (2014). The plant microbiome at work. *Mol. Plant-Microbe Interact.* 28 (3): 212–217.

90 Nemergut, D.R., Schmidt, S.K., Fukami, T. et al. (2013). Patterns and processes of microbial community assembly. *Microbiol. Mol. Biol. Rev.* 77 (3): 342–356. https://doi.org/10.1128/MMBR.00051-12 PMID: 24006468.

91 Agler, M.T., Ruhe, J., Kroll, S. et al. (2016). Microbial hub taxa link host and abiotic factors to plant microbiome variation. *PLoS Biol.* 14 (1).

92 Bulgarelli, D., Rott, M., Schlaeppi, K. et al. (2012). Revealing structure and assembly cues for arabidopsis root-inhabiting bacterial microbiota. *Nature* 488 (7409): 91–95. https://doi.org/10.1038/nature11336. PMID: 22859207.

93 Lundberg, D.S., Lebeis, S.L., Paredes, S.H. et al. (2012). Defining the core Arabidopsis thaliana root microbiome. *Nature* 488 (7409): 86–90. https://doi.org/10.1038/nature11237. PMID: 22859206.

94 Bulgarelli, D., Schlaeppi, K., Spaepen, S. et al. (2013). Structure and functions of the bacterial microbiota of plants. In: *Annual Review of Plant Biology*, vol. 64 (ed. S.S. Merchant). Palo Alto: Annual Reviews, 807–838. https://doi.org/10.1146/annurev-arplant-050312-120106. PMID: 23373698.

95 Hacquard, S., Garrido-Oter, R., Gonzalez, A. et al. (2015). Microbiota and host nutrition across plant and animal kingdoms. *Cell Host Microbe* 17 (5): 603–616.

96 Vandenkoornhuyse, P., Quaiser, A., Duhamel, M. et al. (2015). The importance of the microbiome of the plant holobiont. *New Phytol.* 206 (4): 1196–1206. https://doi.org/10.1111/nph. 13312. PMID: 25655016.

97 Louca, S., Jacques, S.M.S., Pires, A.P.F. et al. (2016). High taxonomic variability despite stable functional structure across microbial communities. *Nat. Ecol. Evol.* 1: 15.

98 Truyens, S., Weyens, N., Cuypers, A., and Vangronsveld, J. (2015). Bacterial seed endophytes: genera, vertical transmission and interaction with plants. *Environ. Microbiol. Rep.* 7 (1): 40–50.

99 Ottesen, A.R., Gorham, S., Reed, E. et al. (2016). Using a control to better understand phyllosphere microbiota. *PLoS One* 11 (9): e0163482. https://doi.org/10.1371/journal.pone.0163482. PMID: 27669159.

100 Wagner, M.R., Lundberg, D.S., Coleman-Derr, D. et al. (2014). Natural soil microbes alter flowering phenology and the intensity of selection on flowering time in a wild arabidopsis relative. *Ecol. Lett.* 17 (6): 717–726. https://doi.org/10.1111/ele.12276. PMID: 24698177.

101 Mueller, U.G. and Sachs, J.L. (2015). Engineering microbiomes to improve plant and animal health. *Trends Microbiol.* 23 (10): 606–617. https://doi.org/10.1016/j.tim.2015.07.009. PMID: 26422463.

102 Ritpitakphong, U., Falquet, L., Vimoltust, A. et al. (2016). The microbiome of the leaf surface of arabidopsis protects against a fungal pathogen. *New Phytol.* 210 (3): 1033–1043.

103 Lakshmanan, V. (2015). Root microbiome assemblage is modulated by plant host factors. In: *Plant Microbe Interactions*, Advances in Botanical Research, vol. 75 (ed. H. Bais and J. Sherrier), 57–79.

104 Berg, G., Rybakova, D., Grube, M., and Koberl, M. (2016). The plant microbiome explored: implications for experimental botany. *J. Exp. Bot.* 67 (4): 995–1002. https://doi.org/10.1093/jxb/erv466. PMID: 26547794.

105 Lebeis, S.L., Paredes, S.H., Lundberg, D.S. et al. (2015). Salicylic acid modulates colonization of the root microbiome by specific bacterial taxa. *Science* 349 (6250): 860–864. https://doi.org/10.1126/science.aaa8764. PMID: 26184915.

106 Werner, G.D. and Kiers, E.T. (2015). Order of arrival structures arbuscular mycorrhizal colonization of plants. *New Phytol.* 205 (4): 1515–1524. https://doi.org/10.1111/nph.13092. PMID: 25298030.

107 Yuan, Z., Druzhinina, I.S., Labbe, J. et al. (2016). Specialized microbiome of a halophyte and its role in helping non-host plants to withstand salinity. *Sci. Rep.* 6: 32467.

108 Mendes, R. and Raaijmakers, J.M. (2015). Cross-kingdom similarities in microbiome functions. *ISME J.* 9: 1905–1907.

109 Lakshmanan, V., Selvaraj, G., and Bais, H.P. (2014). Functional soil microbiome: belowground solutions to an above ground problem. *Plant Physiol.* 166: 689–700.

110 Stuart, L.M., Paquette, N., and Boyer, L. (2013). Effector-triggered versus pattern-triggered immunity: how animals sense pathogens. *Nat. Rev. Immunol.* 13: 199–206.

111 Haldar, S. and Sengupta, S. (2015). Plant-microbe cross-talk in the rhizosphere: insight and biotechnological potential. *Open Microbiol. J.* 9: 1–7.
112 Chaparro, J.M., Badri, D.V., Bakker, M.G. et al. (2013). Root exudation of phytochemicals in Arabidopsis follows specific patterns that are developmentally programmed and correlate with soil microbial functions. *PLoS One* 8: e55731.
113 Akiyama, K., Matsuzaki, K., and Hayashi, H. (2005). Plant sesquiterpenes induce hyphal branching in arbuscular mycorrhizal fungi. *Nature* 435: 824–827.
114 Reinhold-Hurek, B., Bunger, W., Burbano, C.S. et al. (2015). Roots shaping their microbiome: global hotspots for microbial activity. *Annu. Rev. Phytopathol.* 53: 403–424.
115 Ottesen, A.R., Pena, A.G., and White, J.R. (2013). Baseline survey of the anatomical microbial ecology of an important food plant: Solanum lycopersicum (tomato). *BMC Microbiol.* 13: 114.
116 Mendes, R., Kruijt, M., and de Bruijn, I. (2011). Deciphering the rhizosphere microbiome for disease-suppressive bacteria. *Science* 332: 1097–1100.
117 Cordovez, V., Carrion, V.J., and Etalo, D.W. (2015). Diversity and functions of volatile organic compounds produced by Streptomyces from a disease-suppressive soil. *Front. Microbiol.* 6: 1081.
118 Ardanov, P., Sessitsch, A., Haggman, H. et al. (2012). Methylobacterium-induced endophyte community changes correspond with protection of plants against pathogen attack. *PLoS One* 7: e46802.
119 Mendes, R., Garbeva, P., and Raaijmakers, J.M. (2013). The rhizosphere microbiome: significance of plant beneficial, plant pathogenic, and human pathogenic microorganisms. *FEMS Microbiol. Rev.* 37: 634–663.
120 Jimenez, P.N., Koch, G., Thompson, J.A. et al. (2012). The multiple signaling systems regulating virulence in Pseudomonas aeruginosa. *Microbiol. Mol. Biol. Rev.* 76: 46–65.
121 Hardoim, P.R., van Overbeek, L.S., and Berg, G. (2015). The hidden world within plants: ecological and evolutionary considerations for defining functioning of microbial endophytes. *Microbiol. Mol. Biol. Rev.* 79: 293–320.
122 Berg, G., Krause, R., and Mendes, R. (2015). Cross-kingdom similarities in microbiome ecology and biocontrol of pathogens. *Front. Microbiol.* 6: 1311.
123 Keller, N.P., Turner, G., and Bennett, J.W. (2005). Fungal secondary metabolism – from biochemistry to genomics. *Nat. Rev. Microbiol.* 3: 937–947.
124 Yin, W. and Keller, N.P. (2011). Transcriptional regulatory elements in fungal secondary metabolism. *J. Microbiol.* 49: 329–339.
125 Demain, A.L. and Fang, A. (2000). The natural functions of secondary metabolites. *Adv. Biochem. Eng. Biotechnol.* 69: 1–39.
126 Bilyk, O. and Luzhetskyy, A. (2016). Metabolic engineering of natural product biosynthesis in actinobacteria. *Curr. Opin. Biotechnol.* 42: 98–107.

127 Tata, A., Perez, C., Campos, M.L. et al. (2015). Imprint desorption electrospray ionization mass spectrometry imaging for monitoring secondary metabolites production during antagonistic interaction of fungi. *Anal. Chem.* 87: 12298–12304.

128 Johnstone, T.C. and Nolan, E.M. (2015). Beyond iron: non-classical biological functions of bacterial siderophores. *Dalton Trans.* 44: 6320–6339.

129 Schmerk, C.L., Welander, P.V., and Hamad, M.A. (2015). Elucidation of the Burkholderia cenocepacia hopanoid biosynthesis pathway uncovers functions for conserved proteins in hopanoid-producing bacteria. *Environ. Microbiol.* 17: 735–750.

130 Malott, R.J., Steen-Kinnaird, B.R., Lee, T.D., and Speert, D.P. (2012). Identification of hopanoid biosynthesis genes involved in polymyxin resistance in Burkholderia multivorans. *Antimicrob. Agents Chemother.* 56: 464–471.

131 Lopez-Lara, I.M., Sohlenkamp, C., and Geiger, O. (2003). Membrane lipids in plant-associated bacteria: their biosyntheses and possible functions. *Mol. Plant-Microbe Interact.* 16: 567–579.

132 Chamoun, R., Aliferis, K.A., and Jabaji, S. (2015). Identification of signatory secondary metabolites during mycoparasitism of Rhizoctonia solani by Stachybotrys elegans. *Front. Microbiol.* 6: 353.

133 Prakash, V. (2015). Endophytic fungi as resource of bioactive compounds. *Int J Pharm. Bio. Sci* 6: 887–898.

134 Suryanarayanan, T.S. and Shaanker, R.U. (2015). Fungal endophytes – biology and bioprospecting preface. *Curr. Sci.* 109: 37–38.

135 Schulz, B. and Boyle, C. (2005). The endophytic continuum. *Mycol. Res.* 109: 661–686.

136 Strobel, G.A. (2003). Endophytes as sources of bioactive products. *Microbes Infect.* 5: 535–544.

137 Abe, N., Sugimoto, O., Arakawa, T. et al. (2001). Sorbicillinol, a key intermediate of bisorbicillinoid biosynthesis in Trichoderma sp USF-2690. *Biosci. Biotechnol. Biochem.* 65: 2271–2279.

138 Salas-Marina, M.A., Silva-Flores, M.A., Uresti-Rivera, E.E. et al. (2011). Colonization of Arabidopsis roots by Trichoderma atroviride promotes growth and enhances systemic disease resistance through jasmonic acid/ethylene and salicylic acid pathways. *Eur. J. Plant Pathol.* 131: 15–26.

139 Araujo, W.L., Marcon, J., Maccheroni, W. Jr. et al. (2002). Diversity of endophytic bacterial populations and their interaction with Xylella fastidiosa in citrus plants. *Appl. Environ. Microbiol.* 68: 4906–4914.

140 Lacava, P.T., Araujo, W.L., Marcon, J. et al. (2004). Interaction between endophytic bacteria from citrus plants and the phytopathogenic bacteria Xylella fastidiosa, causal agent of citrus-variegated chlorosis. *Lett. Appl. Microbiol.* 39: 55–59.

141 Dourado, M.N., Santos, D.S., Nunes, L.R. et al. (2015). Differential gene expression in Xylella fastidiosa 9a5c during co-cultivation with the endophytic bacterium Methylobacterium mesophilicum SR1.6/6. *J. Basic Microbiol.* 55: 1357–1366.

142 Araujo, W.L., Creason, A.L., and Mano, E.T. (2016). Genome sequencing and transposon mutagenesis of Burkholderia seminalis TC3.4.2R3 identify genes contributing to suppression of orchid necrosis caused by B. gladioli. *Mol. Plant-Microbe Interact.* 29: 435–446.

143 Kim, H.S., Schell, M.A., and Yu, Y. (2005). Bacterial genome adaptation to niches: divergence of the potential virulence genes in three Burkholderia species of different survival strategies. *BMC Genomics* 6: 174.

144 Sim, B.M., Chantratita, N., and Ooi, W.F. (2010). Genomic acquisition of a capsular polysaccharide virulence cluster by non-pathogenic Burkholderia isolates. *Genome Biol.* 11: R89.

145 Bradley, A.S., Pearson, A., Sáenz, J.P., and Marx, C.J. (2010). Adenosylhopane: the first intermediate in hopanoid side chain biosynthesis. *Org. Geochem.* 41: 1075–1081.

146 Welander, P.V., Hunter, R.C., Zhang, L. et al. (2009). Hopanoids play a role in membrane integrity and pH homeostasis in Rhodopseudomonas palustris TIE-1. *J. Bacteriol.* 191: 6145–6156.

147 Seipke, R.F. and Loria, R. (2009). Hopanoids are not essential for growth of Streptomyces scabies 87-22. *J. Bacteriol.* 191: 5216–5223.

148 Jones, D.S., Albrecht, H.L., and Dawson, K.S. (2012). Community genomic analysis of an extremely acidophilic sulfur-oxidizing biofilm. *ISME J.* 6: 158–170.

149 Muller, E.E., Hourcade, E., Louhichi-Jelail, Y. et al. (2011). Functional genomics of dichloromethane utilization in Methylobacterium extorquens DM4. *Environ. Microbiol.* 13: 2518–2535.

150 Nalin, R., Putra, S.R., Domenach, A.M. et al. (2000). High hopanoid/total lipids ratio in Frankia mycelia is not related to the nitrogen status. *Microbiology* 146: 3013–3019.

151 Kulkarni, G., Busset, N., and Molinaro, A. (2015). Specific hopanoid classes differentially affect free-living and symbiotic states of Bradyrhizobium diazoefficiens. *MBio* 6: e01251–e01315.

152 Morissette, D.C., Driscoll, B.T., and Jabaji-Hare, S. (2003). Molecular cloning, characterization, and expression of a cDNA encoding an endochitinase gene from the mycoparasite Stachybotrys elegans. *Fungal Genet. Biol.* 39: 276–285.

153 Morissette, D.C., Dauch, A., Beech, R. et al. (2008). Isolation of mycoparasitic-related transcripts by SSH during interaction of the mycoparasite Stachybotrys elegans with its host Rhizoctonia solani. *Curr. Genet.* 53: 67–80.

154 Martins, M.B. and Carvalho, I. (2007). Diketopiperazines: biological activity and synthesis. *Tetrahedron* 63: 9923–9932.

155 Traxler, M.F., Watrous, J.D., Alexandrov, T. et al. (2013). Interspecies interactions stimulate diversification of the Streptomyces coelicolor secreted metabolome. *MBio* 4.

156 Nutzmann, H.W., Reyes-Dominguez, Y., and Scherlach, K. (2011). Bacteria-induced natural product formation in the fungus Aspergillus nidulans requires Saga/Ada-mediated histone acetylation. *Proc. Natl. Acad. Sci. U. S. A.* 108: 14282–14287.

157 Schroeckh, V., Scherlach, K., and Nutzmann, H.W. (2009). Intimate bacterial–fungal interaction triggers biosynthesis of archetypal polyketides in Aspergillus nidulans. *Proc. Natl. Acad. Sci. U. S. A.* 106: 14558–14563.

158 Faraldo-Gomez, J.D. and Sansom, M.S. (2003). Acquisition of siderophores in gram-negative bacteria. *Nat. Rev. Mol. Cell Biol.* 4: 105–116.

159 Visca, P., Imperi, F., and Lamont, I.L. (2007). Pyoverdine siderophores: from biogenesis to biosignificance. *Trends Microbiol.* 15: 22–30.

160 Lamont, I.L., Beare, P.A., Ochsner, U. et al. (2002). Siderophore-mediated signaling regulates virulence factor production in Pseudomonas aeruginosa. *Proc. Natl. Acad. Sci. U. S. A.* 99: 7072–7077.

161 Guan, L.L., Kanoh, K., and Kamino, K. (2001). Effect of exogenous siderophores on iron uptake activity of marine bacteria under iron-limited conditions. *Appl. Environ. Microbiol.* 67: 1710–1717.

162 Partida-Martinez, L.P. and Hertweck, C. (2005). Pathogenic fungus harbours endosymbiotic bacteria for toxin production. *Nature* 437: 884–888.

163 Partida-Martinez, L.P., Monajembashi, S., Greulich, K.O., and Hertweck, C. (2007). Endosymbiont-dependent host reproduction maintains bacterial-fungal mutualism. *Curr. Biol.* 17: 773–777.

164 Uzum, Z., Silipo, A., Lackner, G. et al. (2015). Structure, genetics and function of an exopolysaccharide produced by a bacterium living within fungal hyphae. *Chembiochem* 16: 387–392.

165 Ross, C., Opel, V., Scherlach, K., and Hertweck, C. (2014). Biosynthesis of antifungal and antibacterial polyketides by Burkholderia gladioli in coculture with Rhizopus microsporus. *Mycoses* 57: 48–55.

166 Blow, F. and Douglas, A.E. (2019). The hemolymph microbiome of insects. *J. Insect Physiol.* 115: 33–39. https://doi.org/10.1016/j.jinsphys.2019.04.002.

167 Douglas, A.E. (1998). Nutritional interactions in insect-microbial symbioses: aphids and their symbiotic bacteria buchnera. *Annu. Rev. Entomol.* 43: 17–37. https://doi.org/10.1146/annurev.ento.43.1.17.

168 Thao, M.L. and Baumann, P. (2004). Evolutionary relationships of primary prokaryotic endosymbionts of whiteflies and their hosts. *Appl. Environ. Microbiol.* 70: 3401–3406. https://doi.org/10.1128/AEM.70.6.3401-3406.2004.

169 Thao, M.L., Moran, N.A., Abbot, P. et al. (2000). Cospeciation of psyllids and their primary prokaryotic endosymbionts. *Appl. Environ. Microbiol.* 66: 2898–2905. https://doi.org/10.1128/AEM.66.7.2898-2905.2000.

170 Luan, J.B., Chen, W., Hasegawa, D.K. et al. (2015). Metabolic coevolution in the bacterial symbiosis of whiteflies and related plant sap-feeding insects. *Genome Biol. Evol.* 7: 2635–2647. https://doi.org/10.1093/gbe/evv170.

171 Raychoudhury, R., Sen, R., Cai, Y. et al. (2013). Comparative metatranscriptomic signatures of wood and paper feeding in the gut of the termite reticulitermes flavipes (*Isoptera: Rhinotermitidae*). *Insect Mol. Biol.* 22: 155–171. https://doi.org/10.1111/imb.12011.

172 Brune, A. (2014). Symbiotic digestion of lignocellulose in termite guts. *Nat. Rev. Microbiol.* 12: 168–180. https://doi.org/10.1038/nrmicro3182.

173 Berlanga, M., Llorens, C., Comas, J., and Guerrero, R. (2016). Gut bacterial community of the *Xylophagous cockroaches Cryptocercus punctulatus* and *Parasphaeria boleiriana*. *PLoS One* 11: e0152400. https://doi.org/10.1371/journal.pone.0152400.

174 Kroiss, J., Kaltenpoth, M., Schneider, B. et al. (2010). Symbiotic streptomycetes provide antibiotic combination prophylaxis for wasp offspring. *Nat. Chem. Biol.* 6: 261–263. https://doi.org/10.1038/nchembio.331.

175 Shao, Y., Chen, B., Sun, C. et al. (2017). Symbiont-derived antimicrobials contribute to the control of the lepidopteran gut microbiota. *Cell Chem. Biol.* 24: 66–75. https://doi.org/10.1016/j.chembiol.2016.11.015.

176 Oliver, K.M., Moran, N.A., and Hunter, M.S. (2005). Variation in resistance to parasitism in aphids is due to symbionts not host genotype. *Proc. Natl. Acad. Sci. U. S. A.* 102: 12795–12800. https://doi.org/10.1073/pnas.0506131102.

177 Oliver, K.M., Russell, J.A., Moran, N.A., and Hunter, M.S. (2003). Facultative bacterial symbionts in aphids confer resistance to parasitic wasps. *Proc. Natl. Acad. Sci. U. S. A.* 100: 1803–1807. https://doi.org/10.1073/pnas.0335320100.

178 Vorburger, C., Sandrock, C., Gouskov, A. et al. (2009). Genotypic variation and the role of defensive endosymbionts in an all-parthenogenetic host–parasitoid interaction. *Evol. Int. J. Organ. Evol.* 63: 1439–1450. https://doi.org/10.1111/j.1558-5646.2009.00660.x.

179 Schmid, M., Sieber, R., Zimmermann, Y.-S., and Vorburger, C. (2012). Development, specificity and sublethal effects of symbiont-conferred resistance to parasitoids in aphids. *Funct. Ecol.* 26: 207–215. https://doi.org/10.1111/j.1365-2435.2011.01904.x.

180 Brandt, J.W., Chevignon, G., Oliver, K.M., and Strand, M.R. (2017). Culture of an aphid heritable symbiont demonstrates its direct role in defence against parasitoids. *Proc. R. Soc. B Biol. Sci.* 284: 20171925. https://doi.org/10.1098/rspb.2017.1925.

181 Frago, E., Dicke, M., and Godfray, H.C.J. (2012). Insect symbionts as hidden players in insect–plant interactions. *Trends Ecol. Evol.* 27: 705–711. https://doi.org/10.1016/j.tree.2012.08.013.

182 Chung, S.H., Rosa, C., Scully, E.D. et al. (2013). Herbivore exploits orally secreted bacteria to suppress plant defenses. *Proc. Natl. Acad. Sci. U. S. A.* 110: 15728–15733. https://doi.org/10.1073/pnas.1308867110.

183 Su, Q., Oliver, K.M., Xie, W. et al. (2015). The whitefly-associated facultative symbiont *Hamiltonella defensa* suppresses induced plant defences in tomato. *Funct. Ecol.* 29: 1007–1018. https://doi.org/10.1111/1365-2435.12405.

184 Xu, H., Zheng, X., Liu, S. et al. (2009). The role of endosymbionts in insect host resistance against adverse factors. *Chin. Bull. Entomol.* 46: 350–354.

185 Libbrecht, R., Gwynn, D.M., and Fellowes, M.D.E. (2007). *Aphidius ervi* preferentially attacks the green morph of the pea aphid, acyrthosiphon pisum. *J. Insect Behav.* 20: 25–32. https://doi.org/10.1007/s10905-006-9055-y.

186 Tsuchida, T., Koga, R., Fujiwara, A., and Fukatsu, T. (2014). Phenotypic effect of "Candidatus Rickettsiella viridis," a facultative symbiont of the pea aphid (*Acyrthosiphon pisum*), and its interaction with a coexisting symbiont. *Appl. Environ. Microbiol.* 80: 525–533. https://doi.org/10.1128/AEM.03049-13.

187 Tsuchida, T., Koga, R., Horikawa, M. et al. (2010). Symbiotic bacterium modifies aphid body color. *Science* 330: 1102–1104. https://doi.org/10.1126/science.1195463.

188 Moran, N.A., Russell, J.A., Koga, R., and Fukatsu, T. (2005). Evolutionary relationships of three new species of enterobacteriaceae living as symbionts of aphids and other insects. *Appl. Environ. Microbiol.* 71: 3302–3310. https://doi.org/10.1128/AEM.71.6.3302-3310.2005.

189 Xia, X., Sun, B., Gurr, G.M. et al. (2018). Gut microbiota mediate insecticide resistance in the diamondback moth, *Plutella xylostella* (L.). *Front. Microbiol.* 9 (25). https://doi.org/10.3389/fmicb.2018.00025.

190 Xia, X., Zheng, D., Zhong, H. et al. (2013). DNA sequencing reveals the midgut microbiota of diamondback moth, *Plutella xylostella* (L.) and a possible relationship with insecticide resistance. *PLoS One* 8: e68852. https://doi.org/10.1371/journal.pone.0068852.

191 Barnard, K., Jeanrenaud, A.C.S.N., Brooke, B.D., and Oliver, S. (2019). The contribution of gut bacteria to insecticide resistance and the life histories of the major malaria vector *Anopheles arabiensis* (Diptera: Culicidae). *Sci. Rep.* 9: 9117. https://doi.org/10.1038/s41598-019-45499-z.

192 McFall-Ngai, M., Hadfield, M.G., Bosch, T.C. et al. (2013). Animals in a bacterial world, a new imperative for the life sciences. *Proc. Natl. Acad. Sci. U. S. A.* 110: 3229–3236. https://doi.org/10.1073/pnas.1218525110.

193 Engel, P. and Moran, N.A. (2013). The gut microbiota of insects – diversity in structure and function. *FEMS Microbiol. Rev.* 37: 699–735. https://doi.org/10.1111/1574-6976.12025.

194 Adamo, S.A. (2013). Parasites: evolution's neurobiologists. *J. Exp. Biol.* 216: 3–10. https://doi.org/10.1242/jeb.073601.

195 Hemarajata, P. and Versalovic, J. (2013). Effects of probiotics on gut microbiota: mechanisms of intestinal immunomodulation and neuromodulation. *Ther. Adv. Gastroenterol.* 6: 39–51. https://doi.org/10.1177/1756283X12459294.

196 Wang, Y. and Kasper, L.H. (2014). The role of microbiome in central nervous system disorders. *Brain Behav. Immun.* 38: 1–12. https://doi.org/10.1016/j.bbi.2013.12.015.

197 Wong, A.C.-N., Wang, Q.-P., Morimoto, J. et al. (2017). Gut microbiota modifies olfactory-guided microbial preferences and foraging decisions in *Drosophila*. *Curr. Biol.* 27: 2397–2404.e4. https://doi.org/10.1016/j.cub.2017.07.022.

198 Farine, J.-P., Habbachi, W., Cortot, J. et al. (2017). Maternally-transmitted microbiota affects odor emission and preference in drosophila larva. *Sci. Rep.* 7: 6062. https://doi.org/10.1038/s41598-017-04922-z.

199 Qiao, H., Keesey, I.W., Hansson, B.S., and Knaden, M. (2019). Gut microbiota affects development and olfactory behavior in drosophila melanogaster. *J. Exp. Biol.* 222: jeb192500. https://doi.org/10.1242/jeb.192500.

200 Behar, A., Yuval, B., and Jurkevitch, E. (2008). Gut bacterial communities in the mediterranean fruit fly (*Ceratitis capitata*) and their impact on host longevity. *J. Insect Physiol.* 54: 1377–1383. https://doi.org/10.1016/j.jinsphys.2008.07.011.

201 Wong, C.N.A., Ng, P., and Douglas, A.E. (2011). Low-diversity bacterial community in the gut of the fruitfly *Drosophila melanogaster*. *Environ. Microbiol.* 13: 1889–1900. https://doi.org/10.1111/j.1462-2920.2011.02511.x.

202 Ras, E., Beukeboom, L.W., Cáceres, C., and Bourtzis, K. (2017). Review of the role of gut microbiota in mass rearing of the olive fruit fly, bactrocera oleae, and its parasitoids. *Entomol. Exp. Appl.* 164: 237–256. https://doi.org/10.1111/eea.12609.

203 Ramalho, M., Bueno, O., and Saux Moreau, C. (2017). Species-specific signatures of the microbiome from camponotus and colobopsis ants across developmental stages. *PLoS One* 12: e0187461. https://doi.org/10.1371/journal.pone.0187461.

204 Shin, B., Park, S.H., Kim, B.-Y. et al. (2017). Deinococcucins A–D, aminoglycolipids from *Deinococcus* sp., a gut bacterium of the carpenter ant camponotus japonicus. *J. Nat. Prod.* 80: 2910–2916. https://doi.org/10.1021/acs.jnatprod.7b00426.

205 Engel, P., Kwong, W.K., McFrederick, Q. et al. (2016). The bee microbiome: impact on bee health and model for evolution and ecology of host-microbe interactions. *MBio* 7: e02164–15. https://doi.org/10.1128/mBio.02164-15.

206 Minard, G., Mavingui, P., and Moro, C.V. (2013). Diversity and function of bacterial microbiota in the mosquito holobiont. *Parasit. Vectors* 6: 146. https://doi.org/10.1186/1756-3305-6-146.

207 Novakova, E., Woodhams, D.C., Rodríguez-Ruano, S.M. et al. (2017). Mosquito microbiome dynamics, a background for prevalence and seasonality of west nile virus. *Front. Microbiol.* 8: 526. https://doi.org/10.3389/fmicb.2017.00526.

208 Narasimhan, S. and Fikrig, E. (2015). Tick microbiome: the force within. *Trends Parasitol.* 31: 315–323. https://doi.org/10.1016/j.pt.2015.03.010.

209 Hulcr, J., Rountree, N.R., Diamond, S.E. et al. (2012). Mycangia of ambrosia beetles host communities of bacteria. *Microb. Ecol.* 64: 784–793. https://doi.org/10.1007/s00248-012-0055-5.

210 Halpern, M. and Senderovich, Y. (2015). Chironomid microbiome. *Microb. Ecol.* 70: 1–8. https://doi.org/10.1007/s00248-014-0536-9.

211 Himler, A.G., Adachi-Hagimori, T., Bergen, J.E. et al. (2011). Rapid spread of a bacterial symbiont in an invasive whitefly is driven by fitness benefits and female bias. *Science* 332: 254–256. https://doi.org/10.1126/science.1199410.

212 Ridley, E.V., Wong, A.C.-N., Westmiller, S., and Douglas, A.E. (2012). Impact of the resident microbiota on the nutritional phenotype of *Drosophila melanogaster*. *PLoS One* 7: e36765. https://doi.org/10.1371/journal.pone.0036765.

213 Ridley, E.V., Wong, A.C.N., and Douglas, A.E. (2013). Microbe-dependent and nonspecific effects of procedures to eliminate the resident microbiota from *Drosophila melanogaster*. *Appl. Environ. Microbiol.* 79: 3209–3214. https://doi.org/10.1128/AEM.00206-13.

214 Morimoto, J., Simpson, S.J., and Ponton, F. (2017). Direct and trans-generational effects of male and female gut microbiota in drosophila melanogaster. *Biol. Lett.* 13: 20160966. https://doi.org/10.1098/rsbl.2016.0966.

215 Bing, X., Gerlach, J., Loeb, G., and Buchon, N. (2018). Nutrient-dependent impact of microbes on drosophila suzukii development. *MBio* 9: e02199-17. https://doi.org/10.1128/mBio.02199-17.

216 Petri, L. (1910). Untersuchung uber die darmbakterien der olivenfliege. *Zentralblatt Bakteriol. Parasitenkd. Infekt. Hyg.* 26: 357–367.

217 Capuzzo, C., Firrao, G., Mazzon, L. et al. (2005). 'Candidatus erwinia dacicola', a coevolved symbiotic bacterium of the olive fly bactrocera oleae (Gmelin). *Int. J. Syst. Evol. Microbiol.* 55: 1641–1647. https://doi.org/10.1099/ijs.0.63653-0.

218 Ben-Yosef, M., Pasternak, Z., Jurkevitch, E., and Yuval, B. (2014). Symbiotic bacteria enable olive flies (*Bactrocera oleae*) to exploit intractable sources of nitrogen. *J. Evol. Biol.* 27: 2695–2705. https://doi.org/10.1111/jeb.12527.

219 Estes, A.M., Hearn, D.J., Bronstein, J.L., and Pierson, E.A. (2009). The olive fly endosymbiont, "Candidatus erwinia dacicola," switches from an intracellular existence to an extracellular existence during host insect development. *Appl. Environ. Microbiol.* 75: 7097–7106. https://doi.org/10.1128/AEM.00778-09.

220 Estes, A.M., Hearn, D.J., Burrack, H.J. et al. (2012). Prevalence of *Candidatus erwinia dacicola* in wild and laboratory olive fruit fly populations and across developmental stages. *Environ. Entomol.* 41: 265–274. https://doi.org/10.1603/EN11245.

221 Behar, A., Yuval, B., and Jurkevitch, E. (2005). Enterobacteria-mediated nitrogen fixation in natural populations of the fruit fly ceratitis capitata. *Mol. Ecol.* 14: 2637–2643. https://doi.org/10.1111/j.1365-294X.2005.02615.x.

222 Malacrinò, A., Campolo, O., Medina, R.F., and Palmeri, V. (2018). Instar- and host-associated differentiation of bacterial communities in the mediterranean fruit fly *Ceratitis capitata*. *PLoS One* 13: e0194131. https://doi.org/10.1371/journal.pone.0194131.

223 Borm, S., van Buschinger, A., Boomsma, J.J., and Billen, J. (2002). Tetraponera ants have gut symbionts related to nitrogen–fixing root–nodule bacteria. *Proc. R. Soc. Lond. B Biol. Sci.* 269: 2023–2027. https://doi.org/10.1098/rspb.2002.2101.

224 Mason, C.J., Couture, J.J., and Raffa, K.F. (2014). Plant-associated bacteria degrade defense chemicals and reduce their adverse effects on an insect defoliator. *Oecologia* 175: 901–910. https://doi.org/10.1007/s00442-014-2950-6.

225 Behar, A., Ben-Yosef, M., Lauzon, C.R. et al. (2009). Structure and function of the bacterial community associated with the mediterranean fruit fly. *Insect Symbiosis* 3: 251–271. https://doi.org/10.1201/9781420064117.ch11.

226 Lauzon, C. (2003). Symbiotic relationships of tephritids. In: *Insect Symbiosis*, vol. 2 (ed. K. Bourtzis and T. Miller), 115–129. Boca Raton: CRC. https://doi.org/10.1201/9780203009918.ch8.

227 Baumann, P. (2005). Biology bacteriocyte-associated endosymbionts of plant sap-sucking insects. *Annu. Rev. Microbiol.* 59: 155–189. https://doi.org/10.1146/annurev.micro.59.030804.121041.

228 Zhong, J., Jasinskas, A., and Barbour, A.G. (2007). Antibiotic treatment of the tick vector *Amblyomma americanum* reduced reproductive fitness. *PLoS One* 2: e405. https://doi.org/10.1371/journal.pone.0000405.

229 Carter, V., Underhill, A., Baber, I. et al. (2013). Killer bee molecules: antimicrobial peptides as effector molecules to target sporogonic stages of plasmodium. *PLoS Pathog.* 9: e1003790. https://doi.org/10.1371/journal.ppat.1003790.

230 Hultmark, D., Steiner, H., Rasmuson, T., and Boman, H.G. (1980). Insect immunity. Purification and properties of three inducible bactericidal proteins from hemolymph of immunized pupae of hyalophora cecropia. *Eur. J. Biochem.* 106: 7–16. https://doi.org/10.1111/j.1432-1033.1980.tb05991.x.

231 Steiner, H., Hultmark, D., Engström, A. et al. (1981). Sequence and specificity of two antibacterial proteins involved in insect immunity. *Nature* 292: 246–248. https://doi.org/10.1038/292246a0.

232 Yi, H.-Y., Chowdhury, M., Huang, Y.-D., and Yu, X.-Q. (2014). Insect antimicrobial peptides and their applications. *Appl. Microbiol. Biotechnol.* 98: 5807–5822. https://doi.org/10.1007/s00253-014-5792-6.

233 Wu, Q., Patočka, J., and Kuča, K. (2018). Insect antimicrobial peptides, a mini review. *Toxins* 10: 461. https://doi.org/10.3390/toxins10110461.

234 Mitsuhara, I., Matsufuru, H., Ohshima, M. et al. (2000). Induced expression of sarcotoxin IA enhanced host resistance against both bacterial and fungal pathogens in transgenic tobacco. *MPMI* 13: 860–868. https://doi.org/10.1094/MPMI.2000.13.8.860.

235 Sharma, A., Sharma, R., Imamura, M. et al. (2000). Transgenic expression of cecropin B, an antibacterial peptide from *Bombyx mori*, confers enhanced resistance to bacterial leaf blight in rice. *FEBS Lett.* 484: 7–11. https://doi.org/10.1016/S0014-5793(00)02106-2.

236 Jan, P.-S., Huang, H.-Y., and Chen, H.-M. (2010). Expression of a Synthesized gene encoding cationic peptide cecropin B in transgenic tomato plants protects against bacterial diseases. *Appl. Environ. Microbiol.* 76: 769–775. https://doi.org/10.1128/AEM.00698-09.

237 Rodriguez, M.D., Zamudio, F., Torres, J.A. et al. (1995). Effect of a cecropin-like synthetic peptide (shiva-3) on the sporogonic development of *Plasmodium berghei*. *Exp. Parasitol.* 80: 596–604. https://doi.org/10.1006/expr.1995.1075.

238 Fieck, A., Hurwitz, I., Kang, A.S., and Durvasula, R. (2010). *Trypanosoma cruzi*: synergistic cytotoxicity of multiple amphipathic anti-microbial peptides to *T. cruzi* and potential bacterial hosts. *Exp. Parasitol.* 125: 342–347. https://doi.org/10.1016/j.exppara.2010.02.016.

239 Kim, W., Koo, H., Richman, A.M. et al. (2004). Ectopic expression of a cecropin transgene in the human malaria vector mosquito anopheles gambiae (*Diptera: Culicidae*): effects on susceptibility to plasmodium. *J. Med. Entomol.* 41: 447–455. https://doi.org/10.1603/0022-2585-41.3.447.

240 Kokoza, V., Ahmed, A., Woon Shin, S. et al. (2010). Blocking of *Plasmodium* transmission by cooperative action of cecropin A and defensin A in transgenic *Aedes aegypti* mosquitoes. *Proc. Natl. Acad. Sci. U. S. A.* 107: 8111–8116. https://doi.org/10.1073/pnas.1003056107.

241 Shen, W., He, P., Xiao, C., and Chen, X. (2018). From antimicrobial peptides to antimicrobial poly(α-amino acid)s. *Adv. Healthc. Mater.* 7: e1800354. https://doi.org/10.1002/adhm.201800354.

242 Peschen, D., Li, H.-P., Fischer, R. et al. (2004). Fusion proteins comprising a fusarium -specific antibody linked to antifungal peptides protect plants against a fungal pathogen. *Nat. Biotechnol.* 22: 732–738. https://doi.org/10.1038/nbt970.

243 Cao, J., de la Fuente-Nunez, C., Ou, R.W. et al. (2018). Yeast-based synthetic biology platform for antimicrobial peptide production. *ACS Synth. Biol.* 7: 896–902. https://doi.org/10.1021/acssynbio.7b00396.

244 Biswaro, L.S., da Costa Sousa, M.G., Rezende, T.M.B. et al. (2018). Antimicrobial peptides and nanotechnology, recent advances and challenges. *Front. Microbiol.* 9: 855. https://doi.org/10.3389/fmicb.2018.00855.

245 Hosokawa, T., Kikuchi, Y., Shimada, M., and Fukatsu, T. (2007). Obligate symbiont involved in pest status of host insect. *Proc. Biol. Sci.* 274: 1979–1984. https://doi.org/10.1098/rspb.2007.0620.

246 Moran, N.A. and Yun, Y. (2015). Experimental replacement of an obligate insect symbiont. *Proc. Natl. Acad. Sci. U. S. A.* 112: 2093–2096. https://doi.org/10.1073/pnas.1420037112.

247 Coon, K.L., Vogel, K.J., Brown, M.R., and Strand, M. (2014). Mosquitoes rely on their gut microbiota for development. *Mol. Ecol.* 23: 2727–2739. https://doi.org/10.1111/mec.12771.

248 Kešnerová, L., Mars, R.A.T., Ellegaard, K.M. et al. (2017). Disentangling metabolic functions of bacteria in the honey bee gut. *PLoS Biol.* 15: e2003467. https://doi.org/10.1371/journal.pbio.2003467.

249 Wong, A.C.-N., Chaston, J.M., and Douglas, A.E. (2013). The inconstant gut microbiota of drosophila species revealed by 16S rRNA gene analysis. *ISME J.* 7: 1922–1932. https://doi.org/10.1038/ismej.2013.86.

250 Hilgenboecker, K., Hammerstein, P., Schlattmann, P. et al. (2008). How many species are infected with wolbachia? – a statistical analysis of current data. *FEMS Microbiol. Lett.* 281: 215–220. https://doi.org/10.1111/j.1574-6968.2008.01110.x.

251 Stouthamer, R., Breeuwer, J.A., and Hurst, G.D. (1999). *Wolbachia pipientis*: microbial manipulator of arthropod reproduction. *Annu. Rev. Microbiol.* 53: 71–102. https://doi.org/10.1146/annurev.micro.53.1.71.

252 Werren, J.H. (1997). Biology of wolbachia. *Annu. Rev. Entomol.* 42: 587–609. https://doi.org/10.1146/annurev.ento.42.1.587.

253 Bourtzis, K., Dobson, S.L., Braig, H.R., and O'Neill, S.L. (1998). Rescuing wolbachia have been overlooked. *Nature* 391: 852–853. https://doi.org/10.1038/36017.

254 Saridaki, A. and Bourtzis, K. (2010). Wolbachia: more than just a bug in insects genitals. *Curr. Opin. Microbiol.* 13: 67–72. https://doi.org/10.1016/j.mib.2009.11.005.

255 Segoli, M., Hoffmann, A.A., Lloyd, J. et al. (2014). The effect of virus-blocking wolbachia on male competitiveness of the dengue vector mosquito, *Aedes aegypti*. *PLoS Negl. Trop. Dis.* 8: e3294. https://doi.org/10.1371/journal.pntd.0003294.

256 Alam, U., Medlock, J., Brelsfoard, C. et al. (2011). Wolbachia symbiont infections induce strong cytoplasmic incompatibility in the tsetse fly glossina morsitans. *PLoS Pathog.* 7: e1002415. https://doi.org/10.1371/journal.ppat.1002415.

257 Berasategui, A., Shukla, S., Salem, H., and Kaltenpoth, M. (2016). Potential applications of insect symbionts in biotechnology. *Appl. Microbiol. Biotechnol.* 100: 1567–1577. https://doi.org/10.1007/s00253-015-7186-9.

258 Zabalou, S., Apostolaki, A., Pattas, S. et al. (2008). Multiple rescue factors within a wolbachia strain. *Genetics* 178: 2145–2160. https://doi.org/10.1534/genetics.107.086488.

259 Zhang, L., Zhang, X., Zhang, Y. et al. (2016). A new formulation of *Bacillus thuringiensis*: UV protection and sustained release mosquito larvae studies. *Sci. Rep.* 6: 39425. https://doi.org/10.1038/srep39425.

260 O'Connor, L., Plichart, C., Sang, A.C. et al. (2012). Open release of male mosquitoes infected with a wolbachia biopesticide: field performance and infection containment. *PLoS Negl. Trop. Dis.* 6: e1797. https://doi.org/10.1371/journal.pntd.0001797.

261 Cattel, J., Kaur, R., Gibert, P. et al. (2016). Wolbachia in European populations of the invasive pest drosophila suzukii: regional variation in infection frequencies. *PLoS One* 11: e0147766. https://doi.org/10.1371/journal.pone.0147766.

262 Ometto, L., Cestaro, A., Ramasamy, S. et al. (2013). Linking genomics and ecology to investigate the complex evolution of an invasive drosophila pest. *Genome Biol. Evol.* 5: 745–757. https://doi.org/10.1093/gbe/evt034.

263 Siozios, S., Cestaro, A., Kaur, R. et al. (2013). Draft genome sequence of the wolbachia endosymbiont of *Drosophila suzukii*. *Genome Announc.* 1: e00032-13. https://doi.org/10.1128/genomeA.00032-13.

264 Cattel, J., Nikolouli, K., Andrieux, T. et al. (2018). Back and forth wolbachia transfers reveal efficient strains to control spotted wing drosophila populations. *J. Appl. Ecol.* 55: 2408–2418. https://doi.org/10.1111/1365-2664.13101.

265 Hamm, C.A., Begun, D.J., Vo, A. et al. (2014). Wolbachia do not live by reproductive manipulation alone: infection polymorphism in *Drosophila suzukii* and *D. subpulchrella*. *Mol. Ecol.* 23: 4871–4885. https://doi.org/10.1111/mec.12901.

266 Mouton, L., Henri, H., Charif, D. et al. (2007). Interaction between host genotype and environmental conditions affects bacterial density in wolbachia symbiosis. *Biol. Lett.* 3: 210–213. https://doi.org/10.1098/rsbl.2006.0590.

267 LePage, D.P., Metcalf, J.A., Bordenstein, S.R. et al. (2017). Prophage WO genes recapitulate and enhance wolbachia-induced cytoplasmic incompatibility. *Nature* 543: 243–247. https://doi.org/10.1038/nature21391.

268 Shropshire, J.D., On, J., Layton, E.M. et al. (2018). One prophage WO gene rescues cytoplasmic incompatibility in *Drosophila melanogaster*. *Proc. Natl. Acad. Sci. U. S. A.* 115: 4987–4991. https://doi.org/10.1073/pnas.1800650115.

269 Beckmann, J.F., Ronau, J.A., and Hochstrasser, M. (2017). A wolbachia deubiquitylating enzyme induces cytoplasmic incompatibility. *Nat. Microbiol.* 2: 17007. https://doi.org/10.1038/nmicrobiol.2017.7.

270 Bian, G., Xu, Y., Lu, P. et al. (2010). The endosymbiotic bacterium wolbachia induces resistance to dengue virus in aedes aegypti. *PLoS Pathog.* 6: e1000833. https://doi.org/10.1371/journal.ppat.1000833.

271 Van den Hurk, A.F., Hall-Mendelin, S., Pyke, A. et al. (2012). Impact of wolbachia on infection with chikungunya and yellow fever viruses in the mosquito vector aedes aegypti. *PLoS Negl. Trop. Dis.* 6: e1892. https://doi.org/10.1371/journal.pntd.0001892.

272 Aliota, M.T., Peinado, S.A., Velez, I.D., and Osorio, J. (2016). The *w*Mel strain of *Wolbachia* reduces transmission of zika virus by *Aedes aegypti*. *Sci. Rep.* 6: 28792. https://doi.org/10.1038/srep28792.

273 Carrington, L.B., Tran, B.C.N., Le, N.T.H. et al. (2018). Field- and clinically derived estimates of wolbachia-mediated blocking of dengue virus transmission potential in *Aedes aegypti* mosquitoes. *Proc. Natl. Acad. Sci. U. S. A.* 115: 361–366. https://doi.org/10.1073/pnas.1715788115.

274 Kambris, Z., Cook, P.E., Phuc, H.K., and Sinkins, S.P. (2009). Immune activation by life-shortening wolbachia and reduced filarial competence in mosquitoes. *Science* 326: 134–136. https://doi.org/10.1126/science.1177531.

275 Moreira, L.A., Iturbe-Ormaetxe, I., Jeffery, J.A. et al. (2009). A wolbachia symbiont in *Aedes aegypti* limits infection with dengue, chikungunya, and plasmodium. *Cell* 139: 1268–1278. https://doi.org/10.1016/j.cell.2009.11.042.

276 Geoghegan, V., Stainton, K., Rainey, S.M. et al. (2017). Perturbed cholesterol and vesicular trafficking associated with dengue blocking in wolbachia-infected aedes aegypti cells. *Nat. Commun.* 8: 526. https://doi.org/10.1038/s41467-017-00610-8.

277 Thomas, S., Verma, J., Woolfit, M., and O'Neill, S.L. (2018). Wolbachia-mediated virus blocking in mosquito cells is dependent on XRN1-mediated viral RNA degradation and influenced by viral replication rate. *PLoS Pathog.* 14: e1006879. https://doi.org/10.1371/journal.ppat.1006879.

278 Hoffmann, A.A., Montgomery, B.L., Popovici, J. et al. (2011). Successful establishment of wolbachia in aedes populations to suppress dengue transmission. *Nature* 476: 454–457. https://doi.org/10.1038/nature10356.

279 Nazni, W.A., Hoffmann, A.A., NoorAfizah, A. et al. (2019). Establishment of wolbachia strain wAlbB in malaysian populations of *Aedes aegypti* for dengue control. *Curr. Biol.* 29: 4241–4248.e5. https://doi.org/10.1016/j.cub.2019.11.007.

280 Ryan, P.A., Turley, A.P., Wilson, G. et al. (2020). Establishment of wMel wolbachia in aedes aegypti mosquitoes and reduction of local dengue transmission in cairns and surrounding locations in northern Queensland, Australia. *Gates Open Res.* 3: 1547. https://doi.org/10.12688/gatesopenres.13061.2.

281 Tantowijoyo, W., Andari, B., Arguni, E. et al. (2020). Stable establishment of wMel wolbachia in *Aedes aegypti* populations in yogyakarta, Indonesia. *PLoS Negl. Trop. Dis.* 14: e0008157. https://doi.org/10.1371/journal.pntd.0008157.

282 Aksoy, S., Weiss, B., and Attardo, G. (2008). Paratransgenesis applied for control of tsetse transmitted sleeping sickness. In: *Transgenesis and the Management of Vector-Borne Disease. Advances in Experimental Medicine and Biology*, vol. 627 (ed. S. Aksoy), 35–48. New York: Springer. https://doi.org/10.1007/978-0-387-78225-6_3.

283 Coutinho-Abreu, I.V., Zhu, K.Y., and Ramalho-Ortigao, M. (2010). Transgenesis and paratransgenesis to control insect-borne diseases: current status and future challenges. *Parasitol. Int.* 59: 1–8. https://doi.org/10.1016/j.parint.2009.10.002.

284 Caragata, E.P. and Walker, T. (2012). Using bacteria to treat diseases. *Expert. Opin. Biol. Ther.* 12: 701–712. https://doi.org/10.1517/14712598.2012.677429.

285 Beard, C.B., Durvasula, R.V., and Richards, F.F. (2000). Bacterial symbiont transformation in Chagas disease vectors. In: *Insect Transgenesis: Methods and Applications* (ed. A.M. Handler and A.A. James), 289–303. Boca Raton: CRC Press. https://doi.org/10.1201/9781420039399.ch16.

286 Beard, C., O'Neill, S.L., Tesh, R. et al. (1993). Modification of arthropod vector competence via symbiotic bacteria. *Parasitol. Today* 9: 179–183. https://doi.org/10.1016/0169-4758(93)90142-3.

287 Durvasula, R.V., Kroger, A., Goodwin, M. et al. (1999). Strategy for introduction of foreign genes into field populations of chagas disease vectors. *Ann. Entomol. Soc. Am.* 92: 937–943. https://doi.org/10.1093/aesa/92.6.937.

288 Hurwitz, I., Hillesland, H., Fieck, A. et al. (2011). The paratransgenic sand fly: a platform for control of Leishmania transmission. *Parasit. Vectors* 4: 82. https://doi.org/10.1186/1756-3305-4-82.

289 Taracena, M.L., Oliveira, P.L., Almendares, O. et al. (2015). Genetically modifying the insect gut microbiota to control chagas disease vectors through systemic RNAi. *PLoS Negl. Trop. Dis.* 9: e0003358. https://doi.org/10.1371/journal.pntd.0003358.

290 De Vooght, L., Caljon, G., De Ridder, K., and Van Den Abbeele, J. (2014). Delivery of a functional anti-trypanosome nanobody in different tsetse fly tissues via a bacterial symbiont, *Sodalis glossinidius*. *Microb. Cell Factories* 13: 156. https://doi.org/10.1186/s12934-014-0156-6.

291 De Vooght, L., Caljon, G., Stijlemans, B. et al. (2012). Expression and extracellular release of a functional anti-trypanosome nanobody® in *Sodalis glossinidius*, a bacterial symbiont of the tsetse fly. *Microb. Cell Factories* 11: 23. https://doi.org/10.1186/1475-2859-11-23.

292 Bisi, D.C. and Lampe, D.J. (2011). Secretion of anti-plasmodium effector proteins from a natural pantoea agglomerans isolate by using PelB and HlyA secretion signals. *Appl. Environ. Microbiol.* 77: 4669–4675. https://doi.org/10.1128/AEM.00514-11.

293 Dinparast Djadid, N., Jazayeri, H., Raz, A. et al. (2011). Identification of the midgut microbiota of an. stephensi and an. maculipennis for their application as a paratransgenic tool against malaria. *PLoS One* 6: e0028484. https://doi.org/10.1371/journal.pone.0028484.

294 Bongio, N.J. and Lampe, D.J. (2015). Inhibition of plasmodium berghei development in mosquitoes by effector proteins secreted from *Asaia* sp. Bacteria using a novel native secretion signal. *PLoS One* 10: e0143541. https://doi.org/10.1371/journal.pone.0143541.

295 Fang, W., Vega-Rodríguez, J., Ghosh, A.K. et al. (2011). Development of transgenic fungi that kill human malaria parasites in mosquitoes. *Science* 331: 1074–1077. https://doi.org/10.1126/science.1199115.

296 Hopkins, D. and Purcell, A. (2002). *Xylella fastidiosa*: cause of pierce's disease of grapevine and other emergent diseases. *Plant Dis.* 86: 1056–1066. https://doi.org/10.1094/PDIS.2002.86.10.1056.

297 Hackett, K.J. (2003). Investigating invasives: it takes a team. *Agric. Res.* 51: 2.

298 Bextine, B.R., Harshman, D., Johnson, M.C., and Miller, T.A. (2004). Impact of pymetrozine on glassy-winged sharpshooter feeding behavior and rate of *Xylella fastidiosa* transmission. *J. Insect Sci.* 4: 34. https://doi.org/10.1673/031.004.3401.

299 Killiny, N., Martinez, R.H., Dumenyo, C.K. et al. (2013). The exopolysaccharide of *Xylella fastidiosa* is essential for biofilm formation, plant virulence, and vector transmission. *Mol. Plant-Microbe Interact.* 26: 10444–10453. https://doi.org/10.1094/MPMI-09-12-0211-R.

300 Saiman, L., Chen, Y., Tabibi, S. et al. (2001). Identification and antimicrobial susceptibility of *Alcaligenes xylosoxidans* isolated from patients with cystic fibrosis. *J. Clin. Microbiol.* 39: 3942–3945. https://doi.org/10.1128/JCM.39.11.3942-3945.2001.

301 Arora, A.K., Forshaw, A., Miller, T.A., and Durvasula, R. (2015). A delivery system for field application of paratransgenic control. *BMC Biotechnol.* 15: 59. https://doi.org/10.1186/s12896-015-0175-3.

302 Leonard, S.P., Powell, J.E., Perutka, J. et al. (2020). Engineered symbionts activate honey bee immunity and limit pathogens. *Science* 367: 573–576. https://doi.org/10.1126/science.aax9039.

303 USEPAO (2015). *What are Biopesticides? US EPA*. https://www.epa.gov/pesticides/biopesticides#what (accessed February 2020).

304 Hajek, A.E. and Delalibera, I. (2010). Fungal pathogens as classical biological control agents against arthropods. *BioControl* 55: 147–158. https://doi.org/10.1007/s10526-009-9253-6.

305 Lacey, L.A. and Georgis, R. (2012). Entomopathogenic nematodes for control of insect pests above and below ground with comments on commercial production. *J. Nematol.* 44: 218–225.

306 Ibrahim, M.A., Griko, N., Junker, M., and Bulla, L.A. (2010). *Bacillus thuringiensis*. *Bioeng. Bugs* 1: 31–50. https://doi.org/10.4161/bbug.1.1.10519.

307 Lacey, L.A., Frutos, R., Kaya, H.K., and Vail, P. (2015). Insect pathogens as biological control agents: back to the future. *J. Invertebr. Pathol.* 132: 1–41. https://doi.org/10.1006/bcon.2001.0938.

308 Hurst, M.R., Glare, T.R., Jackson, T.A., and Ronson, C.W. (2000). Plasmid-located pathogenicity determinants of Serratia entomophila, the causal agent of amber disease of grass grub, show similarity to the insecticidal toxins of Photorhabdus luminescens. *J. Bacteriol.* 182: 5127–5138. https://doi.org/10.1128/jb.182.18.5127-5138.2000.

309 Rippere, K.E., Tran, M.T., Yousten, A.A. et al. (1998). Bacillus popilliae and Bacillus lentimorbus, bacteria causing milky disease in Japanese beetles and related scarab larvae. *Int. J. Syst. Bacteriol.* 48: 395–402. https://doi.org/10.1099/00207713-48-2-395.

310 Sinha, K.K., Choudhary, A.K., and Kumari, P. (2016). Entomopathogenic fungi. In: *Ecofriendly Pest Management for Food Security*, 475–505. Amsterdam: Elsevier Inc.

311 Solter, L.F., Becnel, J.J., and Oi, D.H. (2012). Chapter 7: microsporidian entomopathogens. In: *Insect Pathology*, 2nde (ed. F.E. Vega and H.K. Kaya), 1–45. San Diego, CA: Elsevier.

312 Chiu, E., Coulibaly, F., and Metcalf, P. (2012). Insect virus polyhedra, infectious protein crystals that contain virus particles. *Curr. Opin. Struct. Biol.* 22: 234–240. https://doi.org/10.1016/j.sbi.2012.02.003.

313 Bravo, A., Gill, S.S., and Soberón, M. (2007). Mode of action of bacillus thuringiensis cry and cyt toxins and their potential for insect control. *Toxicon* 49: 423–435. https://doi.org/10.1016/j.toxicon.2006.11.022.

314 Schnepf, E., Crickmore, N., Van Rie, J. et al. (1998). *Bacillus thuringiensis* and its pesticidal crystal proteins. *Microbiol. Mol. Biol. Rev.* 62: 775–806. https://doi.org/10.1128/MMBR.62.3.775-806.1998.

315 Sanchis, V. (2011). From microbial sprays to insect-resistant transgenic plants: history of the biospesticide *Bacillus thuringiensis:* a review. *Agron. Sustain. Dev.* 31: 217–231. https://doi.org/10.1051/agro/2010027.

316 Mertz, F.P. and Yao, R.C. (1990). *Saccharopolyspora spinosa* sp. nov. isolated from soil collected in a sugar mill rum still. *Int. J. Syst. Evol. Microbiol.* 40: 34–39. https://doi.org/10.1099/00207713-40-1-34.

317 Wechsler, S. (2018). *USDA ERS - Recent Trends in GE Adoption. USDA*. https://www.ers.usda.gov/data-products/adoption-of-genetically-engineered-crops-in-the-us/recent-trends-in-ge-adoption.aspx (accessed July 2020).

318 Racke, K.D. (2006). A reduced risk insecticide for organic agriculture: spinosad case study. In: *Crop Protection Products for Organic Agriculture: Environmental, Health and Efficacy Assessment, ACS Symposium Series*, vol. 947 (ed. A.S. Felsot and K.D. Racke), 92–108. Washington, DC: American Chemical Society. https://doi.org/10.1021/bk-2007-0947.ch007.

319 Katz, M., Hover, B.M., and Brady, S.F. (2016). Culture-independent discovery of natural products from soil metagenomes. *J. Ind. Microbiol. Biotechnol.* 43: 129–141. https://doi.org/10.1007/s10295-015-1706-6.

320 Gabor, E.M., Alkema, W.B.L., and Janssen, D.B. (2004). Quantifying the accessibility of the metagenome by random expression cloning techniques. *Environ. Microbiol.* 6: 879–886. https://doi.org/10.1111/j.1462-2920.2004.00640.x.

321 Hofer, U. (2018). The majority is uncultured. *Nat. Rev. Microbiol.* 16: 716–717. https://doi.org/10.1038/s41579-018-0097-x.

322 Witzgall, P., Kirsch, P., and Cork, A. (2010). Sex pheromones and their impact on pest management. *J. Chem. Ecol.* 36: 80–100. https://doi.org/10.1007/s10886-009-9737-y.

323 Cardé, R.T. and Minks, A.K. (1995). Control of moth pests by mating disruption: successes and constraints. *Annu. Rev. Entomol.* 40: 559–585. https://doi.org/10.1146/annurev.en.40.010195.003015.

324 Davis, T.S., Crippen, T.L., Hofstetter, R.W., and Tomberlin, J. (2013). Microbial volatile emissions as insect semiochemicals. *J. Chem. Ecol.* 39: 840–859. https://doi.org/10.1007/s10886-013-0306-z.

325 Lemfack, M.C., Gohlke, B.-O., Toguem, S.M.T. et al. (2017). mVOC 2.0: a database of microbial volatiles. *Nucleic Acids Res.* 46: D1261–D1265. https://doi.org/10.1093/nar/gkx1016.

326 Choudoir, M., Rossabi, S., Gebert, M. et al. (2019). A phylogenetic and functional perspective on volatile organic compound production by actinobacteria. *MSystems* 4: e00295-18. https://doi.org/10.1128/mSystems.00295-18.

327 Medema, M.H., Blin, K., Cimermancic, P. et al. (2011). antiSMASH: rapid identification, annotation and analysis of secondary metabolite biosynthesis gene clusters in bacterial and fungal genome sequences. *Nucleic Acids Res.* 39: W339–W346. https://doi.org/10.1093/nar/gkr466.

328 Weber, T., Blin, K., Duddela, S. et al. (2015). antiSMASH 3.0 – a comprehensive resource for the genome mining of biosynthetic gene clusters. *Nucleic Acids Res.* 43: W237–W243. https://doi.org/10.1093/nar/gkv437.

329 Rappé, M.S. and Giovannoni, S.J. (2003). The uncultured microbial majority. *Annu. Rev. Microbiol.* 57: 369–394. https://doi.org/10.1146/annurev.micro.57.030502.090759.

330 Ballhorn, D.J., Kautz, S., and Schädler, M. (2013). Induced plant defense via volatile production is dependent on rhizobial symbiosis. *Oecologia* 172: 833–846. https://doi.org/10.1007/s00442-012-2539-x.

331 Clardy, J., Fischbach, M.A., and Walsh, C.T. (2006). New antibiotics from bacterial natural products. *Nat. Biotechnol.* 24: 1541–1550. https://doi.org/10.1038/nbt1266.

332 Fontana, A., Reichelt, M., Hempel, S. et al. (2009). The effects of arbuscular mycorrhizal fungi on direct and indirect defense metabolites of plantago lanceolata L. *J. Chem. Ecol.* 35: 833–843. https://doi.org/10.1007/s10886-009-9654-0.

333 Li, L., Chen, J., Yang, Z. et al. (2015). The effect of nano-Mg $(OH)_2$ on insecticidal activity and UV resistance of bacillus thuringiensis protein. *J. Agric. Biotechnol.* 23: 1452–1457.

334 Sharifi, R., Lee, S.-M., and Ryu, C.-M. (2018). Microbe-induced plant volatiles. *New Phytol.* 220: 684–691. https://doi.org/10.1111/nph.14955.

335 Keesey, I.W., Knaden, M., and Hansson, B.S. (2015). Olfactory specialization in *Drosophila suzukii* supports an ecological shift in host preference from rotten to fresh fruit. *J. Chem. Ecol.* 41: 121–128. https://doi.org/10.1007/s10886-015-0544-3.

336 Lasa, R., Tadeo, E., Toledo-Hérnandez, R.A. et al. (2017). Improved capture of drosophila suzukii by a trap baited with two attractants in the same device. *PLoS One* 12: e0188350. https://doi.org/10.1371/journal.pone.0188350.

337 Ishii, Y., Akasaka, N., Goda, I. et al. (2015). Effective trapping of fruit flies with cultures of metabolically modified acetic acid bacteria. *Appl. Environ. Microbiol.* 81: 2265–2273. https://doi.org/10.1128/AEM.03678-14.

338 Stensmyr, M.C., Dweck, H.K.M., Farhan, A. et al. (2012). A conserved dedicated olfactory circuit for detecting harmful microbes in *Drosophila*. *Cell* 151: 1345–1357. https://doi.org/10.1016/j.cell.2012.09.046.

339 Huang, B., Chen, F., Shen, Y. et al. (2018). Advances in targeted pesticides with environmentally responsive controlled release by nanotechnology. *Nanomaterials* 8: 102. https://doi.org/10.3390/nano8020102.

340 Epsky, N.D., Heath, R.R., Dueben, B.D. et al. (1998). Attraction of 3-methyl-1-butanol and ammonia identified from *Enterobacter agglomerans* to *Anastrepha suspensa*. *J. Chem. Ecol.* 24: 1867–1880. https://doi.org/10.1023/A:1022363718193.

341 Wallingford, A.K., Hesler, S.P., Cha, D.H., and Loeb, G. (2016). Behavioral response of spotted-wing drosophila, *Drosophila suzukii* matsumura, to aversive odors and a potential oviposition deterrent in the field. *Pest Manag. Sci.* 72: 701–706. https://doi.org/10.1002/ps.4040.

342 Lam, K., Tsang, M., Labrie, A. et al. (2010). Semiochemical-mediated oviposition avoidance by female house flies, musca domestica, on animal feces colonized with harmful fungi. *J. Chem. Ecol.* 36: 141–147. https://doi.org/10.1007/s10886-010-9741-2.

343 Daisy, B.H., Strobel, G.A., Castillo, U. et al. (2002). Naphthalene, an insect repellent, is produced by muscodor vitigenus, a novel endophytic fungus. *Microbiology* 148: 3737–3741. https://doi.org/10.1099/00221287-148-11-3737.

344 Gliszczyńska, A., Gladkowski, W., Dancewicz, K., and Gabryś, B. (2015). Enantioselective microbial hydroxylation as a useful tool in the production of jasmonate derivatives with aphid deterrent activity. *Curr. Microbiol.* 71: 83–94. https://doi.org/10.1007/s00284-015-0831-9.

345 Skrobiszewski, A., Gladkowski, W., Mazur, M. et al. (2018). Microbial hydrolysis of racemic β-Aryl-γ-ethylidene-γ-lactones and antifeedant activity of the products against alphitobius diaperinus panzer. *Molecules* 23: 1516. https://doi.org/10.3390/molecules23071516.

346 Kajla, M.K., Barrett-Wilt, G.A., and Paskewitz, S.M. (2019). Bacteria: a novel source for potent mosquito feeding-deterrents. *Sci. Adv.* 5: eaau6141. https://doi.org/10.1126/sciadv.aau6141.

347 Leroy, P.D., Sabri, A., Verheggen, F.J. et al. (2011). The semiochemically mediated interactions between bacteria and Insects. *Chemoecology* 21: 113–122. https://doi.org/10.1007/s00049-011-0074-6.

348 Schulz-Bohm, K., Zweers, H., de Boer, W., and Garbeva, P. (2015). A fragrant neighborhood: volatile mediated bacterial interactions in soil. *Front. Microbiol.* 6: 1212. https://doi.org/10.3389/fmicb.2015.01212.

349 Cook, S.M., Khan, Z.R., and Pickett, J.A. (2006). The use of push-pull strategies in integrated pest management. *Annu. Rev. Entomol.* 52: 375–400. https://doi.org/10.1146/annurev.ento.52.110405.091407.

350 Kim, J.S. and Je, Y.H. (2012). Milling effect on the control efficacy of spray-dried *Bacillus thuringiensis* technical powder against diamondback moths. *Pest Manag. Sci.* 68: 321–323. https://doi.org/10.1002/ps.2330.

351 Vassilev, N., Vassileva, M., Martos, V. et al. (2020). Formulation of microbial inoculants by encapsulation in natural polysaccharides: focus on beneficial properties of carrier additives and derivatives. *Front. Plant Sci.* 11: 270. https://doi.org/10.3389/fpls.2020.00270.

352 Khot, L.R., Sankaran, S., Maja, J.M. et al. (2012). Applications of nanomaterials in agricultural production and crop protection: a review. *Crop Prot.* 35: 64–70. https://doi.org/10.1016/j.cropro.2012.01.007.

353 Kumar, S., Nehra, M., Dilbaghi, N. et al. (2019). Nano-based smart pesticide formulations: emerging opportunities for agriculture. *J. Control. Release* 294: 131–153. https://doi.org/10.1016/j.jconrel.2018.12.012.

354 Dyck, V.A., Hendrichs, J., and Robinson, A.S. (ed.) (2006). *Sterile Insect Technique: Principles and Practice in Area-Wide Integrated Pest Management*, 39–68. Dordrecht: Springer https://doi.org/10.1007/1-4020-4051-2.

355 Vreysen, M.J., Saleh, K., Mramba, F. et al. (2014). Sterile insects to enhance agricultural development: the case of sustainable tsetse eradication on Unguja Island, Zanzibar, using an area-wide integrated pest management approach. *PLoS Negl. Trop. Dis.* 8: e2857. https://doi.org/10.1371/journal.pntd.0002857.

356 Munhenga, G., Brooke, B.D., Gilles, J.R. et al. (2016). Mating competitiveness of sterile genetic sexing strain males (GAMA) under laboratory and semi-field conditions: steps towards the use of the sterile insect technique to control the major malaria vector anopheles arabiensis in South Africa. *Parasit. Vectors* 9: 122. https://doi.org/10.1186/s13071-016-1385-9.

357 Harris, A.F., McKemey, A.R., Nimmo, D. et al. (2012). Successful suppression of a field mosquito population by sustained release of engineered male mosquitoes. *Nat. Biotechnol.* 30: 828–830. https://doi.org/10.1038/nbt.2350.

358 Raza, M.F., Yao, Z., Bai, S. et al. (2020). Tephritidae fruit fly gut microbiome diversity, function and potential for applications. *Bull. Entomol. Res.* 110: 423–437. https://doi.org/10.1017/S0007485319000853.

359 Drew, R.A. and Yuval, B. (2000). The evolution of fruit fly feeding behavior. In: *Fruit Flies (Tephritidae) Phylogeny and Evolution of Behavior* (ed. M. Aluja and A.L. Norrbom), 749–768. Boca Raton: CRC Press. https://doi.org/10.1201/9781420074468.ch27.

360 Ami, E.B., Yuval, B., and Jurkevitch, E. (2010). Manipulation of the microbiota of mass-reared mediterranean fruit flies ceratitis capitata (diptera: tephritidae) improves sterile male sexual performance. *ISME J.* 4: 28–37. https://doi.org/10.1038/ismej.2009.82.

361 Stathopoulou, P., Asimakis, E.D., Khan, M. et al. (2019). Irradiation effect on the structure of bacterial communities associated with the oriental fruit fly, bactrocera dorsalis. *Entomol. Exp. Appl.* 167: 209–219. https://doi.org/10.1111/eea.12770.

2

Engineering and Management of Agricultural Microbiomes for Improving Crop Health

CONTENTS

Microbiomics and Sustainable Crop Production, First Edition. Mohammad Yaseen Mir and Saima Hamid.

2.1 Why to Modify Plant Microbiome?

There are two distinct definitions for the term microbiome as it has been described that microbiome is the collection of genomes of microorganisms in a specific environment, while the microbiota is the set of microorganisms in a specific habitat. As a result, they use the name microbiota referring to microbial communities. Others use the term microbiome to refer to all of a community's bacteria, and in particular the plant microbiome, which submits to the microbial species associated with the plant that can survive and communicate with various tissues, such as shoots and roots [1, 2]. Bacteria, fungi, protozoa, archaea, and viruses make up the microbiome, which is made up of a variety of species [3]. Plant microbiomes can shield plants from diseases while also enhancing growth, health, and productivity. They can also provide plants with an adaptive advantage [4, 5]. In comparison to other classes of species, the bacterial microbiome, which performs beneficial interactions, is particularly important for sustainable agriculture. According to Glick [6] and Santoyo and coworkers [7], plant growth is regulated by either direct or indirect mechanisms of plant growth-promoting bacteria (PGPB). In a nutshell, a bacterium promotes plant growth by facilitating the acquisition of vital nutrients or modulating the amount of hormones within a plant. PGPB facilitates the acquisition of nutrients such as nitrogen, phosphorus, and iron, which are necessary for plant growth [8]. In the synthesis of one or more phytohormones, auxins, cytokinins, and gibberellins, PGPB can modulate hormone levels [9]. Furthermore, certain PGPB can reduce ethylene levels by producing an enzyme called 1-aminocyclopropane-1-carboxylate (ACC) deaminase, which cleaves the compound ACC, which is the immediate precursor of ethylene in all higher plants [6, 10]. When a PGPB reduces plant damage after infection with a plant pathogen, such as harmful fungi or bacteria, this is known as indirect plant growth promotion. This is generally attributed to the pathogens being inhibited by PGPB [11]. Antibiotics (such as 2,4-diacetylphloroglucinol), proteases, chitinases, bacteriocins, siderophores, lipopeptides (such as iturin A, bacillomycin D, and mycosubtilin), and volatile organic compounds, to name a few, are synthesized and excreted by these pathways [12]. The microbiome's capacities make it an important component for the plant to carry out physiological functions, and, in certain situations, the microbiome is a vital part of basic processes like plant production or the growth of crucial organs like the root, for improved nutrient and water absorption [13]. Despite the fact that the plant microbiome is made up of a variety of organisms that all play important roles, the emphasis of this analysis is on bacterial species, especially those that have biocontrol and plant growth-promoting properties. The plant microbiome, as well as the regions or zones where the microbes can interact with the plant, either they live outside (rhizosphere), inside (endosphere) or on (phyllosphere) the plant. The rhizosphere is the soil portion influenced by plant roots. The rhizosphere is the part of the soil where plant roots have an influence [14]. Chemical communications and the exchange of compounds and nutrients between plant and soil microorganisms take place in this microecosystem [15]. The collection and alteration of the rhizosphere microbiome is part of a strategy to enhance plant health and growth [16]. Germida et al. [17] and Márquez-Santacruz et al. [18] suggested that the endospheric microbiome is a branch of the rhizosphere microbiome. Bai et al. [19] discovered a significant overlap between the *Arabidopsis* leaf and root microbiomes, revealing the constantly changing state of microbes communicating with the plant host. The endosphere is a plant's internal ecosystem where endophytes such as bacteria, archaea, fungi, and viruses live, colonize, and thrive without harming the plant [20]. The colonization of internal plant tissues by endophytes has been studied extensively. Santoyo et al. [21] analysed available evidence on different entry points, with the rhizosphere being a key area for microbe entry into the plant due to its nutrient-rich existence, supplied by root exudates, and hence its high microbe concentration [22]. Lenticels, stomas, cuts, ruptures, and nodules are examples of such points of entry. Endophytes may also be passed down through the generations through seeds [21]. Some scientists suggest that the inner microbiome of plants has a benefit over external bacteria (phyllosphere or rhizosphere) and that it has no effects, including bacterial degradation, upon soil conditions. Furthermore, (chemical) contact inside tissues of the plant was suggested to become more efficient [23], as they become more effective for plant than those secreted in open environments, such as the rhizosphere, as lower concentrations of metabolite secreted by bacterial endophytes will mitigate and limit their effects by biotic and abiotic influences [24]. For example, other species such as fungi, which have non-specific transporters of siderophore–Fe complexes, collect and internalize siderophores of bacterial origin excreted in the rhizosphere [25]. Various experiments have shown that rhizosphere composition is complex rather than endosphere composition. An increased number of operational taxonomical units were identified in a library of clones from the rhizosphere in the sample of rhizosphere and endosphere 16S rRNA genes from Mexican husk tomato plants (*Physalis ixocarpa*), relative to the root endosphere. The number of clones collected spontaneously from the endosphere against the rhizosphere is statistically comparable, which should be noted. *Stenotrophomonas*, *Burkholderia*, *Bacillus*, and *Pseudomonas* are major group of rhizospheric bacteria that are present in the plant's internal tissues [18]. Various findings revealed the presence of a core microbiome, which is transmitted from generation to generation with positive functions in *Phy. ixocarpa* crops. Phyllosphere consists of leaf tissue surface and apoplast [26]. It has been suggested for years that the phyllosphere microbiome is of major significance for biocontrol and the promotion of plant growth. Indeed, several authors have suggested the exercise of beneficial activity for foliar microbiota, comparable with that for endophytic and rhizospheric microbes [27]. Many foliar microbiotas, for example, will fix nitrogen, which is the primary source of nitrogen for plants in tropical humid environments [28]. Indirect pathogen resistance and the development of plant hormones are two other beneficial functions of the microbiota on the plant surface. As a result, foliar microbiota can be used to minimize the use of agrochemicals in the management of leaf pathogens and to stimulate plant growth [27]. However, further research is required to get a deeper understanding of how these microorganisms function. The aforementioned mechanisms have been thoroughly investigated, and several studies have shown that PGPB has the ability to inhibit phytopathogen development while promoting plant growth. Any plant that has ever been researched has been linked to microorganisms, to the best of our understanding. Furthermore, as part of their internal microbiome, all plants are believed to contain endophytic cells [11]. We know that vigorous organisms that are part of plant microbiomes modify the physiology, overall development, tolerance range against abiotic factors like salinity stress, and enhance disease resistance [29]. These roles are not performed by the entire microbiome; however, these are performed by one or a few species of microbials and often due to additional or synergistic effects of two or more strains [30]. As per findings of Mueller and Sachs [3] and Yuan and coworkers [31], not all the microbial population play aforementioned role in plants but only few bacterial species have the capacity to perform such functions of metabolic pathways in order to increase fruit yield. At various locales, plants are exposed to different environmental factors like rainfall or nutrient availability, which enables the plant varieties to have different tolerance range, but plant microbiome can help them to cope up with changing biotic and abiotic stress conditions [24]. Hence the priority is to transfer these traits, which will help to fight against pathogens in those plant species that are lacking this regulatory mechanism [32]. As opposed to single gene transfer, transferring one or more microbial organisms, mostly bacteria, allows for the transfer of a larger volume of genetic material or gene numbers as in *Bacillus thuringiensis*, which contains cry toxins due to embedded genes to perform some other functions as well [33]. Moreover, *Pseudomonas fluorescens* UM270 or *Arthrobacter agilis* UMCV2 are the PGPB that can play essential as well as other activities that are secondary mechanisms to enhance overall plant development [34]. However, in-depth studies are the need of the hour to get details of core microbiomes whether present in host plant or inoculated to some other plant species [24, 35].

2.2 Methods for Detecting Endophytes Within the Plant

Characterization of endophytic microorganisms is a critical phase in understanding their phylogeny, complexity, relationships with plants, and potential uses as bio-inoculants to increase agricultural plant fitness and investigate origins of biologically active molecules that are holding significance in the field of industries and medicinal values [36, 37]. Multiple endophytes should be recovered, but isolation protocols should still be rigorous enough to remove saprophytes and other plant associates [36]. Endophytic microbes, including fungi and bacteria, are isolated in a similar way. Surface sterilization, plating tissue parts or crushed plant tissues on various media, and purification of microbes that emerge from tissues are all steps in the isolation process [38]. In view of the fact that endophytes are being isolated from all the different plant parts like shoots and roots, as the isolation is always associated with surface microbial pathogens and saprobic species, fresh and young plant parts like twigs, leaves, stem, and fruits are preferred. However, the isolated endophytes took bit longer to grow than the saprobes thus hampering their growth [39]. After the collection of plant parts, the minimum storage should consist of 24 hours or less that too kept in refrigerator, followed by cutting of obtained plant parts to proceed with surface sterilization methods. As the surface sterilization choice is not much so far because the removal efficiency of various available methods is not fully known yet as their demerits more importantly. Every method includes dipping of plant tissue in CH_3OH (75–90% for 1–2 minutes), NaClO (4% for 5–10 minutes), followed by washing with CH_3OH (90% for 30 seconds), and final washing with distilled water that too multiple times to wash off all the sterilizing agents [40]. Having an imprint of a sterilized tissue sample may be used to test the effectiveness of surface sterilization [41]. Since the nature, thickness, roughness, and permeability of various plant tissues differ, the steps and concentrations of sterilizing agents can differ from plant to plant and tissue to tissue. As a result, some work is needed to improve the surface sterilization protocol prior to endophyte isolation [42]. The type of endophytes (fungi, bacteria, and actinomycetes) to be isolated determines the best isolation media.

2.2.1 Media for Isolation of Fungal Endophytes

Various researchers analysed different prepared media for the isolation in order to get pure cultures for the endophytic fungi as well as bacteria. Bills and Polishook [43] obtained some of the isolates by utilizing malt-yeast extract media (1% malt extract, 0.2% yeast extract) for fungal diversity from the root and fruits of *Azadirachta indica*, in 2011, while Verma et al. [44] utilized malt-yeast extract agar (MYA), potato dextrose agar (PDA), mycological agar (MCA: 1% papaic digest of soybean meal, 1% dextrose, 1.5% agar), and nutrient agar (0.5% peptone, 0.3% beef extract, 0.5% NaCl, 1.5% agar) in order to obtain fungi with better efficacy for PDA and MCA. Some of the medium used for the isolation of endophytic fungi are shown in Table 2.1.

Table 2.1 Some of the common media and antimicrobial agents used for the isolation of endophytic fungi.

Endophytic fungi medium	Composition	Antimicrobials	References
Czapek-Dox agar (CzDA)	$MgSO_4.7H_2O$ 0.5 g; KCl 0.5 g; $NaNO_3$ 2.0 g; K_2HPO_4 1.0 g; $FeSO_4$ 0.01 g; sucrose 30.0 g; agar 15.0 g; distilled water	Streptomycin and/or ciprofloxacin (100–150 μg mL^{-1})	[45, 46]
Potato dextrose agar (PDA)	Potato 200.0 g; dextrose 20.0 g; agar 20.0 g; distilled water	Streptomycin and/or ciprofloxacin (100–150 μg mL^{-1})	[45, 46]
Malt extract agar	Malt extract 30.0 g; K_2HPO_4 1.0 g; NH_4Cl 1.0 g; citric acid 1N 15.0 mL; agar 20.0 g; distilled water	Streptomycin and/or ciprofloxacin (100–150 μg mL^{-1})	[45, 46]

2.2.2 Media for Isolation of Bacterial Endophytes

Hallmann et al. [36] provided detailed instructions on how to isolate bacterial endophytes. Various forms of solid agar media may be applied with crushed tissue samples in distilled water. Luria-Bertani agar (LBA), tryptic soy agar (TSA: 10%), yeast extract sucrose agar (YESA), and nutrient agar are some of the most popular media used to isolate bacterial endophytes. Antifungals such as cycloheximide can also be used in media for bacterial endophyte isolation (Table 2.2). For actinomycete isolation, starch casein agar (SCA) with minor ingredient changes

Table 2.2 Some of the common media and antimicrobial agents used for the isolation of endophytic bacteria.

Endophytic bacteria	Composition	Antimicrobials	References
Tryptic soy agar (10%)	Tryptic soy agar 4 g; agar 14 g; distilled water	Cycloheximide or nalidixic acid (100–150 μg mL^{-1})	[47]
Yeast extract sucrose agar	Yeast extract 10 g; sucrose 10 g; agar 15 g; distilled water	Cycloheximide or nalidixic acid (100–150 μg mL^{-1})	[47]
Luria-Bertani agar (LBA)	LBA 15 g; distilled water	Cycloheximide or nalidixic acid (100–150 μg mL^{-1})	[47]

is most widely used [40, 48]. To prevent fungal growth, antifungal agents such as cycloheximide or nalidixic acid should be added to the media [49].

2.2.3 Identification of Endophytes

Following the isolation and purification of endophytes, the identification of microbes and assessment of species diversity are critical next steps [50]. Endophytes may be classified using culture-dependent methods or culture-independent methods directly from the tissues [50]. The cultivation-dependent technique, which is commonly used in determining endophyte species diversity, is the most popular tool for identifying fungal endophytes [45, 46] in the bioprospection of endophytes for bioactive metabolite production [51, 52] and as mediators of host defence against biotic and abiotic stresses [47, 53]. Endophytic microbes are isolated and purified from healthy tissues onto culture media (as previously described), and identified. Development techniques, colony colour, texture, and fruiting structures, such as spores, conidia, conidiophores, sporangiophores, or sexual spores, can all be used to identify endophytic fungi in media. This necessitates taxonomic knowledge as well as references that relate fungi structural features to their identities. *Illustrated Genera of Imperfect Fungi* by Barnett and Hunter [54], *The Genera of Fungi Sporulating in Pure Culture* by Von Arx [55], *More Dematiaceous Hyphomycetes* by Ellis [56], and *The Fungi: An Advanced Treatise* by Ainsworth et al. [57] are some of the most commonly used standard taxonomic manuals for fungi. The appearance of mycelia sterilia (fungi that do not produce sexual or asexual spores) is a common problem in fungal endophyte research, making identification difficult [50]. Taxonomic protocols also lack sufficient information on spore-producing fungi to enable species identification. Investigators were able to identify 28 species by morphological characters from 40 different morphotypes in a study of fungal endophytes of Mahua plants (*Morina longifolia*) from India [46], 14 species out of 17 morphotypes in *Adenocalymma alliaceum* [38], 24 out of 29 morphotypes in *Tinospora cordifolia* [45], and 17 out of 19 morphotypes in *Nyctanthes arbor-tristis* have been identified based upon morphological characters [58]. Bacterial species cannot be differentiated solely on the basis of morphology due to the lack of distinguishing morphological characteristics.

2.2.4 Molecular Tools to Identify Endophytes

Many endophytes would benefit from the morphological method of endophyte identification [46, 59]. Due to inability to sporulate in culture, a significant number of isolates are classified as mycelia sterilia. In a study of endophytes in palm (*Trachycarpus fortunei*), 11% were reported as mycelia sterilia, while 54% were reported as mycelia sterilia in a study of endophytes in *Quercus ilex* in Switzerland [60–62]. Few researchers attempted, but failed, to enhance sporulation in mycelia sterilia in culture by altering the nutritional composition of the medium [61]. Due to a lack of literature, even certain spore-producing fungi are difficult to classify into proper taxonomic units [46]. Furthermore, some taxonomists conclude that morphological characters are insufficient to differentiate some taxa or are not accurate taxonomic features [50]. As a result, molecular approaches could be the best option for identifying obtained and cultured endophytic microbes. The taxonomic status of fungi isolated from various environments has been determined using molecular methods [63]. Isolation of total genomic DNA from all recovered isolates from the host plant, polymerase chain reaction (PCR) amplification of conserved DNA (ribosomal DNA [rDNA]) sequences using specific or universal primers, sequencing of PCR products, BLAST options in GenBank or other DNA databanks, retrieval of related sequences from the databank, and construction of phylogenetic trees using molecular software are all needed for molecular identification [64]. Since certain sections of nucleic acid sequences are extremely stable and conserved across a broad variety of species, genotype-based microbe recognition is thought to be more accurate [65]. Certain rDNA sequence positions have been shown to have the best specificity for identifying bacteria and fungi up to species level, and they can also be used in phylogenetic studies [66].

2.2.5 Markers and Primers for Endophyte Identification

The 18S, 28s, 5.8S, ITS1, and ITS2 subunit regions make up the fungal nuclear rRNA operon (rDNA), which is present in several copies in fungal genomes [67]. The smaller subunit gene (SSU:18S) is isolated from the larger subunit gene (LSU:28S) by the internal transcribed spacer (ITS1 and ITS2) regions. Another 5.8S gene intercalates ITS1 and ITS2, which are strongly variable spacers. Two intergeneric spacers (IGS1 and IGS2) split each rRNA gene, with an additional 5S gene intercalated. Highly conserved rDNA regions and less variation across related taxa are needed for a successful marker, but a useful marker must also be able to distinguish between closely related taxa [65]. Nuclear rDNA genes have been used as an ideal marker for fungi recognition because of their high interspecies variance and low intraspecies variation [65]. For accurate analysis, amplification of fragment gene via primer is essential for phylogenetic studies. In the detection of fungal endophytes, the primers ITS1, ITS2, and ITS4 are commonly used [45]. The primers ITS1 (F) and ITS4 (R) amplify partial 18S, complete ITS1, ITS2, and 5.8S, as well as partial 28S rRNA gene sequences. Some researchers contend that using NL1 and NL2 primers to amplify the D1 and D2 regions of 28S rRNA provides better resolution than using ITS primers [68]. Similarly, since the 16S rDNA gene is heavily conserved and provides strong species resolution in most bacterial classes, many bacterial endophytes can be classified using it. 8F, UnF, 11F, 27F, 6R, UnR, 907R, 1492R, and 1525R are some of the primers used to amplify 16S rDNA genes, each of which amplifies about 1000–1500 base pairs of rDNA [69]. While 16S rDNA gene sequences allow for more stable, repeatable, and reliable bacterial identification than phenotypic testing, it is often difficult to classify species in some bacterial groups due to limited diversity of the 16S rRNA gene in some groups and limited coverage in some databases [70].

2.2.6 Techniques to Evaluate Endophyte Distribution in Plants

Tissue blotting techniques were developed to ensure surface sterilization while isolating endophytes from the interior of healthy plant tissues [41]. This method may also be used to see whether the hypothesized endophytes are present on plant tissues. Microdissection and histological visualization of plant tissues using laser, confocal, or electron microscopy can be used to better assess endophytes [50]. Histological investigation of tissue is a method to observe all types of endophytes, including bacteria and fungi, within living host tissues. This enables us to see endophyte growth patterns, e.g. intra- or intercellular nature, and endophyte density in tissue and identify the part of tissue in which the endophyte is more concentrated. Most endophytes within plant tissues cannot be identified to any taxonomic category because they do not produce any identifying characters in vivo. The tissues must be stained with a fungal-specific or stain mixture and placed for microscopic examination of endophytes. Stone et al. [71] present a step-by-step procedure for clearing tissues to enable direct visualization of endophytic fungi in host tissues. The following staining techniques must be advanced for plant tissue and endophytic microbes [72].

2.2.6.1 Hood and Shew Staining Protocol

Surface sterilization techniques involve washing of stem or leaf tissues (0.5 cm^2) under triple distilled water followed by potassium hydroxide (2.5%) with many times repetition for several days and with heating at 40–60 °C with changing this solution three to four times a day for the period of 10 days. The leaf tissues are rinsed three times in distilled water before being bleached for three minutes in 4% NaOCl and rinsed again in distilled water. Solution of 0.05% aniline blue dye in 0.067 M K_2HPO_4 at pH 9.0 is prepared at least two to three hours before use for staining KOH-aniline blue. Leaf portions are placed on glass slides and visualized under a compound light microscope with tiny drops of stain solution.

2.2.6.2 Fluorescent Probes for Localization of Bacterial and Fungal Endophytes

In the analysis of endophyte–plant interactions, fluorescent probing to establish internal endophyte distribution (intra- or intercellular) and density of endophytic microbes in different plant tissues has become a common technique. Green fluorescent protein (GFP) that autofluoresces is one kind of fluorescent probe, while chemically related fluorescent compounds that bind to proteins and nucleic acids or react with enzymes are another [73, 74]. To monitor the position of endophytic bacteria within plant tissues, tagging bacteria with GFP using compatible vector has become common. The most important move, however, is to tag with GFP as Mousa et al. [75] used confocal scanning microscopy to image endophytic bacteria within root tissues of maize, wheat, and millets after successfully tagging the endophyte *Enterobacter* sp. with GFP using vector (pDSKGFPuv). Fluorescein diacetate (FDA), carboxyfluorescein diacetate (CFDA), chloromethylfluorescein diacetate (CMFDA), SYTO9 (S9), propidium iodide (PI), diamidino-2-phenylindole (DAPI), and other fluorescein-based compounds or dyes have been used to observe in vivo tissue colonization by endophytic bacteria and fungi. A cellular esterase enzyme present in plants and microbes activates the fluorescence of FDA, CFDA, and CMFDA. Fescue grass belonging to Pooideae family was used to investigate *Epichloë coenophiala* using CFDA, CMFDA, FDA, and calcofluor white fluorescent compounds [74] with preference over choice of method for staining being CMFDA by producing dissimilarity among endophytes and host tissues. SYTO9, SYTO13, PI, and DAPI are nucleic acid-staining dyes [76]. SYTO9 and SYTO13 have been shown to be very useful to stain and visualize living bacterial endophytes in different plant tissues [77]. SYTO9 stained live bacteria in banana shoot tops, while PI and DAPI stained dead bacteria or bacteria inside infected plant tissues, according to Thomas and Reddy [76]. SYTO9 and SYTO13 have also been used by White and his colleagues to find the bacterial endophyte in a variety of plant tissues, including vanilla, cotton, turf grasses, and amaranthus. Excitation/emission bands for DNA and RNA are 485/498 nm and 486/501 nm, respectively. Surface-sterilized plant tissues were cut into fine sections and stained with SYTO9 or SYTO13 (10–20 M) in water for 5–10 minutes in a fluorescent microscope to observe endophytic bacteria in vitro. Staining without sectioning will reveal endophytic bacteria around seedling root cells and root hairs [77].

2.2.6.3 ROS Staining to Study Bacterial Endophytes

In plants, reactive oxygen species such as H_2O_2, $\cdot O^2$, and $\cdot OH$ are known to be essential signalling molecules in a variety of metabolic and developmental processes. H_2O_2 is the most soluble reactive oxygen species in tissue, and it is shown to be elevated in response to biotic and abiotic stresses [78]. Many bacterial endophytes have been found to invade plant cells and produce hydrogen peroxide in cells near intracellular bacteria [79]. In case of bacterial endophytes, 3,30-diaminobenzidine tetrachloride (DAB) was found to be the preferred choice to stain tissues and organs along with counterstain aniline blue/lactophenol in order to visualize both intra- and intercellular endophytic bacteria; moreover, using cellular staining techniques including H_2O_2 can perform well for vascular plant species [79, 80]. The protocol to stain reactive oxygen species includes thorough washing of roots with distilled water followed by overnight soaking of root seedlings in DAB solution (2.5 mM), then washing with deionized water, counterstaining with aniline blue, and finally slide mounting. Surface bacteria are spotted with aniline blue, while endophytic bacteria stained with DAB are dark brown. This method of staining is selective in order to stain bacteria in and outside of the cells. The combination of DAB–aniline–blue stain allows the visualization of bacterial deformity and lysis in plant root cells, too. Many root cells of many plants were found with different types and swollen endophytic bacteria (L-forms) [79, 80]. White and his team are currently focusing on this staining process in order to envision the endophysical bacterial population of numerous vascular plants like *Cynodon* sp., *Phragmites* sp., *Daucus carota*, *Oryza sativa*, *Triticum aestivum*, etc. The staining process is therefore not successful in the visualization of endophytes in shoot systems, as waxy cuticle and pigmentation fail to penetrate the stem.

2.2.7 Analysis of Endophyte Diversity

Non-molecular approaches may be used to investigate the diversity of isolated endophytic microbes on various culture media. There are various methods for evaluating the diversity of isolated bacteria, such as frequency of colonization, frequency of separation, alpha diversity indices (Simpson, Shannon-Wiener diversity indices), species diversity evenness, etc. [45, 46, 81]. The extent of colonization (percent cf) of endophytic fungi can be measured manually with Hata and Futai formulation [81]: CF 1/4 percent CF/Nt 100, where the number of segments of Ncol 1/4 was analysed and the number of Nt 1/4 of the total segments colonized by fungal population. Verma and coworkers [46] and Orduña et al. [82] concluded that biodiversity Pro, Origin software, and PAST are programmed softwares to analyse the diversity indices like Simpson's and Shannon-Wiener by using formula of Simpson's diversity, ¼ 1 Σ(pi) 2, and Shannon-Wiener diversity, ¼ Σ s (pi log pi), as well as evenness (E) ¼ H/log(S), where H ¼ is Shannon-Wiener diversity and S ¼ is species richness (i.e. total number of species). To measure significant differences among mean diversity indices of samples obtained from different locations and tissues, statistical verification can be performed using analysis of variance (ANOVA) with SPSS 16.0 or a newer version of software (to check the temporal and spatial variation in diversity). Kharwar and his colleagues used the same method to examine the variety of endophytic fungi in over 10 medicinal plants from northern India. Principal component analysis (PCA) can be used to assess the interrelationship and tissue and site specificity of endophytic microbial communities among different tissue samples and sites, as well as ecological interactions between endophytes and different tissues and locations of the host [82].

2.2.8 Non-Culture Methods

The number of microbes isolated into pure culture and the number of endophytes found in plants vary (as little as 1–5% of microbes can be isolated from the environments of the natural world) [83]. Many microbes are unlikely to be isolated, and cultivation-based approaches are likely to skip the majority of endophytic diversity, as well as other microbial habitats [84]. Often, bacteria are used to isolate them because they do not thrive in the current conditions. Molecular methods have been used to uncover the diversity of microbes inside host tissues in recent years [85]. To elucidate fungal endophyte diversity, Guo et al. [85] amplified the 18S rRNA gene from complete DNA isolated from *Livistona chinensis* frond tissues, then cloned, sequenced, and performed phylogenetic analysis. Some new species of fungal endophytes have been obtained that have not before been isolated by cultivation methods. The analysis of molecular diversity and endophyte phylogeny has been advanced by technology [85]. Denaturing gradient gel electrophoresis (DGGE) has been successfully used to research the direct variety of endophytes in *Magnolia liliifera's* leaves [85].

2.2.9 Metagenomics and Pyrosequencing

Studying the genetics of recovered samples collected from environment like water, soil, plants, and animals comes under the category of metagenomics or environmental genomics. As per the various scientific reports like from Guo et al. [85], Duong et al. [86], and Marco [87], the metagenomics offers flexible study of genes beyond the phylogenetic studies that rely upon 16S or 18S RNA as the techniques have been employed for plethora of endophytes in order to study its functional genes along with diversity. According to reports of Romero et al. [88] and Manter et al. [89], researchers employed the technique of pyrosequencing in order to get idea of diversities for endophytic fungal and bacterial communities in various host plants. As next-generation sequencing, pyrosequencing can yield advanced results for sequencing without cloning process, which is cost effective. Shotgun metagenomics and pyrosequencing reveal the study of theoretical metabolic aspects of organism and its diversity in host plants [90]. In the study of functional endophytism, this may be very interesting. Metagenomics and pyrosequencing technologies have been recently successfully used to examine the functional metabolic role of endophytes and their structural diversity in several host plants [90]. Endophytes are active in fostering plant growth and improving stress resistance to biotic and abiotic stresses, according to Sessitsch et al. [91], who found that endophytes play a role in the functional functions of the endophytic population of rice roots. Endophytes may also play a role in the whole nitrogen cycle, including nitrogen fixation, denitrification, and nitrification, according to the researchers. Tian et al. [90] looked at the functional genomics of stable and nematode-infected tomato root-endophyte microbiomes and compared them. It has been demonstrated that in nematode pathogenesis, the bacterial colonies were involved. Jumpponen and Jones [92] used 454 pyrosequencing to examine fungal communities in the phyllosphere of *Quercus macrocarpa* and compare fungal diversity and distribution among trees from various locations. Toju et al. [93] identified the association of endophytic and mycorrhizal fungi in the roots of various plant species using 454 pyrosequencing techniques. In recent decades, metagenomics and next-generation sequencing (NGS) technology have been widely used in biological research, but the use of the technology to disclose endophytic functional diversity and its structure is limited to a very small number of studies. Any technical shortcomings include very complicated and detailed databases, which require innovative instruments, mathematical applications, extensive storage, and viewing and analysing current data sets [94]. In addition, the annotation of the presented sequences in public libraries is incomplete and this may be the one massive restriction of metagenomic approaches.

2.2.10 Microarray: Gene Chips to Study the Expression and Mechanisms of Interaction

One of the fastest developing modern molecular techniques is a microarray of DNA (usually known as the DNA chip or the gene chip) along with real-time PCR (RT-PCR) analysis. The RT-PCR is now very popular technique used in modulating the gene expression of host plants [95]. But it is restricted at a time to a few genetic experiments; however, the DNA microarray has the advantage of studying whole genetic expression in the organism and is being used in biological systems for the study of differential gene expression. The molecular dynamics of plant–endophyte interactions could be studied using this essential technology [96]. Since little is known about the molecular mechanisms underlying plant–endophyte interactions, a microarray analysis followed by RT-PCR may shed more light on the topic. Using DNA chip techniques, several experiments have attempted to compare gene expression in endophyte-free and endophyte-containing plants [97, 98]. Many crops and medicinal plants have commercially available gene chips. For at least those plants for which microarray platforms are available or for their close relatives, researchers have the ability to analyse and discern the relative expression of a wide number of genes simultaneously [96]. Many genes involved in essential metabolism, signal transduction, and stress control were found to be unregulated in *Arabidopsis–Ps. fluorescens* (endophyte) interactions, according to microarray research. Auxin- and nodulin-related genes were significantly unregulated, while ethylene-responsive genes were significantly downregulated [99]. Dinkins et al. [96] used the Affymetrix Wheat Genome Array GeneChip and Barley1 Genome Array GeneChip to investigate the differential gene expression in tall fescue with their obligate endophyte

Ep. coenophiala, with some results. Irizarry and White have used Affymetrix cotton Gene Chip to investigate the function of *Bacillus* sp. (endophyte) in cotton seedling gene expression differentials. Endophytes were discovered to upregulate certain primary genes involved in nitrogen metabolism and auxin synthesis (not published). They have also discovered that the same bacteria have a phenotypic impact on the growth of stronger root architecture than control bacteria [100]. Development promotion, stress tolerance, increased mineral acquisition from soils, and improved nitrogen utilization are all phenotypic effects of endophytes on their hosts [101]. The molecular mechanisms of endophytes' beneficial effects on plants could be studied using microarray analysis. The lack of gene chip platforms for individual plants, as well as compatible reference data sets for statistical analysis, is a significant weakness of microarray analysis. There are a variety of methods for inoculating strains to introduce and/or engineer the plant microbiome in plants. Inoculation into bulk dirt, rhizosphere, seeds, or seedlings; atomization into tissues such as roots, leaves, and bulbs; and direct injection into tissues or wounds are all examples of host-mediated and multi-generation microbiome range. The following are some recent related research concerning the aforementioned techniques.

2.3 Engineering of the Plant Microbiome

2.3.1 Host-Mediated and Multi-Generation Microbiome Selection

According to Mueller and Sachs [3], host-mediated microbiome selection is an engineering strategy that selects microbial communities indirectly via the host and takes advantage of host characteristics that have evolved to affect microbiomes. Swenson et al. [102] conducted groundbreaking work in which they attempted 'artificial selection of the environment' by choosing the maximum (high selection lines) or lowest (low selection lines) plant biomass of *Arabidopsis thaliana* that had previously been changed by microbial contact (bacteria, algae, protozoa, and fungi) with the plant over 16 generations. An aquatic microcosm has been designed in order to deteriorate 3-chloroaniline, which works at low or high pH; hence the applicability of bioengineering can be used to analyse microbiota at ecosystem levels, which would have given a clearer image of the microecosystem's biotic evolution. Panke-Buisse et al. [103] recently used a methodology that resembles with Swenson et al. [102]; a multi-generation experiment was conducted on *Ar. thaliana* Col-0 plants to select a microbiota that affected their life events, such as late and early flowering. Three related *Arabidopsis* (Ler, Be, RLD) and *Brassica rapa* genotypes were inoculated with the microbiome. The variations in the microbiome over various generations through the sequence of the 16S rRNA genes were analysed by various researchers, showing successive rises in inflorescence biomass of the three genetic forms of *Arabidopsis* and a rise in overall *B. rapa* biomass of the inocular microbiome correlated with late blooming. The authors suggested that their selection findings based on different microbial species had significant potential to aid the use of microbiomes to solve complex agronomic and environmental issues, based on the finding that increased biomass coincided with increased activity of microbial extracellular enzymes associated with nitrogen mineralization in soils.

2.3.2 Inoculation into the Soil and Rhizosphere

External strains inoculated from bulk or rhizospheric soils can alter the microbial community's structure. Chihaoui et al. [104] investigated the effects of inoculating *Phaseolus vulgaris*, a nodule-inhabiting endophyte, with *Agrobacterium* sp. 10-C2 on nodulation, plant formation, and the rhizosphere microbiome in semi-natural conditions. This strain does not form nodules and is non-pathogenic. The 10-C2 strain encouraged a large rise in the amount of nodules and plant biomass after inoculation. The amount of phosphorus, polyphenols, and flavonoids in the beans provided increased significantly, as did their total antioxidant ability. The terminal restriction fragment (TRF) length polymorphism technique of PCR-amplified 16S rRNA genes was used to assess the impact of the 10-C2 strain on the microbiome. The TRF richness remained constant 15 days after the 10-C2 strain was inoculated, but the bacterial population composition had improved. When bean plants were grown in these soils for a much longer period of time, 75 days, the TRF richness and composition were also affected by the 10-C2 strain. The TRF richness in the rhizosphere soil improved, but not in the soil without the vine. *Bacillus licheniformis*, *Bacillus pumilus*, *Paenibacillus koreensis*, and the genera *Arthrobacter*, *Microbacterium*, *Actinomyces*, and *Brevibacterium* have all been linked to the inoculation of the 10-C2 strain [104]. According to reports of Glick [105], and Santoyo et al. [7], bacterial species were remarked as plant growth promoters. The authors of this study concluded that the 10-C2 strain affected the rhizosphere positively by stimulating bacterial species known to promote plant development.

2.3.3 Inoculation into Seeds or Seedlings

There have been few reports on the interaction of rhizosphere and endosphere inhabitants of slow-growing plants like orchids. However, because natural environments such as forests have been degraded, these plants' natural habitat has been limited, and several attempts to save them have been made. Orchids do, without a doubt, have unique relationships with mycorrhizal fungi, as evidenced by numerous studies. Inoculation of bacterial species such as *Ps. fluorescens* and *Klebsiella oxytoca* into *Dendrobium nobile* Lindl. increased the adaptive, germination, and growth potential of orchids, according to a recent report of Pavlova et al. [106]. The authors of this research looked at the GFP colonization patterns of *Ps. fluorescens* and *K. oxytoca* in epiphytic orchid plants, seedlings, and roots. The effect of inoculating these strains on the endophyte microbiota of orchids would have been useful to investigate. A beneficial additive effect of two endophyte strains, *Pseudomonas stutzeri* E25 and *Stenotrophomonas maltophilia* CR71, when both strains were co-inoculated into the rhizosphere of tomato seedlings (*Lycopersicon esculentum* cv Saladette) relative to single inoculation, was discovered when the effect of encouraging plant growth was improved. The volatile cocktails released by strains E25 and CR71 were strikingly close, indicating that sulphur-containing compounds such as the antimicrobial volatile dimethyl disulphide (DMDS) are generated [107].

2.3.4 Tissue Atomization

The impact of the microbiome on plant phenotypic and physiological characteristics can be attributed to the numerous pathways enacted by different PGPB [108]. Plant growth promotion (e.g. phytohormones or organic volatile compounds) as well as the biocontrol activity against phytopathogens (e.g. the development of antibiotic compounds) are commonly identified [108]. The positive effects of the microbiota, on the other hand, may be due to interactions between human species and functions. Timm et al. [30] discovered that two strains of *Pseudomonas* and *Burkholderia* isolated from *Populus deltoides* had a beneficial additive effect, as inoculating *Populus* species with a mixture of these strains enhanced root biomass and photosynthetic ability more than individual inoculation or no inoculation. Through the transcriptomics, metabolic pathways along with response were observed after each strain activated genes unique for *Pseudomonas* and *Burkholderia*, as well as genes common to both strains, including genes encoding for ethylene responses and stress responses to temperature and salinity. The mixed inoculum also influenced genes involved in the synthesis of thiamine, sulphate, and lipids. In comparison to individual inoculation, the investigators looked at the effect of mixed inoculum on the leaf's metabolic profile. Mitter et al. [109] used a novel method called atomization to introduce the endophyte bacterium *Paraburkholderia phytofirmans* PsJN into the flowers of monocotyledonous and dicotyledonous plants. The authors were able to change the seed microbiome as well as the growth characteristics of wheat plants by vertical inheritance. Seeds that inherited the PsJN strain demonstrated substantial variations in plant growth as compared to controls in both greenhouse and field studies (plants with an unmodified microbiome). This research also looked into how the endogenous microbiota of seeds evolved after the endophyte PsJN was introduced, and discovered that some classes had a major difference in abundance. The population of α- (alpha) and γ- (gamma) proteobacteria decreased dramatically, while the population of β- (beta) proteobacteria increased. One of the challenges of this approach, according to the scientists, is that the modified microbiome must be passed on for more than one generation, so the PsJN strain is not passed on to the second generation of plants. This innovative approach, on the other hand, raises a lot of hopes for plant microbiome bioengineering without affecting the plant's genetics.

2.3.5 Direct Injection into Tissues or Wounds

As we have wide varieties of plants that are very fragile towards pathogens, hence few plants like Mānuka (*Leptospermum scoparium*) have capability to produce antimicrobial oils that possess antibacterial property against pathogenic strain of *Pseudomonas syringae* and this plant is endemically found in New Zealand. As a result, it is quite wondering that this quality of biocontrol could be inoculating endophytic bacteria namely *L. scoparium* into *Actinidia deliciosa* (Kiwi fruit) [32]. Three endophytic strains of the genus *Pseudomonas* were inoculated into wounds of *Ac. deliciosa* in this experiment; they were able to live in this new host and impart resistance to *Ps. syringae* infection. Various reports came from many investigators, including Avilés-García et al. [34], Orozco-Mosqueda et al. [110], and Velázquez-Becerra et al. [111], regarding direct injection of PGPB *A. agilis* UMCV2 into stems of *Medicago truncatula* plants where bacteria flourish, colonize, and survive within the plant. According to Avilés-García et al. [34], thus colonization of this kind can prove more effective in bioavailability of nutrients (especially Fe) of the plant. In other plant species, such as the teocinte ancestor of maize (TeA Mays), guided injection techniques involving bacterial endophytes were applied [112]. It is interesting that the authors noted that the endophyte atmosphere was left behind and the rhizosphere with two *Enterobacter asburiae* isolates was inoculated by stem injection. There are two factors in the colonization: either the plant is able, as is the case for many low-molecular-weight compounds exuded from the roots, to eject bacteria into the rhizospheric or surrounding soil, or the endophytes can have its own opening and closing of the entrance door to the plant tissues. Both theories have not yet been clarified and further research is warranted. Understanding this kind of feature could allow one to discover the entry and exit pathways for endophytes and pathogenic species into and out of plant tissue.

2.4 In Situ Harnessing of Agricultural Microbiome

2.4.1 Recent Advancement in Plant Microbiome Studies

A new scientific frontier in agriculture is the modification of plant-associated microbiomes in situ to increase plant production. This is happening in tandem with the advancement of innovative technology to incorporate desirable functions into single or multiple microbial organisms [3]. Indeed, plant microbes in plant phyllosphere (top layer), rhizosphere (lower ground compartments), and endosphere (in tissue) play important roles in increasing plant survival in restricted conditions [1]. Nearly about more than 95% of microbes that are present in an environment are non-culturable as we are bound to culture certain genera that are beneficial and suitable for genetic engineering. By using the inherent capacities of a wide proportion of the endogenous plant microbiome, new and enhanced microbial functions can be selected. A mounting literature indicates that plants harbour microbial communities that are known as typical microbial assemblies in various plant species [35]. Members of the so-called plant 'core' microbiome reliably and routinely related to one specific plant species under differing environmental conditions [113]. A number of studies, including maize, rice, and sugar cane, characterizing the composition of the core microbiota on various plant species have identified hundreds or more core microbial taxa occurring with each plant species [114], with compositional variations related to plant function [113]. Within the core microbiota of plants, the 'hub' of microbiota is characterized as the plant micronutrients, which are capable of forming solid, facilitative, and mutualistic interactions. Recent research has emphasized the importance of such 'hub' bacteria, which are thought to play a key role in orchestrating the assembly of other plant-associated microbiomes inside and around host plants [115]. Hub microbiota is thought to be obtained in two ways: transmission and recruiting. Plant seeds contain a substantial number of microorganisms, with a considerable proportion of them having been transported from the parent plant, similar to human microbiota, where the capacity of microbial species to be passed from mother to offspring before, during, and after (via milk) birth has been widely demonstrated [116]. In addition to the inherited microbiota, the release of targeted signals and root exudates can become main factors for plants to regulate the recruitment of associated microbes starting at the seed germination level. The formation of the core microbiota and overall assembly can be aided by the initial hub microbiota formed by plant collection and filtering. The intrinsic abilities of a significant proportion of the indigenous plant microbiome to recruit beneficial microbes can be harnessed for the collection of novel and enhanced microbial functions, given the close association between plants and their related hub microbiota. Hub microbes are typically classified based on the degree of their associations, and their relationships and personalities can be decoded using network analysis [115], making the task of designing these essential members of the microbiota in situ much easier. Network maps will illuminate microbiota centre and its related participants with unique roles. These microbes can be used to classify their functional capacity for isolation and genome sequence. The identification of hub microbiota and their influence in a plant microbiome reveal target microorganisms responsible for essential interactions between host–microbe and microbes. The artificial application of natural ecological processes using microbial, biochemical, and molecular tooling will modify the indigenous microbiomes of plants in situ. Given next are several interesting methods that have the potential to accelerate and enhance our ability to harness these critical relationships in order to boost agricultural productivity.

2.4.2 Microbial-Based Strategies

A hierarchical organization, where these central group members are important and have a role as regulators in controlling the development and operation of the other microbial assemblies, suggests the idea of the experimental preliminary evidence [117]. Hub microbiota can thus represent an appealing target for microbial manipulation in agricultural settings, where management of the order and timing of hub microbial arrival (i.e. priority effects) can be harnessed to enhance the recruitment of other beneficial members of the plant microbiome during early stages of plant development. Based on an approach of Wei and Jousset [118], in which the plant hub microbiome is engineered in situ using a microbial-based plant breeding process, plant microbiota can be updated in this technique by using a step-wise approach to inoculate vertically transmitted microbiota to the next generation. Plant microbiomes are first profiled using next-generation sequencing technology, and core and hub microbes are classified using statistical and network analysis. Following that, metagenomic sequencing is used to identify microbes with essential functions (e.g. benefit to plant growth or pathogen defence). These closely linked hub microbiota can then be extracted from the plant and sprayed on parental flowers as a microbial mixture, resulting in seeds that are enriched with complex hub microbiota until a seedling is formed. Current efforts to implement related floral inoculation techniques show that seed microbes are likely to survive and move on to the next generation [109]. This technique may be an alternative to manipulate priority effects to produce plants with 'improved' plant microbiota (i.e. seeds with an enhanced capacity to attract useful microbiome).

2.4.3 Biochemical Strategies

To attract and preserve hub microbiota or other beneficial microbiota in situ, biochemical techniques to engineer indigenous microbiomes include the use of chemical compounds naturally formed by both plants and microbes. Root exudates, for example, are well known for attracting beneficial microbes or reshaping microbiome assembly in the plant rhizosphere with exudates' activity being favoured under environmental stress [119]. One of the recently published reports has shown the ability of organic plant volatility to attract soil bacteria from long distances, which suppress the pathogens, suggesting the application of soil microbiome engineering to improve plant protection using certain organic volatile substances [120]. In order for the intended root compounds to be distilled or synthesized and used for enhancing the capacity of plants to attract and retain beneficial microbes and their behaviours in the rhizosphere, the root metabolites need to be associated with a specific proliferation of rhizosphere microbials. The use of microbial structures analogous to microbial quorum sensing processes results in an alternative method for enriching beneficial members of the plant microbiome. Quorum sensing is a population-dependent gene expression control in microbes, and has an acknowledged role in modulating collective microbial behaviour by releasing complex contact (signal) molecular arrays [103]. Significant ecological microbial roles may be influenced by these signal molecules, such as promoting nutrient or niche acquisition, modulating mutual defence against rivals, and facilitating group escape in the face of population extinction. Plants can also sense and be stimulated positively by microbial signals such as heat shock proteins and reactive oxygen species that are found during the stress response in the rhizosphere [121]. Signal molecules could be used to facilitate hub or beneficial bacteria, improve microbial-mediated nutrient supply (e.g. fixation, mineralization, and mobilization), or evoke a microbial-mediated reaction to pathogens in the sense of microbiome in situ manipulation. As a result, signal molecules may be a useful method for controlling plant–microbe interactions in order to maximize resource supply and plant safety. Our knowledge of the identification, roles, and mechanistic interactions of signal molecules used by plants or microbes for these communications, on the other hand, is exceedingly restricted, limiting our ability to completely exploit these promising resources. To further characterize their variety and precision, spectroscopy-based detection technologies would need to be made more responsive, along with the convergence of metagenomics, metatranscriptomics, and metabolomics approaches.

2.4.4 Molecular Strategies

Multiple plant genes and functions have been associated with heterogeneity in the plant microbiota through environmental conditions indicating that host genetics plays a significant role in deciding total microbiome structure, abundance, or function in plants [122]. Given the close relationship between the host genome and the composition of the microbiome, harnessing the interactions between the microbiome and the whole host genome could provide the foundation for creating plants with a better microbiome. In this regard, quantitative genetic tools such as quantitative trait loci (QTL) mapping can be especially useful, as they enable the detection of genes or genetic loci underlying essential biological traits (phenotypes) in any organism of interest [123]. This method has been widely used in mouse and human microbiome research to classify genetic loci that affect particular microbial taxa or pathways, as well as to relate environmental factors to changes in microbial composition. A significant opportunity has been shown for this technology, either by genetic modification or by conventional breeder methods, to produce better plant varieties. When the QTL or genetic features of cultures that mediate crop-beneficial microbiome interactions are determined, they can produce new and enhanced plant varieties that can draw beneficial indigenous microbiota and exploit them. With the introduction of clustered regularly interspaced short palindromic repeats (CRISPR)/Cas9, such methods are able to either enhance or encourage the plants to recruit useful microbiota [124]. The initial colonization of the hub microbiota, that ultimately can predictably form the core and entire plant microbiome, could be manipulated by a similar approach. If microbial signalling genetic pathways or biochemical signalling pathways are identifiable, these genes can be passed to crops to produce beneficial reactions from plant microbiomes.

2.5 Future Perspective of Agricultural Microbiome Engineering

With advancement of microbial science, the influence and contribution of microbiomes on the ecology of hosts and the features of the environments where they live has been widely known in recent years. Microbiome engineering has shown promise in the treatment of diseases and the improvement of agricultural production, thanks to new knowledge about the microbiome. Microbiome biology, on the other hand, is still in its infancy and has a lot of promise that is yet to be realized. With the rise of synthetic biology in recent decades, advancements in microbiome engineering may be greatly aided. Smart microbes with novel functionalities can be engineered using synthetic biology [125]. Probiotic *Escherichia coli* that can detect, seek, and kill pathogenic *Pseudomonas aeruginosa*, for example, has been demonstrated [126]. This method could be used to exclude strains from the microbiota selectively for microbiome engineering. *Es. coli* has also been modified to disrupt microbe–microbe signalling in order to change the makeup of the microbiota. The Firmicutes/Bacteroidetes ratio in the mouse intestine was rebalanced after streptomycin-induced dysbiosis [127]. In addition, *Es. coli* over-expressing both AI-2 and the genus unique autoinducer-1 (CAI-1) has been used to successfully vaccinate against *Vibrio cholerae* infection in the gut [128]. These examples show how smart microbes can be used to engineer the microbiota for medicinal purposes [128]. The recent discovery of genetic parts and tools for the human commensal bacterium *Bacteroides thetaiotaomicron* [129] could significantly speed up efforts to engineer therapeutic smart microbes for microbiome engineering. Detailed research into the human gut microbiome, and research into the phageome that is connected with it, has only recently started. While studies have suggested that the gut phageome is essential for preserving and protecting the structure and function of the gut microbiome, our knowledge of microbiota ecosystem responses and phageome functions is still restricted. Despite this, phages have been used for antibacterial purposes in agriculture and humans and they can be engineered to change host specificity and learn new functions [130]. As a result, controlling phageomes, such as using modified phages to selectively remove pathogenic microbes, may be a promising avenue for modifying the bacterial population in the animal or human gut to affect microbiome engineering. Despite the fact that advances in HMP and the advancement of omics methods have greatly aided in assessing microbial composition and learning about microbiomes, there is still much to learn about microbiomes. The complexities of the microbiome, as well as the interactions between microbes that determine microbial composition, are still poorly understood. Recently, *Es. coli* was designed to produce readable signals in response to quorum sensing molecules used in microbiota cell-to-cell communication, effectively enabling us to 'listen in' on the bacteria. This strategy may be used to supplement existing omics techniques for high-throughput profiling of signalling inside and changes to the microbial community over time, advancing our understanding of the microbiome and accelerating microbiome engineering development. More studies can also be carried out on other bodily microbiomes such as skin microbiomes, oral cavity, and the urogenital tract in addition to the intestinal microbiome. In order for the applicability of medicinal microbiome technology to be improved, the distal consequences of microbiomes beyond their local area need to be better understood. The composition of intestinal microbiota affects host immune responses, for example, and vaccine growth may be benefiting from the use of microbiome [131]. Advances in these fields are important to understand the maximum potential for therapeutic use in the field of microbiome engineering.

References

1 Bulgarelli, D., Schlaeppi, K., Spaepen, S. et al. (2013). Structure and functions of the bacterial microbiota of plants. *Annu. Rev. Plant Biol.* 64: 807–838.

2 Nelson, E.B. (2017). The seed microbiome: origins, interactions, and impacts. *Plant Soil*. https://doi.org/10.1007/s11104-017-3289-7.

3 Mueller, U.G. and Sachs, J.L. (2015). Engineering microbiomes to improve plant and animal health. *Trends Microbiol.* 23: 606–617.

4 Berg, G., Rybakova, D., Grube, M., and Köberl, M. (2016). The plant microbiome explored: implications for experimental botany. *J. Exp. Bot.* 67: 995–1002.

5 Haney, C.H., Samuel, B.S., Bush, J., and Ausubel, F.M. (2015). Associations with rhizosphere bacteria can confer an adaptive advantage to plants. *Nat. Plants* 1: 15051.

6 Glick, B.R. (2012). Plant growth-promoting bacteria: mechanisms and applications. *Scientifica* 2012: 963401.

7 Santoyo, G., del Carman Orozco-Mosqueda, M., and Govindappa, M. (2012). Mechanisms of biocontrol and plant growth-promoting activity in soil bacterial species of Bacillus and Pseudomonas: a review. *Biocontrol Sci. Tech.* 22: 855–872.

8 Calvo, P., Watts, D.B., Kloepper, J.W., and Torbert, H.A. (2017). Effect of microbial-based inoculants on nutrient concentrations and early root morphology of corn [Zea mays]. *J. Plant Nutr. Soil Sci.* 180: 56–70.

9 Pérez-Flores, P., Valencia-Cantero, E., Altamirano-Hernández, J. et al. (2017). Bacillus methylotrophicus M4-96 isolated from maize [Zea mays] rhizoplane increases growth and auxin content in *Arabidopsis thaliana* via emission of volatiles. *Protoplasma* 254 (6): 2201–2213. https://doi.org/10.1007/s00709-017-1109-9.

10 Glick, B.R. (1995). The enhancement of plant growth by free-living bacteria. *Can. J. Microbiol.* 41: 109–117.

11 Ryan, R.P., Germaine, K., Franks, A. et al. (2008). Bacterial endophytes: recent developments and applications. *FEMS Microbiol. Lett.* 278: 1–9.

12 Martínez-Absalón, S., Rojas-Solís, D., Hernández-León, R. et al. (2014). Potential use and mode of action of the new strain *Bacillus thuringiensis* UM96 for the biological control of the grey mould phytopathogen Botrytis cinerea. *Biocontrol Sci. Tech.* 24: 1349–1362.

13 Gutiérrez-Luna, F.M., López-Bucio, J., Altamirano-Hernández, J. et al. (2010). Plant growth-promoting rhizobacteria modulate root-system architecture i: *Arabidopsis thaliana* through volatile organic compound emission. *Symbiosis* 51: 75–83.

14 Hartmann, A., Rothballer, M., and Schmid, M. (2008). Lorenz Hiltner, a pioneer in rhizosphere microbial ecology and soil bacteriology research. *Plant Soil* 312: 7–14.

15 Pieterse, C.M., de Jonge, R., and Berendsen, B.L. (2016). The soil-borne supremacy. *Trends Plant Sci.* 21: 171–173.

16 Chaparro, J.M., Sheflin, A.M., Manter, D.K., and Vivanco, J.M. (2012). Manipulating the soil microbiome to increase soil health and plant fertility. *Biol. Fertil. Soils* 48: 489–499.

17 Germida, J.J., Siciliano, S.D., Freitas, J.R., and Seib, A.M. (1998). Diversity of root-associated bacteria associated with fieldgrown canola [*Brassica napus* L.] and wheat [*Triticum aestivum* L.]. *FEMS Microbiol. Ecol.* 26: 43–50.

18 Márquez-Santacruz, H.A., Hernandez-Leon, R., del Carmen Orozco-Mosqueda, M. et al. (2010). Diversity of bacterial endophytes in roots of Mexican husk tomato plants [*Physalis ixocarpa*] and their detection in the rhizosphere. *Genet. Mol. Res.* 9: 2372–2380.

19 Bai, Y., Müller, D.B., Srinivas, G. et al. (2015). Functional overlap of the Arabidopsis leaf and root microbiota. *Nature* 528: 364.

20 Compant, S., Clément, C., and Sessitsch, A. (2010). Plant growth-promoting bacteria in the rhizo-and endosphere of plants: their role, colonization, mechanisms involved and prospects for utilization. *Soil Biol. Biochem.* 42: 669–678.

21 Santoyo, G., Moreno-Hagelsieb, G., del Carmen Orozco-Mosqueda, M., and Glick, B.R. (2016). Plant growth-promoting bacterial endophytes. *Microbiol. Res.* 183: 92–99.

22 Badri, D.V. and Vivanco, J.M. (2009). Regulation and function of root exudates. *Plant Cell Environ.* 32: 666–681.

23 Ali, S., Charles, T.C., and Glick, B.R. (2012). Delay of flower senescence by bacterial endophytes expressing 1-aminocyclopropane-1-carboxylate deaminase. *J. Appl. Microbiol.* 113: 1139–1144.

24 Santoyo, G., Hernández-Pacheco, C., Hernández-Salmerón, J., and Hernández-León, R. (2017). The role of abiotic factors modulating the plant-microbe-soil interactions: toward sustainable agriculture: a review. *Spanish J. Agric. Res.* 15: e03R01.

25 Philpott, C.C. (2006). Iron uptake in fungi: a system for every source. *Biochim. Biophys. Acta-Mol. Cell Res.* 1763: 636–645.

26 Vorholt, J.A. (2012). Microbial life in the phyllosphere. *Nat. Rev. Microbiol.* 10: 828.

27 Peñuelas, J. and Terradas, J. (2014). The foliar microbiome. *Trends Plant Sci.* 19: 278–280.

28 Abril, A.B., Torres, P.A., and Bucher, E.H. (2005). The importance of phyllosphere microbial populations in nitrogen cycling in the Chaco semi-arid woodland. *J. Trop. Ecol.* 21: 103–107.

29 Yaish, W.M., Al-Harrasi, I., Alansari, A.S. et al. (2017). The use of high throughput DNA sequence analysis to assess the endophytic microbiome of date palm roots grown under different levels of salt stress. *Int. Microbiol.* 19: 143–155.

30 Timm, C.M., Pelletier, D.A., Jawdy, S.S. et al. (2016). Two poplar-associated bacterial isolates induce additive favorable responses in a constructed plant-microbiome system. *Front. Plant Sci.* 7: 497.

31 Yuan, Z., Druzhinina, I.S., Labbé, J. et al. (2016). Specialized microbiome of a halophyte and its role in helping non-host plants to withstand salinity. *Sci. Rep.* 6: 32467.

32 Wicaksono, W.A., Jones, E.E., Casonato, S. et al. (2017). Biological control of *Pseudomonas syringae* pv. actinidiae [Psa], the causal agent of bacterial canker of kiwifruit, using endophytic bacteria recovered from a medicinal plant. *Biol. Control* 116: 103–112. https://doi.org/10.1016/j.biocontrol.2017.03.003.

33 Romeis, J., Meissle, M., and Bigler, F. (2006). Transgenic crops expressing *Bacillus thuringiensis* toxins and biological control. *Nat. Biotechnol.* 24: 63–71.

34 Avilés-García, M.E., Flores-Cortez, I., Hernández-Soberano, C. et al. (2016). La rizobacteria promotora del crecimiento vegeta: *Arthrobacter agilis* UMCV2 coloniza endofíticamente a *Medicago truncatula*. *Rev. Argent. Microbiol.* 48: 342–346.

35 Mendes, R., Garbeva, P., and Raaijmakers, J.M. (2013). The rhizosphere microbiome: significance of plant beneficial, plant pathogenic, and human pathogenic microorganisms. *FEMS Microbiol. Rev.* 37: 634–663.

36 Hallmann, J., Schulz, B., and Berg, G. (2006). Isolation procedures for endophytic microorganisms. In: *Microbial Root Endophytes*, Soil Biology, vol. 9 (ed. S. BJE, B. CJC and T.N. Sieber), 299–319. Springer.

37 Kharwar, R.N., Mishra, A., Gond, S.K. et al. (2011). Anticancer compounds derived from fungal endophytes: their importance and future challenges. *Nat. Prod. Rep.* 28: 1208–1228.

38 Kharwar, R.N., Verma, S.K., Mishra, A. et al. (2011). Assessment of diversity, distribution and antibacterial activity of endophytic fungi isolated from a medicinal plant *Adenocalyma alliaceum* Miers. *Symbiosis* 55: 39–46.

39 Verma, V.C., Gond, S.K., Kumar, A. et al. (2007). Endophytic mycoflora from leaf, bark, and stem of *Azadirachta indica* A Juss. from Varanasi, India. *Microb. Ecol.* 54: 119–125.

40 Kumar, J., Sharma, V.K., Singh, D.K. et al. (2016). Epigenetic activation of antibacterial property of an endophytic Streptomyces coelicolor Strain AZRA 37 and identification of the induced protein using MALDI TOF MS/MS. *PLoS One* 11 (2): e0147876. https://doi.org/10.1371/journal.pone.0147876.

41 Schulz, B., Guske, S., Dammann, U., and Boyle, C. (1998). Endophyte–host interaction II. Defining symbiosis of the endophyte–host interaction. *Symbiosis* 25: 213–227.

42 Hyde, K.D. and Soytong, K. (2008). The fungal endophyte dilemma. *Fungal Divers.* 33: 163–173.

43 Bills, G.F. and Polishook, J.D. (1992). Recovery of endophytic fungi from Chamaecyparis thyoides. *Sydowia* 44: 1–12.

44 Verma, V.C., Gond, S.K., Kumar, A. et al. (2011). Endophytic fungal flora from roots and fruits of an Indian neem plant *Azadirachta indica* A. Juss., and impact of culture media on their isolation. *Indian J. Microbiol.* 51 (4): 469–476.

45 Mishra, A., Gond, S.K., Kumar, A. et al. (2012). Season and tissue type affect fungal endophyte communities of the Indian medicinal plant *Tinospora cordifolia* more strongly than geographic location. *Microb. Ecol.* 64: 3288–3398.

46 Verma, S.K., Gond, S.K., Mishra, A. et al. (2014). Impact of environmental variables on the isolation, diversity and antibacterial activity of endophytic fungal communities from Madhuca indica Gmel. at different locations in India. *Ann. Microbiol.* 64 (2): 721–734.

47 Gond, S.K., Bergen, M.S., Torres, M.S., and White, J.F. (2015). Endophytic Bacillus spp. Produce antifungal lipopeptides and induce host defence gene expression in maize. *Microbiol. Res.* 172: 79–87.

48 Kutser, E. and Williams, S.T. (1964). Selection of media for the isolation of Streptomyces. *Nature* 202: 928–929.

49 Verma, V.C., Gond, S.K., Kumar, A. et al. (2009). Endophytic actinomycetes from *Azadirachta indica* A. Juss.: isolation, diversity, and anti-microbial activity. *Microb. Ecol.* 57: 749–756.

50 Sun, X. and Guo, L.-D. (2012). Endophytic fungal diversity: review of traditional and molecular techniques. *Mycology* 3 (1): 65–76.

51 Xu, J., Yang, X., and Lin, Q. (2014). Chemistry and biology of Pestalotiopsis-derived natural products. *Fungal Divers.* 66: 37–68.

52 Li, E., Tian, R., Liu, S. et al. (2008). Pestalotheols A-D, bioactive metabolites from the plant endophytic fungus Pestalotiopsis theae. *J. Nat. Prod.* 71 (4): 664–668.

53 Omacini, M., Chaneton, E.J., Ghersa, C.M., and Müller, C.B. (2001). Symbiotic fungal endophytes control insect host–parasite interaction webs. *Nature* 409: 78–81.

54 Barnett, H.L. and Hunter, B.B. (1998). *Illustrated Genera of Imperfect Fungi*, 4e. St. Paul, MN: The American Phytopathological Society.

55 Von Arx, J.A. (1978). The genera of fungi sporulating in pure culture. In: (ed. A.R. Gantner and K.G. Verlag). FL-9490 Vaduz, Liechtenstein.

56 Ellis, M.B. (1976). *More Dematiaceous Hyphomycetes. Commonwealth Mycological Institute*, 507. Surrey, England: Kew. Obtainable from Commonwealth Agricultural Bureaux, Central Sales Branch, Farnham Royal, Slough SL2 3BN, England.

57 Ainsworth, G.C., Sparrow, F.K., and Sussman, A.S. (1973). *The Fungi: An Advanced Treatise*, vol. 4A. New York: Academic Press.

58 Gond, S.K., Mishra, A., Sharma, V.K. et al. (2011). Diversity and antimicrobial activity of endophytic fungi isolated from Nyctanthes arbor-tristis, a wellknown medicinal plant of India. *Mycoscience* 53: 113–121.

59 Kumar, D.S.S. and Hyde, K.D. (2004). Biodiversity and tissue-recurrence of endophytic fungi in Tripterygium wilfordii. *Fungal Divers.* 17: 69–90.

60 Fisher, P.J., Pertini, O., Petrini, L.E., and Sutton, B.C. (1994). Fungal endophytes from leaves and twigs of *Quercus ilex* L. from England, Majorca and Switzerland. *New Phytol.* 127: 133–137.

61 Fröhlich, J., Hyde, K.D., and Petrini, O. (2000). Endophytic fungi associated with palms. *Mycol. Res.* 104: 1202–1212.

62 Ghimire, S.R. and Hyde, K.D. (2004). Fungal endophytes. In: *Plant Surface Microbiology* (ed. A. Varma, L. Abbott, D. Werner and R. Hampp), 281–288. Berlin: Springer.

63 Ranghoo, V.M., Hyde, K.D., Liew, E.C.Y., and Spatafora, J.W. (1999). Family placement of Ascotaiwanian and Ascolacicola based on DNA sequences from the large subunit rRNA gene. *Fungal Divers.* 2: 159–168.

64 Saitou, N. and Nei, M. (1987). The neighbor-joining method: a new method for reconstructing phylogenetic trees. *Mol. Biol. Evol.* 4 (4): 406–425.

65 Lindahl, B.D., Nilsson, R.H., Tedersoo, L. et al. (2013). Fungal community analysis by high-throughput sequencing of amplified markers – a user's guide. *New Phytol.* 199: 288–299.

66 Sugita, T. and Nishikawa, A. (2003). Fungal identification method based on DNA sequence analyisis: reassessment of the method of Pharmaceutical society of Japan and the Japanese Pharmacopoeia. *J. Health Sci.* 49 (6): 531–533.

67 Iwen, P.C., Hinrichs, S.H., and Rupp, M.E. (2002). Utilization of the internal transcribed spacer regions as molecular targets to detect and identify human fungal pathogens. *Med. Mycol.* 40: 87–109.

68 Khot, P.D., Ko, D.L., and Fredricks, D.N. (2009). Sequencing and analysis of fungal rRNA operons for development of broad-range fungal PCR assays. *Appl. Environ. Microbiol.* 75: 1559–1565.

69 Barghouthi, S.A. (2011). A universal method for the identification of bacteria based on general PCR primers. *Indian J. Microbiol.* 51 (4): 430–444.

70 Clarridge, J.E. (2004). Impact of 16S rRNA gene sequence analysis for identification of bacteria on clinical microbiology and infectious diseases. *Clin. Microbiol. Rev.* 17: 840–862.

71 Stone, J.K., Polishook, J.D., and White, J.F. (2004). Endophytic fungi. In: *Biodiversity of Fungi: Inventory and Monitoring Methods* (ed. G. Mueller, G.F. Bills and M.S. Foster), 241–270. Burlington, MA: Elsevier.

72 Hood, M.E. and Shew, H.D. (1996). Applications of KOH-aniline blue fluorescence in the study of plantfungal interactions. *Phytopathology* 86: 704–708.

73 Thirugnanasambandam, A., Wright, K.M., Atkins, S.D. et al. (2011). Infection of Rrs1 barley by an incompatible race of the fungus Rhynchosporium secalis expressing the green fluorescent protein. *Plant Pathol.* 60: 513–521.

74 Card, S.D., Tapper, B.A., Loyd-West, C., and Wright, K.M. (2013). Assessment of fluorescein-based fluorescent dyes for tracing Neotyphodium endophytes in planta. *Mycologia* 105 (1): 221–229.

75 Mousa, W.K., Shearer, C., Limay-Rios, V. et al. (2016). Root-hair endophyte stacking in finger millet creates a physicochemical barrier to trap the fungal pathogen Fusarium graminearum. *Nat. Microbiol.* 16167. https://doi.org/10.1038/NMICROBIOL.2016.167.

76 Thomas, P. and Reddy, K.M. (2013). Microscopic elucidation of abundant endophytic bacteria colonizing the cell wall–plasma membrane peri-space in the shoot-tip tissue of banana. *AoB PLANTS* 5: plt011. https://doi.org/10.1093/aobpla/plt011.

77 White, J.F. Jr., Crawford, H., Torres, M.S. et al. (2012). A proposed mechanism for nitrogen acquisition by grass seedlings through oxidation of symbiotic bacteria. *Symbiosis* 57: 161–117.

78 Choudhury, S., Panda, P., Sahoo, L., and Panda, S.K. (2013). Reactive oxygen species signaling in plants under abiotic stress. *Plant Signal. Behav.* 8: e23681. https://doi.org/10.4161/psb.2368.

79 White, J.F. Jr., Torres, M.S., Somu, M.P. et al. (2014). Hydrogen peroxide staining to visualize bacterial infections of seedling root cells. *Microsc. Res. Tech.* 77: 566–573.

80 White, J.F., Chen, Q., Torres, M.S. et al. (2015). Collaboration between grass seedlings and rhizobacteria to scavenge organic nitrogen in soils. *AoB PLANTS* 7 (plu093). https://doi.org/10.1093/aobpla/plu093.

81 Hata, F. and Futai, K. (1995). Endophytic fungi associated with healthy pine needles and needles infested by pine needle gall midge Thecodiplosis japonensis. *Can. J. Bot.* 73: 384–390.

82 Orduña, F.N.R., Sanchez, R.A.S., Bustamante, Z.R.F. et al. (2011). Diversity of endophytic fungi of *Taxus globosa* [Mexican yew]. *Fungal Divers.* 47: 65–74.

83 Stewart, E.J. (2012). Growing unculturable bacteria. *J. Bacteriol.* 194: 4151–4416.

84 Hugenholtz, P., Goebel, B.M., and Pace, N.R. (1998). Impact of culture-independent studies on the emerging phylogenetic view of bacterial diversity. *J. Bacteriol.* 180 (18): 4765–4774.

85 Guo, L.D., Hyde, K.D., and Liew, E.C.Y. (2001). Detection and taxonomic placement of endophytic fungi within frond tissues of *Livistona chinensis* based on rDNA sequences. *Mol. Phylogenet. Evol.* 20: 1–13.

86 Duong, L.M., Jeewon, R., Lumyong, S., and Hyde, K.D. (2006). DGGE coupled with ribosomal DNA gene phylogenies reveal uncharacterized fungal phylotypes. *Fungal Divers.* 23: 121–138.

87 Marco, D. (2011). *Metagenomics: Current Innovations and Future Trends.* Caister Academic Press.

88 Romero, F.M., Marina, M., and Pieckenstain, F.I. (2014). The community of tomato [*Solanum lycopersicum* L.] leaf endophytic bacteria, analyzed by 16s-ribosomal RNA gene pyrosequencing. *FEMS Microbiol. Lett.* 351: 187–194.

89 Manter Daniel, K., Delgado, J.A., Holm, D.J., and Stong, R.A. (2010). Pyrosequencing reveals a highly diverse and cultivar-specific bacterial endophyte community in potato roots. *Microb. Ecol.* 60 (1): 157–166.

90 Tian, B.Y., Cao, Y., and Zhang, K.-Q. (2015). Metagenomic insights into communities, functions of endophytes, and their associates with infection by root-knot nematode, Meloidogyne incognita, in tomato roots. *Sci. Rep.* 5: 17087.

91 Sessitsch, P., Hardoim, J., Doring, A. et al. (2012). Functional characteristics of an endophyte community colonizing rice roots as revealed by metagenomic analysis. *Mol. Plant-Microbe Interact.* 25: 28–36.

92 Jumpponen, A. and Jones, K.L. (2009). Massively parallel 454 sequencing indicate hyperdiverse fungal communities in temperate *Quercus macrocarpa* phyllosphere. *New Phytol.* 18: 438–448.

93 Toju, H., Yamamoto, S., Sato, H. et al. (2013). Community composition of root-associated fungi in a Quercus-dominated temperate forest: 'codominance' of mycorrhizal and root-endophytic fungi. *Ecol. Evol.* 3: 1281–1293. https://doi.org/10.1002/ece3.546.

94 Thomas, T., Gilbert, J., and Meyer, F. (2012). Metagenomics – a guide from sampling to data analysis. *Microb. Inf. Exp.* 2: 3. http://www.microbialinformaticsj.com/content/2/1/3.

95 Gond, S.K., Torres, M.S., Bergen, M.S. et al. (2015). Induction of salt tolerance and up-regulation of aquaporin genes in tropical corn by rhizobacterium Pantoea agglomerans. *Lett. Appl. Microbiol.* 60: 392–399.

96 Dinkins, R.D., Barnes, A., and Waters, W. (2010). Microarray analysis of endophyte-infected and endophyte-free tall fescue. *J. Plant Physiol.* 167: 1197–1203.

97 Feletti, S.A., Shields, K., Ramsperger, M. et al. (2004). Gene discovery and microarray based transcriptome analysis in grass endophytes. In: *Proceedings of the 3rd International Symposium, Molecular Breeding of Forage and Turf* (ed. A. Hopkins et al.), pp. 145–153. Dordrecht: Kluwer Academic

98 Felitti, S., Shields, K., Ramsperger, M. et al. (2006). Transcriptome analysis of Neotyphodium and Epichloe grass endophytes. *Fungal Genet. Biol.* 43: 465–475.

99 Wang, J.-W., Wang, L.-J., Mao, Y.-B. et al. (2005). Control of root cap formation by microRNA-targeted auxin response factors in Arabidopsis. *Plant Cell.* 17: 2204–2216.

100 Irizarry, I. and White, J.F. (2017). Application of bacteria from non-cultivated plants to promote growth, alter root architecture and alleviate salt stress of cotton. *J. Appl. Microbiol.* https://doi.org/10.1111/jam.13414.

101 Malinowski, D.P. and Belesky, D.P. (2000). Adaptations of endophyte-infected cool-season grasses to environmental stresses: mechanisms of drought and mineral stress tolerance. *Crop Sci.* 40: 923–940.

102 Swenson, W., Wilson, D.S., and Elias, R. (2000). Artificial ecosystem selection. *Proc. Natl. Acad. Sci. U. S. A.* 97: 9110–9114.

103 Panke-Buisse, K., Poole, A., Goodrich, J. et al. (2015). Selection on soil microbiomes reveals reproducible impacts on plant function. *ISME J.* 9: 980–989.

104 Chihaoui, S.A., Trabelsi, D., Jdey, A. et al. (2015). Inoculation of *Phaseolus vulgaris* with the nodule-endophyte Agrobacterium sp. 10C2 affects richness and structure of rhizosphere bacterial communities and enhances nodulation and growth. *Arch. Microbiol.* 197: 805–813.

105 Glick, B.R. (2015). *Beneficial Plant-Bacterial Interactions.* Heidelberg: Springer.

106 Pavlova, A.S., Leontieva, M.R., Smirnova, T.A. et al. (2017). Colonization strategy of the endophytic plant growth promoting strains of *Pseudomonas fluorescens* and *Klebsiella oxytoca* on the seeds, seedlings and roots of the epiphytic orchid, *Dendrobium nobile* Lindl. *J. Appl. Microbiol.* 123 (1): 217–232. https://doi.org/10.1111/jam.13481.

107 Rojas-Solís, D., Zetter-Salmón, E., Contreras-Pérez, M. et al. (2018). *Pseudomonas stutzeri* E25 and Stenotrophomonas maltophilia CR71 endophytes produce antifungal volatile organic compounds and exhibit additive plant growth-promoting effects. *Biocatal. Agric. Biotechnol.* 13: 46–52.

108 Hernández-León, R., Rojas-Solís, D., Contreras-Pérez, M. et al. (2015). Characterization of the antifungal and plant growth-promoting effects of

diffusible and volatile organic compounds produced by *Pseudomonas fluorescens* strains. *Biol. Control* 81: 83–92.
109 Mitter, B.N., Pfaffenbichler, R., Flavell, S. et al. (2017). A new approach to modify plant microbiomes and traits by introducing beneficial bacteria at flowering into progeny seeds. *Front. Microbiol.* 8: 11.
110 Orozco-Mosqueda, M.C., Velázquez-Becerra, I., Macías-Rodríguez, L. et al. (2013). *Arthrobacter agilis* UMCV2 induces iron acquisition in *Medicago truncatula* [strategy I plant] in vitro via dimethylhexadecylamine emission. *Plant Soil* 362: 51–66.
111 Velázquez-Becerra, C., Macías-Rodríguez, L.I., López-Bucio, J. et al. (2013). The rhizobacterium *Arthrobacter agilis* produces dimethylhexadecylamine, a compound that inhibits growth of phytopathogenic fungi in vitro. *Protoplasma* 250: 1251–1262.
112 Johnston-Monje, D. and Raizada, M.N. (2011). Conservation and diversity of seed associated endophytes in Zea across boundaries of evolution, ethnography and ecology. *PLoS One* 6: e20396.
113 Lemanceau, P., Blouin, M., Muller, D., and Moenne-Loccoz, Y. (2017). Let the core microbiota be functional. *Trends Plant Sci.* 22: 583–595.
114 Hamonts, K., Trivedi, P., Garg, A. et al. (2018). Field study reveals core plant microbiota and relative importance of their drivers. *Environ. Microbiol.* 20: 124–140.
115 Shade, A. and Handelsman, J. (2012). Beyond the Venn diagram: the hunt for a core microbiome. *Environ. Microbiol.* 14: 4–12.
116 Gundel, P.E., Rudgers, J.A., and Ghersa, C.M. (2011). Incorporating the process of vertical transmission into understanding of host-symbiont dynamics. *Oikos* 120: 1121–1128. https://doi.org/10.1111/j.1600-0706.2011.19299.x.
117 Niu, B., Paulson, J.N., Zheng, X., and Kolter, R. (2017). Simplified and represenative bacterial community of maiz roots. *Proc. Natl. Acad. Sci.* 114: 2450–2459.
118 Wei, Z. and Jousset, A. (2017). Plant breeding goes microbial. *Trends Plant Sci.* 22: 555–558.
119 Berendsen, R.L., Pieterse, C.M., and Bakker, P.A. (2012). The rhizosphere microbiome and plant health. *Trends Plant Sci.* 17: 478–486.
120 Schulz-Bohm, K., Gerards, S., Hundscheid, M. et al. (2018). Calling from distance: attraction of soil bacteria by plant root volatiles. *ISME J.* 12: 1252.
121 Bauer, W.D. and Mathesius, U. (2004). Plant responses to bacterial quorum sensing signals. *Curr. Opin. Plant Biol.* 7: 429–433.
122 Turner, T.R., James, E.K., and Poole, P.S. (2013). The plant microbiome. *Genome Biol.* 14: 209.
123 McCarthy, M.I., Abecasis, G.R., Cardon, L.R. et al. (2008). Genome-wide association studies for complex traits: consensus, uncertainty and challenges. *Nat. Rev. Genet.* 9: 356.
124 Schaeffer, S.M. and Nakata, P.A. (2015). CRISPR/Cas9-mediated genome editing and gene replacement in plants: transitioning from lab to field. *Plant Sci.* 240: 130–142.

125 Hwang, I.Y., Koh, E., Kim, H.R. et al. (2016). Reprogrammable microbial cell-based therapeutics against antibioticresistant bacteria. *Drug Resist. Updat.* 27: 59–71.

126 Hwang, I.Y., Tan, M.H., Koh, E. et al. (2014). Reprogramming microbes to be pathogen-seeking killers. *ACS Synth. Biol.* 3: 228–237.

127 Thompson, J.A., Oliveira, R.A., Djukovic, A. et al. (2015). Manipulation of the quorum sensing signal AI-2 affects the antibiotic-treated gut microbiota. *Cell Rep.* 10: 1861–1871.

128 Sun, Z., Grimm, V., and Riedel, C.U. (2015). AI-2 to the rescue against antibiotic-induced intestinal dysbiosis? *Trends Microbiol.* 23: 327–328.

129 Mimee, M., Tucker, A.C., Voigt, C.A., and Lu, T.K. (2015). Programming a human commensal bacterium, *Bacteroides thetaiotaomicron*, to sense and respond to stimuli in the murine gut microbiota. *Cell Syst.* 1: 62–71.

130 Lu, T.K. and Collins, J.J. (2007). Dispersing biofilms with engineered enzymatic bacteriophage. *Proc. Natl. Acad. Sci. U. S. A.* 104: 11197–11202.

131 Ferreira, R.B., Antunes, L.C., and Finlay, B.B. (2010). Should the human microbiome be considered when developing vaccines? *PLoS Pathog.* 6: e1001190.

3

Approaches and Challenges in Agricultural Microbiome Research

CONTENTS

3.1 Microbiome Research in the Omics Era

Plant microbiomes are classified as rhizospheric, endorhizospheric, or phyllospheric [1]. Aerial surfaces, tissue kinds, and surrounding environmental conditions can define the uniqueness of microbial communities in these ecological niches [2, 3]. The phyllosphere niche is nutrient-poor and subjected to temperature, radiation, and moisture changes on a regular basis [3]. The rhizospheric niche, on the other hand, is located closest to the plant root system and is relatively rich in nutrients produced from and controlled by mucilage and root exudate deposition [4]. These host exudates have been shown to be significant in shaping distinct microbiomes that are altered by chemo-attractants and repellents [5, 6]. The endorhizospheric microbiome is a subset of the rhizobacterial populations that can infiltrate root hair [7]. These niches of microbial populations aid in the growth of the host by limiting the growth of plant diseases and by constructing plant growth hormones [7–9]. In addition, the rhizospheric microbiome contributes to biogeochemical cycles that produce a number of chemical

Microbiomics and Sustainable Crop Production, First Edition. Mohammad Yaseen Mir and Saima Hamid.

components essential for appropriate plant development [10]. High-throughput genome sequencing tools have made previously unavailable data on the taxonomy and functional complexity of these organisms readily available today [11–13]. These niches have been studied using multi-omics technologies, and it is clear that the microbial communities associated with them are taxonomically and functionally distinct throughout the plant host [2, 3, 14–16]. Many different host plants (such as *Arabidopsis* and rice) have had their rhizospheric microbiomes studied using metagenomics to better understand how external stressors influence the microbial communities (2013). The majority of rhizospheric microbial communities consist of Proteobacteria, Firmicutes, and Actinobacteria taxa; however, their composition changes frequently depending on the host's developmental stage, cultivar, and soil type [17, 18]. Phyllospheric microbial communities are more dynamic than rhizospheric ones, yet they often contain species that are comparable to those found in rhizospheric areas [19]. Metaproteogenomics was used by Knief et al. [20] to investigate the microbiome present in the rice phyllosphere and rhizosphere and found that phyllospheric microbial populations are affected by the potential assimilation of simple carbohydrates, ammonium, and amino acids derived from plants. A nutrient-poor phyllosphere is indicated by elevated expression of stress and transport response proteins (porins, Adenosine triphosphate (ATP) binding cassette transporter components, and TonB-dependent receptors). Furthermore, in the rhizospheric microbiome region as opposed to the phyllospheric microbiome, several nitrogen-fixing genes (such as nitrogenase [nifH]) are uniquely expressed. Ryffel et al. [21] performed metabolic footprint experiments and found simple carbohydrates produced from plants, such as sucrose, fructose, and glucose, in the phyllospheric zone. However, changes to these sugars were only noticed in *Arabidopsis thaliana* leaves after they were colonized by the organo-heterotroph *Sphingomonas melonis* or the phytopathogen *Pseudomonas syringoides* tomato, showing that these are most likely the principal carbon sources in the phyllosphere. After filling their surface, phyllospheric microbial populations alter host-derived arginine metabolites and phytoalexin production, according to metabolomic research. As a result, high-throughput proteomics and metabolomics investigations can significantly improve our understanding of the molecular basis of plant–microbe interactions. Endophytic communities were once thought to be pollutants until the discovery of high-throughput technologies, while some were indicated as symbionts of the host [22]. Endophytes of *Paullinia cupana* with asymptomatic and symptomatic anthracnose were studied using culture-dependent and 16S ribosomal RNA (rRNA) clone libraries, with clone libraries revealing more phylogenetic diversity than culture-dependent approaches [23]. A study done by Lundberg and colleagues [24], however, found that more than 600 *A. thaliana* plants grown in various soil types and developmental stages had their 16S rRNA gene pyrosequenced by the same group of researchers. The study found a core endophytic microbiome that is consistent across soil types and developmental stages and is related to Proteobacteria and Actinobacteria. These types of microbial communities are confined to a small area and are likely functionally linked to the host's metabolic processes [13]. A wide range of studies have been conducted to better understand the phylogenetic nature of microbial communities associated with various hosts and at various life stages, as described previously. However, it is critical to identify which microorganisms are active in which areas of the plant and what they are doing at different stages of the host's development. Until recently, the bulk of research has employed biomarkers to explain gene function using various molecular methods, such as Yang et al. [25], who used biomarkers to estimate maize nitrogen status. Genetic tools such as reporter genes make it possible to determine how different rhizosphere microbes perceive chemical, physical, and biological stimuli in their surroundings [26–28]. It is unclear exactly what the plant microbiome does, but some studies have employed multi-omics technologies to gain some insight into how it works as a whole [20, 29, 30]. It was possible to use biomarker assays to identify only a few proteins till now thanks to these investigations. The study of plant microbiome functions could help to improve the sustainability of agriculture [24, 31]. Beneficial microbes could be used as biofertilizers, biocontrol agents, or stress-relieving items to achieve these breakthroughs. To limit the spread of rhizospheric plant diseases, a deeper functional understanding of the plant microbiome may be required. For example, by mixing diseased suppressive soils with infected conducive soils, a succession of plant disease managements has been achieved [32].

3.2 New Efforts and Challenges in Assigning Function to Microbes

Beyond determining microbiome structures under different environments, it is particularly challenging to assign a function to individual microbiome members or groups and we tend to consider each microbial group as a functional group. However, even species within a particular genus can have completely different lifestyles – from pathogen to mutualist depending on the environment [33, 34], or due to horizontal transfer of specific functional genes [35]. This variability can lead to dramatic changes in microbial phenotypes of desired traits, such as phosphorus mobilization [36]. More sensitive methods to characterize the microbiome beyond the genus level are needed, along with a better functional characterization of each species, which would require large-scale/high-throughput techniques [37]. With the rate at which technology advances, combining, e.g. computational/modelling methods are very promising. For instance, there is a transition from metagenomics to metaphenomics, which combine the product of the metagenome (or 'expressed functions encoded in microbial genomes') with the environment, taking all the parameters into account that may influence the dynamics of the interactions within the community and the environment [38]. In this respect, metagenomics can be potentially powerful and provide information about the broad functional capabilities (e.g. secondary metabolite production or carbohydrate utilization) [39] or about specific gene sets (e.g. metabolic pathways) of the microbiome [40, 41]. With new analytical methods, we can also gain deeper insights into the specific taxa responsible for harbouring key functional traits [42]. With the development of easy-to-use commercial kits, extracting deoxyribonucleic acid (DNA) from a given sample is straightforward. However, DNA only gives the functional 'potential' of a microbial community. Furthermore, particularly in soils, the vast majority (>90%) of microbial biomass (or genetic information) is inactive or dormant [43]. Nevertheless, this number is likely to drop significantly in the rhizosphere and root, due to the stepwise selection of microbes by plant-mediated factors meaning more microbes are metabolically active in these niches [14]. Efforts have been made to extract RNA from rhizosphere samples to look at those microbes that are metabolically active and reveal which genetic pathways they are inducing in response to plant and microbial stimuli [13, 44, 45]. Therefore, many studies combine the enrichment of 13C-labelled CO_2 with metatranscriptomics to identify microbes responding to plant exudates to improve our understanding on the interactions between microbiome and plant host [46, 47]. Unlike the shorter turnover times of RNA, which reduces the simplicity and robustness of sampling efficiency [42], proteins, especially exoproteins, are more stable in the environment [48]. Thus, sampling is methodologically easier prone to sampling errors/artefacts. Metaproteomics also provides an exciting opportunity in omics research as it gives a profile of expressed proteins, and hence (metabolic) activities in a given sample [49]. In turn, exoproteomics or exometaproteomics enriches for the more ecologically important proteins that are involved in nutrient acquisition and microbial–microbial and microbial–host interactions (e.g. extracellular hydrolytic enzymes and transporter systems) [36, 48]. However, there are several drawbacks to metaproteomics, particularly a requirement for sufficient starting material (sometimes up to 100 g of soil is needed) [50], as well as the accurate assignment of peptides detected to the correct proteins, which relies on a comprehensive database (metagenome) and sufficient computational power [51–53]. Perhaps this is why meta(exo)proteomics has not been extensively utilized in rhizosphere research, in comparison to studies of other less complex microbial niches, e.g. seawater and anaerobic digesters [49, 54, 55], and enriched chitin-degrading sandy soil samples [50]. Similarly, this approach can be very powerful for identifying the major extracellular enzymes involved in phosphate mobilization within the rhizosphere [36].

3.3 Characterization of Complex Microbial Communities

Since 1990, various molecular methods capable of tracking one to hundreds of biomarkers have been widely used to analyse microbial community structure, such as polymerase chain reaction (PCR) amplification-based gene cloning, sequencing of 16S rRNA genes and functional genes, amplified ribosomal DNA restriction analysis, denaturing gradient gel electrophoresis, terminal restriction fragment length polymorphism, phospholipid fatty acid analysis, and BioLog EcoPlates for measuring carbon and nitrogen metabolisms [51]. Especially in the last decade, high-throughput molecular technologies capable of tracking multiple thousands of biomarkers have been developed for characterizing microbial communities, including high-throughput DNA/RNA sequencing, PhyloChip, GeoChip, mass spectrometry-based proteomics for community analysis, and metabolite analysis [52, 53]. We can group high-throughput molecular microbial detection technologies into two major categories: open and closed formats. 'Open format' refers to technologies whose potential experimental results cannot be anticipated prior to performing the analysis, and thus the experimental outcome is considered open. For instance, when using sequencing to analyse a microbial community, we will not know what types of sequences will be obtained prior to sequencing [52]. The main characteristics of technologies of this type are that they typically do not require a priori sequence information from the community of interest and, overall, they enable discovery of new genes, pathways, and taxa. This category includes a variety of molecular techniques, such as high-throughput sequencing technologies, screening for functional expression, fingerprinting methods, and mass spectrometry-based proteomic and metabolomic approaches. 'Closed format' refers to the detection technologies whose range of potential experimental results is defined prior to performing the analysis, and thus the experimental outcome is considered closed. For example, when a functional gene array containing 10 000 probes is used for analysing a microbial community, the experimental results from this sample cannot go beyond the detection capability of the probes (10 000) fabricated on the array. The main features of technologies of this type are that they require a priori sequence information and they do not provide new molecular information because all molecules used for designing the querying devices are known. DNA arrays, protein arrays, carbohydrate arrays, phenotype arrays, and BioLog EcoPlates, as well as quantitative PCR, are all considered closed-format technologies [52, 54].

Open- and closed-format technologies typically differ in sample preparation and quality control, data processing and analysis, performance, and application, and each presents advantages and limitations. In the following discussion, we compare features important to meaningful applications of each platform by giving special consideration to their usefulness in analysing complex microbial communities like those in soils. Since next-generation sequencing and microarrays are the best and most widely used representatives of open- and closed-format technologies, respectively, our comparison and discussion are primarily focused on these technologies [52, 55].

3.4 Advanced Fundamental Research on Microbe–Microbe and Plant–Microbe Interactions: Bridging the Lab–Field Gap

Besides comprehending the complexity of microbial communities and interactions, there is the need to uncover basic regulatory (communicative) principles of interactions that can inform experimental design. Recently, cytology-based systems have been developed to study microbial interactions. Hennessy et al. [56] established a microplate reader-based system, to quantify the activities and interactions between living microbes. This method represents a potential high-throughput screen by using live imaging of fluorescing metabolites and microbial growth to identify and trace the expression of certain genes in defined microbial communities. In their study, they measured the rate of the production of fluorescent metabolites of *Pseudomonas fluorescens* in response to the presence of *Fusarium graminearum*, as an indicator for their interaction [56]. Alternatively, Massalha et al. [57] published recently a microfluidics-based system for in vivo imaging of plant root–microbe interactions. Using a transparent chamber, they could record root zone preferably colonized by a fluorophore-tagged microbe. By adding a second microbe to the system, they were able to study microbe–microbe interactions in real time. Despite their minimalistic setup, such studies reveal fundamental insights into basic principles that shape microbe–microbe and root–microbe interactions. In this respect, the application of transparent soil represents an innovative approach to study and live image microbes on plant roots in an environment that mimics different soil textures [58]. It allows detecting processes driving the distribution of microbes in bulk substrate along the root [59] and studying the effects of major root pests such as nematodes on microbe/community behaviour [60].

In addition to bacterial and fungal microbiomes, soil and plant processes are directly influenced by other organisms including viruses, archaea, nematodes, and insects. Viruses play a very important role in soil biochemical processes and act as gene reservoirs for horizontal gene transfer, although their function is not completely understood [61]. Similarly, Archaea and nematodes significantly contribute to microbiome diversity and in interaction with other microbes to soil–plant processes and ecosystem functioning [62, 63]. In this respect, Benítez et al. [64] have given a very interesting insight into plant–microbe–insect interactions. They reported that soil microorganisms can affect aboveground interactions between plants and insects, by modulating the release of plant volatiles [64–66]. These studies indicate that we need more comprehensive, holistic studies on multi-trophic interactions in order to understand which edaphic and biotic factors determine the structure and, hence, function of soil and plant microbiome structures. Exploiting the full potential of microbes and microbial communities will depend on expertise from different fields. In addition to improving our understanding of complex plant–microbe and multi-trophic interactions using plant biology and microbiology-based approaches, we need to develop new ecological systems with growing complexity. Most critically, in order for this knowledge to be successfully transferred to agriculture it is essential to understand the impact of various farming practices on the microbiome and how this is translated to plant health and, thus, crop productivity. In addition, it is necessary to test microbial community function in a highly complex and diverse system (e.g. field), bridging the gap between the lab and the farm.

3.4.1 Bridging the Lab–Field Gap

3.4.1.1 Limitations on the Experiments Performed in Controlled Conditions: The Lack of Context

Increasing evidence is showing that plant–microbe interactions can be beneficial or detrimental for either the host or microbial symbiont, depending on the balance of associated biotic and abiotic factors. While experiments involving pairwise interactions under controlled conditions have increased our knowledge about gene and metabolite expression profiles involved in plant–microbe interactions, these experiments give us little information about microbial function in a natural ecosystem [67]. Although this was stated by de Boer [67] for fungal–bacterial interactions, it is applicable to many other interactions (even to those such as plant–rhizobia or plant–mycorrhizal interactions): for example, even some species of plant-growth-promoting arbuscular mycorrhizae fungi (AMF) have been shown to inhibit plant growth under certain conditions, e.g. low light, low temperature, or phosphorous (P) availability [68, 69]. In addition, AMF activity can also be suppressed by the soil microbiota [70] highlighting the practical need for field experiments to fully understand microbe behaviour.

The main reason for the existence of this lab–farm gap is that lab studies generally do not capture the complexity of microbe–microbe interactions that occur in a natural setting. However, it is widely known now that microbial communities and plant–microbe interactions are highly dependent on the entire ecosystem [14, 67, 71]. For instance, host genotypes have been shown to shape plant microbial communities [72] and a genome-wide association study (GWAS) analysis revealed that both bacterial and fungal communities are structured by the same host biological processes (such as defence response or signal transduction). However, different genes seem to be involved in the interaction [72], and microbial communities are further fine-tuned during plant development according to host requirements [44]. Host-dependent control of the microbial community is likely controlled by the flow of organic compounds from the root to the rhizosphere (rhizodeposition) [44, 73], which has been shown to attract beneficial microbes and refrain pathogen attack. The legume–rhizobia symbiosis is an elegant example of rhizodeposits selecting for beneficial microbes. In response to low nitrogen, the host releases flavonoid compounds that initiate the molecular dialog with nitrogen-fixing rhizobia, resulting in root nodulation and nitrogen fixation [74]. In addition, rhizodeposition also functions as a chemical signal for the establishment of inter-root or root–microbe interactions [75]. Therefore, different hosts, holding different gene sets, will trigger different responses to the same inoculant. Moreover, the same host will release different root exudates depending on the soil nutrient and microbial environment. All these examples represent the cyclic feedback between all the components of this ecosystem (plant–soil–microbes). This likely explains why field microbial inoculants fail to persist for long periods [76]. Moreover, the soil ecosystem plays a key role in the establishment of root microbiome [77, 78], which means that even if the inoculant survives within the soil community it is not guaranteed that it would colonize the plant host.

Any benefits mediated by microbes observed under controlled conditions will ultimately need to be operative in the field. This implies their persistence in the field over time and successful plant colonization over a wide range of varying biotic, abiotic, and climate conditions. Therefore, finding single inoculants that can perform in such a variety of scenarios will be highly unlikely, which increases the need for the development of microbial precision agriculture mirroring the concept of human personalized medicine [79]. In fact, the abundance of similarities between human gut and plant root microbiomes is striking and reveals the importance of the root microbiome in controlling plant fitness [11]. Specifically, it has been shown that complex microbial inoculums can improve plant disease resistance and promote growth better than individual inoculums [76], highlighting the synergistic effects of a community. However, these findings still need to consider the soil context to address their potential use as soil amendments. In addition, a deeper understanding of a microbe's function within a community and within a host would require functional studies where the ecosystem is challenged with different conditions (e.g. temperature, light, humidity). Those studies would validate their community interactions and their beneficial or detrimental outcome for plants as a prerequisite to justify for further field experiments.

References

1 Berg, G., Rybakova, D., Grube, M., and Koberl, M. (2016). The plant microbiome explored: implications for experimental botany. *J. Exp. Bot.* 67: 995–1002. https://doi.org/10.1093/jxb/erv466.

2 Gottel, N.R., Castro, H.F., Kerley, M. et al. (2011). Distinct microbial communities within the endosphere and rhizosphere of Populus deltoides roots across contrasting soil types. *Appl. Environ. Microbiol.* 77: 5934–5944. https://doi.org/10.1128/AEM.05255-11.

3 Vorholt, J.A. (2012). Microbial life in the phyllosphere. *Nat. Rev. Microbiol.* 10: 828–840. https://doi.org/10.1038/nrmicro2910.

4 Koranda, M., Schnecker, J., Kaiser, C. et al. (2011). Microbial processes and community composition in the rhizosphere of European beech – the influence of plant C exudates. *Soil Biol. Biochem.* 43: 551–558. https://doi.org/10.1016/j.soilbio.2010.11.022.

5 Bais, H.P., Weir, T.L., Perry, L.G. et al. (2006). The role of root exudates in rhizosphere interactions with plants and other organisms. *Annu. Rev. Plant Biol.* 57: 233–266. https://doi.org/10.1146/annurev.arplant.57.032905.105159.

6 Wu, L., Wang, J., Huang, W. et al. (2015). Plant-microbe rhizosphere interactions mediated by Rehmannia glutino. *Sci. Rep.* 5: 15871.

7 Prieto, P., Schiliro, E., Maldonado-Gonzalez, M.M. et al. (2011). Root hairs play a key role in the endophytic colonization of olive roots by Pseudomonas spp. with biocontrol activity. *Microb. Ecol.* 62: 435–445. https://doi.org/10.1007/s00248-011-9827-6.

8 Tkacz, A. and Poole, P. (2015). Role of root microbiota in plant productivity. *J. Exp. Bot.* 66: 2167–2175. https://doi.org/10.1093/jxb/erv157.

9 Lareen, A., Burton, F., and Schafer, P. (2016). Plant root-microbe communication in shaping root microbiomes. *Plant Mol. Biol.* 90: 575–587. https://doi.org/10.1007/s11103-015-0417-8.

10 Rousk, J. and Bengtson, P. (2014). Microbial regulation of global biogeochemical cycles. *Front. Microbiol.* 5: 103. https://doi.org/10.3389/fmicb.2014.00103.

11 Berendsen, R.L., Pieterse, C.M., and Bakker, P.A. (2012). The rhizosphere microbiome and plant health. *Trends Plant Sci.* 17: 478–486. https://doi.org/10.1016/j.tplants.2012.04.001.

12 Bakker, P.A., Berendsen, R.L., Doornbos, R.F. et al. (2013). The rhizosphere revisited: root microbiomics. *Front. Plant Sci.* 4: 165. https://doi.org/10.3389/fpls.2013.00165.

13 Turner, T.R., James, E.K., and Poole, P.S. (2013). The plant microbiome. *Genome Biol.* 14: 209. https://doi.org/10.1186/gb-2013-14-6-209.

14 Bulgarelli, D., Schlaeppi, K., Spaepen, S. et al. (2013). Structure and functions of the bacterial microbiota of plants. *Annu. Rev. Plant Biol.* 64: 807–838. https://doi.org/10.1146/annurev-arplant-050312-120106.

15 Knief, C. (2014). Analysis of plant microbe interactions in the era of next generation sequencing technologies. *Front. Plant Sci.* 5: 216. https://doi.org/10.3389/fpls.2014.00216.

16 Mendes, L.W., Kuramae, E.E., Navarrete, A.A. et al. (2014). Taxonomical and functional microbial community selection in soybean rhizosphere. *ISME J.* 8: 1577–1587. https://doi.org/10.1038/ismej.2014.17.

17 Mendes, R., Kruijt, M., De Bruijn, I. et al. (2011). Deciphering the rhizosphere microbiome for disease-suppressive bacteria. *Science* 332: 1097–1100. https://doi.org/10.1126/science.1203980.

18 Weinert, N., Piceno, Y., Ding, G.C. et al. (2011). PhyloChip hybridization uncovered an enormous bacterial diversity in the rhizosphere of different potato cultivars: many common and few cultivar-dependent taxa. *FEMS Microbiol. Ecol.* 75: 497–506. https://doi.org/10.1111/j.1574-6941.2010.01025.x.

19 Bodenhausen, N., Horton, M.W., and Bergelson, J. (2013). Bacterial communities associated with the leaves and the roots of *Arabidopsis thaliana*. *PLoS One* 8: e56329. https://doi.org/10.1371/journal.pone.0056329.

20 Knief, C., Delmotte, N., Chaffron, S. et al. (2012). Metaproteogenomic analysis of microbial communities in the phyllosphere and rhizosphere of rice. *ISME J.* 6: 1378–1390.

21 Ryffel, F., Helfrich, E.J., Kiefer, P. et al. (2016). Metabolic footprint of epiphytic bacteria on *Arabidopsis thaliana* leaves. *ISME J.* 10: 632–643. https://doi.org/10.1038/ismej.2015.141.

22 Ryan, R.P., Germaine, K., Franks, A. et al. (2008). Bacterial endophytes: recent developments and applications. *FEMS Microbiol. Lett.* 278: 1–9. https://doi.org/10.1111/j.1574-6968.2007.00918.x.

23 Bogas, A.C., Ferreira, A.J., Araujo, W.L. et al. (2015). Endophytic bacterial diversity in the phyllosphere of Amazon *Paullinia cupana* associated with asymptomatic and symptomatic anthracnose. *Springerplus* 4: 258. https://doi.org/10.1186/s40064-015-1037-0.

24 Lundberg, D.S., Lebeis, S.L., Paredes, S.H. et al. (2012). Defining the core *Arabidopsis thaliana* root microbiome. *Nature* 488: 86–90. https://doi.org/10.1038/nature11237.

25 Yang, X.F.S., Wu, J.R., Ziegler, T.E. et al. (2011). Gene expression biomarkers provide sensitive indicators of in planta nitrogen status in maize. *Plant Physiol.* 157 (4): 1841–1852.

26 Deangelis, K.M., Ji, P., Firestone, M.K., and Lindow, S.E. (2005). Two novel bacterial biosensors for detection of nitrate availability in the rhizosphere. *Appl. Environ. Microbiol.* 71: 8537–8547. https://doi.org/10.1128/AEM.71.12.8537-8547.2005.

27 Steindler, L. and Venturi, V. (2007). Detection of quorum-sensing N-acyl homoserine lactone signal molecules by bacterial biosensors. *FEMS Microbiol. Lett.* 266: 1–9. https://doi.org/10.1111/j.1574-6968.2006.00501.x.

28 De Werra, P., Baehler, E., Huser, A. et al. (2008). Detection of plant-modulated alterations in antifungal gene expression in *Pseudomonas fluorescens* CHA0 on roots by flow cytometry. *Appl. Environ. Microbiol.* 74: 1339–1349. https://doi.org/10.1128/AEM.02126-07.

29 Delmotte, N., Knief, C., Chaffron, S. et al. (2009). Community proteogenomics reveals insights into the physiology of phyllosphere bacteria. *Proc. Natl. Acad. Sci. U. S. A.* 106: 16428–16433. https://doi.org/10.1073/pnas.0905240106.

30 Wang, H.B., Zhang, Z.X., Li, H. et al. (2011). Characterization of metaproteomics in crop rhizospheric soil. *J. Proteome Res.* 10: 932–940. https://doi.org/10.1021/pr100981r.

31 Berg, G. (2009). Plant-microbe interactions promoting plant growth and health: perspectives for controlled use of microorganisms in agriculture. *Appl. Microbiol. Biotechnol.* 84: 11–18. https://doi.org/10.1007/s00253-009-2092-7.

32 Mazzola, M. (2007). Manipulation of rhizosphere bacterial communities to induce suppressive soils. *J. Nematol.* 39: 213–220.

33 Hacquard, S., Kracher, B., Hiruma, K. et al. (2016). Survival trade-offs in plant roots during colonization by closely related beneficial and pathogenic fungi. *Nat. Commun.* 7: 11362. https://doi.org/10.1038/ncomms11362.

34 Hiruma, K., Gerlach, N., Sacristán, S. et al. (2016). Root endophyte colletotrichum tofieldiae confers plant fitness benefits that are phosphate status dependent. *Cell* 165: 464–474. https://doi.org/10.1016/j.cell.2016. 02.028.

35 Qiu, X., Kulasekara, B.R., and Lory, S. (2009). Role of horizontal gene transfer in the evolution of Pseudomonas aeruginosa virulence. *Genome Dyn.* 6: 126–139. https://doi.org/10.1159/000235767.

36 Lidbury, I.D., Murphy, A.R.J., Scanlan, D.J. et al. (2016). Comparative genomic, proteomic and exoproteomic analyses of three Pseudomonas strains reveals novel insights into the phosphorus scavenging capabilities of soil bacteria. *Environ. Microbiol.* 18: 3535–3549. https://doi.org/10.1111/1462-2920.13390.

37 Schlaeppi, K. and Bulgarelli, D. (2015). The plant microbiome at work. *Mol. Plant-Microbe Interact.* 28: 212–217. https://doi.org/10.1094/MPMI-10-14-0334-FI.

38 Jansson, J.K. and Hofmockel, K.S. (2018). The soil microbiome – from metagenomics to metaphenomics. *Curr. Opin. Microbiol.* 43: 162–168. https://doi.org/10.1016/j.mib.2018.01.013.

39 Bulgarelli, D., Garrido-Oter, R., Münch, P.C. et al. (2015). Structure and function of the bacterial root microbiota in wild and domesticated barley. *Cell Host Microbe* 17: 392–403. https://doi.org/10.1016/j.chom. 2015.01.011.

40 Wang, Q., Fish, J.A., Gilman, M. et al. (2015). Xander: employing a novel method for efficient gene-targeted metagenomic assembly. *Microbiome* 3: 32. https://doi.org/10.1186/s40168-015-0093-6.

41 Ofaim, S., Ofek-Lalzar, M., Sela, N. et al. (2017). Analysis of microbial functions in the rhizosphere using a metabolic-network based framework for

metagenomics interpretation. *Front. Microbiol.* 8: 1606. https://doi.org/10.3389/fmicb.2017.01606.

42 Prosser, J.I. (2015). Dispersing misconceptions and identifying opportunities for the use of "omics" in soil microbial ecology. *Nat. Rev. Microbiol.* 13: 439–446. https://doi.org/10.1038/nrmicro3468.

43 Fierer, N. (2017). Embracing the unknown: disentangling the complexities of the soil microbiome. *Nat. Rev. Microbiol.* 15: 579–590. https://doi.org/10.1038/nrmicro.2017. 87.

44 Chaparro, J.M., Badri, D.V., and Vivanco, J.M. (2014). Rhizosphere microbiome assemblage is affected by plant development. *ISME J.* 8: 790–803. https://doi.org/10.1038/ismej.2013.196.

45 Yergeau, E., Sanschagrin, S., Maynard, C. et al. (2014). Microbial expression profiles in the rhizosphere of willows depend on soil contamination. *ISME J.* 8: 344–358. https://doi.org/10.1038/ismej.2013.163.

46 Haichar, F.E.Z., Heulin, T., Guyonnet, J.P., and Achouak, W. (2016). Stable isotope probing of carbon flow in the plant holobiont. *Curr. Opin. Biotechnol.* 41: 9–13. https://doi.org/10.1016/j.copbio.2016.02.023.

47 Lueders, T., Dumont, M.G., Bradford, L., and Manefield, M. (2016). RNAstable isotope probing: from carbon flow within key microbiota to targetedtranscriptomes. *Curr. Opin. Biotechnol.* 41: 83–89. https://doi.org/10.1016/j.copbio.2016. 05.001.

48 Armengaud, J., Christie-Oleza, J.A., Clair, G. et al. (2012). Exoproteomics: exploring the world around biological systems. *Expert Rev. Proteomics* 9: 561–575. https://doi.org/10.1586/epr.12.52.

49 Heyer, R., Kohrs, F., Reichl, U., and Benndorf, D. (2015). Metaproteomics of complex microbial communities in biogas plants. *Microb. Biotechnol.* 8: 749–763. https://doi.org/10.1111/1751-7915.12276.

50 Johnson-Rollings, A.S., Wright, H., Masciandaro, G. et al. (2014). Exploring the functional soil-microbe interface and exoenzymes through soil metaexoproteomics. *ISME J.* 8: 2148–2150. https://doi.org/10.1038/ismej.2014.130.

51 Hadwin, A.M., Del Rio, L.F., Pinto, L.J. et al. (2006). Microbial communities in wetlands of the Athabasca oil sands: genetic and metabolic characterization. *FEMS Microbiol. Ecol.* 55: 68–78. https://doi.org/10.1111/j.1574-6941.2005.00009.x.

52 Zhou, J., He, Z., Yang, Y. et al. (2015). Highthroughput metagenomic technologies for complex microbial community analysis: open and closed formats. *MBio* 6 (1): e02288-14. https://doi.org/10.1128/mBio.02288-14.

53 Loman, N.J., Misra, R.V., Dallman, T.J. et al. (2012). Performance comparison of benchtop highthroughput sequencing platforms. *Nat. Biotechnol.* 30: 434–439. https://doi.org/10.1038/nbt.2198.

54 Frias-Lopez, J., Shi, Y., Tyson, G.W. et al. (2008). Microbial community gene expression in ocean surface waters. *Proc. Natl. Acad. Sci. U. S. A.* 105: 3805–3810. https://doi.org/10.1073/pnas.0708897105.

55 Hazen, T.C., Dubinsky, E.A., DeSantis, T.Z. et al. (2010). Deep-sea oil plume enriches indigenous oil-degrading bacteria. *Science* 330: 204–208. https://doi.org/10.1126/science.1195979.

56 Hennessy, R.C., Stougaard, P., and Olsson, S. (2017). A microplate readerbased system for visualizing transcriptional activity during in vivo microbial interactions in space and time. *Sci. Rep.* 7: 281. https://doi.org/10.1038/s41598-017-00 296-4.

57 Massalha, H., Korenblum, E., Malitsky, S. et al. (2017). Live imaging of root–bacteria interactions in a microfluidics setup. *Proc. Natl. Acad. Sci. U. S. A.* 114: 4549–4554. https://doi.org/10.1073/pnas.1618584114.

58 Downie, H., Holden, N., Otten, W. et al. (2012). Transparent soil for imaging the rhizosphere. *PLoS One* 7: e44276. https://doi.org/10.1371/journal.pone.0044276.

59 Downie, H.F., Valentine, T.A., Otten, W. et al. (2014). Transparent soil microcosms allow 3D spatial quantification of soil microbiological processes in vivo. *Plant Signal. Behav.* 9: e970421. https://doi.org/10.4161/1559231 6.2014.970421.

60 O'Callaghan, F.E., Braga, R.A., Neilson, R. et al. (2018). New live screening of plant-nematode interactions in the rhizosphere. *Sci. Rep.* 8: 1440. https://doi.org/10.1038/s41598-017-18797-7.

61 Pratama, A.A. and van Elsas, J.D. (2018). The "Neglected" soil virome – potential role and impact. *Trends Microbiol.* 26: 649–662. https://doi.org/10.1016/j.tim.2017.12.004.

62 Castillo, J.D., Vivanco, J.M., and Manter, D.K. (2017). Bacterial microbiome and nematode occurrence in different potato agricultural soils. *Microb. Ecol.* 74: 888–900. https://doi.org/10.1007/s00248-017-0990-2.

63 Elhady, A., Giné, A., Topalovic, O. et al. (2017). Microbiomes associated with infective stages of root-knot and lesion nematodes in soil. *PLoS One* 12: e0177145. https://doi.org/10.1371/journal.pone.0177145.

64 Benítez, E., Paredes, D., Rodríguez, E. et al. (2017). Bottom-up effects on herbivore-induced plant defences: a case study based on compositional patterns of rhizosphere microbial communities. *Sci. Rep.* 7: 6251. https://doi.org/10.1038/s41598-017-06714-x.

65 Pineda, A., Soler, R., Pozo, M.J. et al. (2015). Editorial: above-belowground interactions involving plants, microbes and insects. *Front. Plant Sci.* 6: 318. https://doi.org/10.3389/fpls.2015.00318.

66 Beck, J.J. and Vannette, R.L. (2017). Harnessing insect-microbe chemical communications to control insect pests of agricultural systems. *J. Agric. Food Chem.* 65: 23–28. https://doi.org/10.1021/acs.jafc.6b04298.

67 de Boer, W. (2017). Upscaling of fungal–bacterial interactions: from the lab to the field. *Curr. Opin. Microbiol.* 37: 35–41. https://doi.org/10.1016/j.mib.2017. 03.007.

68 Smith, F.A. and Smith, S.E. (1996). Mutualism and parasitism: diversity in function and structure in the "arbuscular" (VA) mycorrhizal symbiosis. *Adv. Bot. Res.* 22: 1–43. https://doi.org/10.1016/S0065-2296(08)60055-5.

69 Johnson, N.C., Graham, J.H., and Smith, F.A. (1997). Functioning of mycorrhizal associations along the mutualism-parasitism continuum. *New Phytol.* 135: 575–585. https://doi.org/10.1046/j.1469-8137.1997.00729.x.

70 Svenningsen, N.B., Watts-Williams, S.J., Joner, E.J. et al. (2018). Suppression of the activity of arbuscular mycorrhizal fungi by the soil microbiota. *ISME J.* 12: 1296–1307. https://doi.org/10.1038/s41396-018-0059-3.

71 Lewis, R., Barth, V., Coffey, T. et al. (2018). Altered bacterial communities in long-term no-till soils associated with stratification of soluble aluminum and soil pH. *Soil* 2: 7. https://doi.org/10.3390/soils2010007.

72 Horton, M.W., Bodenhausen, N., Beilsmith, K. et al. (2014). Genome-wide association study of *Arabidopsis thaliana* leaf microbial community. *Nat. Commun.* 5: 5320. https://doi.org/10.1038/ncomms6320.

73 Baran, R., Brodie, E.L., Mayberry-Lewis, J. et al. (2015). Exometabolite niche partitioning among sympatric soil bacteria. *Nat. Commun.* 6: 8289. https://doi.org/10.1038/ncomms9289.

74 Oldroyd, G.E.D. (2013). Speak, friend, and enter: signalling systems that promote beneficial symbiotic associations in plants. *Nat. Rev. Microbiol.* 11: 252–263. https://doi.org/10.1038/nrmicro2990.

75 Jones, D.L., Nguyen, C., and Finlay, R.D. (2009). Carbon flow in the rhizosphere: carbon trading at the soil-root interface. *Plant Soil* 321: 5–33. https://doi.org/10.1007/s11104-009-9925-0.

76 Finkel, O.M., Castrillo, G., Herrera Paredes, S. et al. (2017). Understanding and exploiting plant beneficial microbes. *Curr. Opin. Plant Biol.* 38: 155–163. https://doi.org/10.1016/j.pbi.2017.04.018.

77 Edwards, J., Johnson, C., Santos-Medellín, C. et al. (2015). Structure, variation, and assembly of the rootassociated microbiomes of rice. *Proc. Natl. Acad. Sci. U. S. A.* 112: E911–E920. https://doi.org/10.1073/pnas.1414592112.

78 Zarraonaindia, I., Owens, S.M., Weisenhorn, P. et al. (2015). The soil microbiome influences grapevine-associated microbiota. *MBio* 6: e02527-14. https://doi.org/10.1128/mBio.02527-14.

79 Hamburg, M.A. and Collins, F.S. (2010). The path to personalized medicineperspective. *N. Engl. J. Med.* 363: 301–304. https://doi.org/10.1056/NEJMp1006304.

4

Perceptive of Rhizosphere Microbiome

CONTENTS

Microbiomics and Sustainable Crop Production, First Edition. Mohammad Yaseen Mir and Saima Hamid.

4.1 Introduction

In recent years, as attention in host–microbe biology has grown, a variety of new and re-established terms have been developed to describe complex heritable associations of species. The definition of the 'holobiont' and its 'hologenome' (i.e. a host and all of its microbial symbionts, and thus the adaptive unit's collective genomes) [1] denotes the ability of natural selection to function not only on the organism, but also on the suite of associations with its microbial members. However, the holobiont–hologenome hypothesis has received a lot of attention; it also applies to plants and their microbiomes.

Co-evolutionary powers acting on plants and their microbial symbionts have been studied extensively in the past. According to research on vesicular-arbuscular mycorrhiza (VAM), the highly beneficial symbiosis between the plant root and the fungal symbiont has driven the diversification of plant root morphology as well as VAM structure and function [2]. Furthermore, decades of legume research indicates that interactions like plant sanctions against nitrogen-fixing rhizobia are largely responsible for stabilizing the easily disrupted legume–rhizobium mutualism [3]. The advent of next-generation sequencing and other -omics tools has sparked interest in better understanding how more complex relationships between plants and their microbiome, dubbed the 'phytobiome' [4], affect plant health, growth, and development. Different compartments of a plant can host functionally and taxonomically distinct sets of microbiomes, similar to animal microbiomes. These include the phyllosphere microbiome, which includes aboveground sections of plants, root, and shoot endophytes (microbiomes colonizing root and shoot tissues), and the rhizosphere microbiome, which inhabits soil surrounding plant roots and adheres to root surface [5]. Despite the fact that the rhizosphere is not a plant organ or a physically intact compartment of the plant, it contains a vast array of microorganisms that are either free-living or intricately bound to their plant hosts [6]. The root zone is a rich source of compounds secreted by both plants and microorganisms that play an important role in maintaining plant–microbe interactions [7]. Sugars, complex polysaccharides, amino acids, proteins, and a number of secondary metabolites are all present in the exudates [7]. Signalling of host–microbe and microbe–microbe interactions is an essential component of these compounds [8]. With hundreds of different microbial taxa living in a plant's rhizosphere, the potential for interactions influencing nutrient dynamics and plant growth is vast. We address the background research on population evolution and genetics that is important to rhizosphere controls on nutrient cycling in this context. Then, we suggest conceptual ideas for experiments that specifically evaluate group-level dynamics in rhizospheres that can be targeted to modify nutrient capture and utilization processes that affect plant efficiency.

4.2 Multiple Levels of Selection in the Plant Rhizosphere

Researchers can use multilevel selection theory to understand how multiple levels of selection shape a plant's microbiome by applying it to rhizosphere processes like nutrient cycling. To better understand the role of complex associations of microorganisms in determining plant growth, development, and fitness, experimental tests of group dynamics in the rhizosphere impacting a plant phenotype are needed. In this portion, we propose three experimental design concepts that could help elucidate the function of the rhizosphere microbiome in influencing plant phenotypes through changes in nutrient cycling processes.

4.2.1 Microbial Experimental Systems and Network Analysis

Researchers have been able to monitor changes in population structure and adaptive changes in species over time using microbial experimental systems [9]. The studies usually involved paired comparisons of cultivated bacterial strains or lower eukaryotes to show the outcomes of interactions. While the greatest utility of microbial experimental systems is the ability to control for environmental and biological variables that mask potential microbe-microbe interactions, hierarchies of complexity can be introduced into the system to reveal more dynamic interactions. A small number of these experimental systems have been developed in recent years to see how changes in complex soil microbiomes can affect a plant's biology after generations of selection for microbiota associated with a plant trait. Swenson et al. [10], for example, showed that artificially selecting for rhizosphere microbiomes known to increase plant biomass production could improve the aboveground biomass of *Arabidopsis thaliana*. Flowering period was changed by soil microbiota in a similar study using *A. thaliana*, which was predicated on the assumption that selection can be conducted on a plant's phenotype while the agents of selection are made up of the rhizosphere microbiome [11]. Although the experiment showed taxonomic and functional differences in the microbiome between early and delayed flowering plants, group-level behaviours were difficult to quantify when studying soil microbiomes at the time. Since then, advances in sequencing and bioinformatic analysis have given the opportunity to examine networks of associations across multiple microbial partners that could demonstrate taxonomic coordination [12]. The testing of microbial taxa associations relevant to group-level processes, like mineralization dynamics in the soil, can be conducted. Since some plant community features may be predictive of compositional Sole microbiome [13], plant soil systems with higher capturing or use efficiencies over time may suggest more associations – including positive, negative, or neutral associations – between microbial taxa that can help to model complex interactions among Soil system microbial partners. In rhizosphere studies, network analysis of soil microbiomes can reveal unique taxa that co-occur consistently in an experimental system, suggesting targets for assembling synthetic microbial communities [14]. The use of microbial experimental systems for plant trait generation of microbiomes and of bottom-up approaches using the resulting sequence and bioinformatics data to design synthetic microbial communities could help identify plant growing consortia through alteration to the cycling of nutrients. These isolate consortia could then be tested in a variety of regulated and field environments, providing insight into the interactions that alter the plasticity of plant host traits.

4.2.2 Observing Microbiome Controls over Observed Phenotypes of the Plant Using -Omics Techniques

Plant traits are influenced by microbiomes or individual strains of microorganisms through changes in plant gene expression, thanks to advancements in sequencing technologies [15]. Although it is well known that soil microorganisms influence plant development [16], sequencing techniques such as transcriptomics enable researchers to observe these changes at previously unexplored scales [17]. When a particular microbial partner or consortia of partners interacts with its plant host, plant transcriptional analysis approaches can be used to detect microbial influences at the gene level, which can then be combined with network analysis to identify individual genes or entire metabolic pathways that are impacted. Experiments can be designed to look at the impact of specific microorganisms on the production of plant traits while tracking changes in the transcriptome of the host over time. These experiments enabled researchers to determine the impact of specific microbial associations in the rhizosphere, including effects related to plant nutrition requirements. Other -omics techniques, in addition to plant transcriptomics, may be applied to plant–soil systems to better understand how rhizosphere microbiomes affect host biology under various conditions. Exometabolomics methods, for example, may be used to describe microbial metabolite use in the rhizosphere [18]. This information could then be used to identify rhizosphere microbial exometabolic niches, which could help relate metabolite composition to microbial community structure [19]. Other methods, such as proteomics, may be used in microbiome research to see how different microbiome functions affect the plant host [20]. Researchers will gain a better understanding of how rhizosphere microorganisms affect soil processes and plant host biology by combining these multiple -omics techniques [20].

4.2.3 Genome-Editing Techniques to Uncover Plant Host Controls over Microbiome Composition and Function

The creation of genome-editing techniques such as clustered regulatory interspaced short palindromic repeats (CRISPR)-Cas may be useful in altering rhizosphere processes by influencing plant structure and physiology [21]. Short RNA sequences direct the Cas enzyme to a target site for modification in the CRISPR-Cas process [22]. Mechanistic studies of how plants regulate microbiome assembly and work in the rhizosphere may benefit from systemic changes in plant architecture or physiology. For example, genetic engineering and gene editing may be used to disrupt mutualistic relationships between microorganisms that help with nutrient capture and plants that provide carbon sources for microbial development [2]. Plants that no longer provide readily available carbon sources to soil microorganisms involved in mineralization dynamics may develop a unique microbial community that specializes in the more complex polysaccharides found in soil organic matter. Furthermore, root exudation processes may be tweaked to generate unique rhizosphere chemical families. Plant selection for particular microbial taxa could then be documented with these changes in root exudate composition, revealing taxonomic shifts in the rhizosphere population.

4.2.4 Rhizosphere Engineering and Sustainable Agriculture

The rhizosphere's physical and chemical environment is the product of many competing and interacting processes that are affected by soil type and water content, root-associated microbial population composition and biological activities, and plant physiology [23]. Agronomic management will alter the rhizosphere for brief periods of time during a plant's life cycle [24]. Any time they irrigate their fields or add fertilizer, farmers change the ecosystem around the roots of their crops and pasture plants. The rhizosphere is acidified by ammonium-based fertilizers, whereas it is alkalined by nitrate-based fertilizers. pH changes can affect the growth and composition of microbial communities by altering the chemistry of the soil around roots. Plant breeding and biotechnology can be used to create longer-lasting improvements in the rhizosphere that continue during the growth cycle. More lasting improvements can be accomplished by choosing plants with desirable rhizosphere characteristics and introgressing these traits into elite breeding lines. This technique is based on the ability to recognize characters that are useful, heritable, and easily detectable, all of which are difficult to come by when dealing with belowground traits. Plant breeders have, however, unwittingly engineered beneficial rhizosphere traits into crops by choosing the best performing lines in field trials and crossing them to other breeding lines on a regular basis. The release of malate anions from the root apices in a rhizosphere process is used in wheat breeding for aluminium (Al^{3+}) resistance. This mechanism of this physiology has been known for over a decade [25]. Despite the fact that the gene responsible for it was only recently discovered [26], breeders have been unwittingly selecting for it for over a century by selecting the best performing lines on acid soils [27]. Microorganisms are an integral part of the rhizosphere, and the overall biomass and composition of rhizosphere microbial communities have a direct effect on plant–soil interactions. The development of methods for encouraging the proliferation of beneficially introduced or indigenous microbial populations that facilitate nutrient uptake (e.g. rhizobia and mycorrhiza), promote direct plant growth, or suppress plant pathogens is of great interest. We may try to establish conditions most conducive to plant growth by amending the soil, breeding or engineering better plants, and manipulating plant/microorganism interactions as the rhizosphere's complexity is unravelled.

4.2.5 Engineering Plants

Plants change the rhizosphere by releasing inorganic and organic compounds from their roots (rhizodeposition). These exudates can help with nutrient acquisition, mineral stress avoidance, and the growth of beneficial microorganisms. Most transport processes across the plasma membrane in plant cells, such as the uptake of essential nutrients, depend on H^+ efflux to produce a membrane potential difference and an electrochemical gradient for H^+. H^+ efflux can lead to nutrient acquisition by acidifying the rhizosphere, in addition to providing a driving force for membrane transport [28]. The supply of Fe^{3+} and phosphorus, which are often fixed to the surface of minerals or present as insoluble complexes, can be increased by local acidification. In dicots and non-graminaceous monocots, organic compounds are released that chelate Fe^{3+} and promote its reduction to Fe^{2+} prior to uptake. Grass species, on the other hand, rely on the release of phytosiderophores, which assist in the mining of the poorly soluble Fe^{3+} and deliver it to the root surface for absorption [29]. Some species can reach poorly soluble organic and inorganic phosphorus by releasing organic anions such as citrate and malate, as well as phytases and phosphatases [30]. Some crop species are also shielded from Al^{3+} toxicity in acid soils by organic anion efflux, which chelates harmful Al^{3+} ions and prevents their damaging interactions with root apices [31]. As a result, the release of organic anions from roots has the potential to have a major effect on plant growth and nutrition. Since many of the genes that control these exudates have now been established, genetic engineering can be used to manipulate rhizosphere conditions by changing their expression. Plant genetic engineering is a relatively new technology, and it must overcome a number of challenges before it can be considered a success. Transgenic plants can show subtle, unexplained, or even non-existent phenotypes. Metabolic feedback, substrate availability, and post-translational regulation can all restrict the behaviour of exogenously expressed proteins. Even if a desired feature, such as root exudate, has been successfully bred into a plant, the compounds may be rapidly degraded or otherwise inactivated in the soil, or the rate of exudation may simply be too slow to affect the rhizosphere as expected. The pleiotropic effects of constitutive high or low expression of certain genes on plant growth, fertility, or function may be detrimental to output. Transgenic plants may also exhibit unexpected phenotypes that have little to do with the engineered protein's operation (loss of function due to random insertion of transgene, somaclonal variation, etc.). While several of these issues are rare, they highlight the importance of careful study of transgenic material and the use of appropriate controls. It is important to examine many independent transgenic lines and compare them to control lines once transgenic plants have been produced. There is no straightforward answer to the question of how many transgenic and control lines should be examined. This can vary depending on how subtle or solid the phenotype is, but 'more is better' in general. The study of many separately transformed lines confirms that transgene expression is the source of the observed phenotype. It is also comforting to see that the severity of the phenotype is related to the degree of transgene expression in different lines. Transgenic lines are often compared to wild-type plants, although other forms of controls may also be used. Many workers use control lines or non-transformed plants that have gone through tissue culture while creating transgenic lines from tissue culture. Since it simulates more of the heterogeneity that could be introduced into transgenic lines during tissue culture, this practice increases the chances of identifying real variations between the transgenic and control lines.

The null segregants produced by selfing the primary transformants and analysing the resulting lines, on the other hand, are the preferred controls. Null plants are excellent monitors because they have been subjected to the same treatments as transgenic plants but are devoid of the transgene. When transgenic lines are produced without tissue culture, such as by the flower-dipping method for transforming *Arabidopsis*, the null segregants are also useful [32].

The case studies that follow describe attempts to engineer the rhizosphere by modifying the efflux of H^+ and organic anions from transgenic plants' roots.

4.2.6 Case Study 1: Manipulating Rhizosphere pH

H^+ ATPase proteins, which use energy from ATP catalysis to pump H^+ across the plasma membrane against a steep electrochemical gradient, are primarily responsible for proton efflux from plant cells. For several years, the genes encoding these proteins have been isolated and studied [33]. As a result, controlling rhizosphere pH by overexpressing these genes appears to be an easy process. However, this has not been proven, and only a few studies have succeeded in increasing H^+ efflux enough to alter root function. For example, despite large increases in protein levels, *Nicotiana tabacum* and *Arabidopsis* plants transformed with cDNA encoding H^+ ATPase genes showed no symptoms [34]. It was discovered that an auto-inhibitory domain on the C-terminal tail of the H^+ ATPase protein regulated its function. By preventing protein phosphorylation and subsequent interactions with 14-3-3 regulatory proteins, this domain suppresses enzyme activity [35]. Only when plants were transformed with a tweaked H^+ ATPase protein that replaced the auto-inhibitory area did phenotypes including increased H^+ efflux from roots, more acidic rhizosphere [36], and improved growth at low pH [37] appear in transgenic lines. Minor developmental phenotypes and better growth on 200 mM NaCl were also correlated with constitutive expression of a modified H^+ ATPase gene, PMA4, in tobacco [34]. The increased potential for Na^+/H^+ exchange across the plasma membrane was due to the enhanced salinity resistance. In plant cells, a vacuolar H^+ ATPase and an H^+ pyrophosphatase catalyse the transfer of H^+ through the tonoplast membrane. There was no reason to believe that modulating these enzymes' activity would affect rhizosphere pH because they pump H^+ into the vacuole rather than across the plasma membrane. Nonetheless, in *Arabidopsis*, overexpression of the AVP1 pyrophosphatase resulted in a more acidic rhizosphere, presumably by increasing the activity of the plasma membrane H^+ ATPase [36]. Since the changes in apoplasmic pH and the concomitant up-regulation of auxin transport proteins altered the distribution of auxin along the roots, these transgenic *Arabidopsis* lines showed a variety of other phenotypes that appeared to be linked to auxin transport [38].

4.2.7 Case Study 2: Enhancing Organic Anion Efflux from Roots

Following the discovery that organic anions play a critical role in several rhizosphere processes, many groups started to look at how efflux could be manipulated [39]. To increase organic anion release from roots, two methods have been tried: (i) engineering plants with a higher capacity for synthesizing organic anions, and (ii) engineering plants with a higher capacity for transporting organic anions out of the cell. Since malate and citrate anions are present in almost all living cells and are known to be released from roots in response to nutrient deficiency and mineral stress, these studies have mainly focused on them. Because of the broad outwardly guided electrochemical gradient for anions across the plasma membrane, some researchers believe that efflux from roots is more likely to be restricted by transport across the plasma membrane than by synthesis [40]. If this is the case, strategies aimed at increasing transport ability would be more effective than those aimed at changing metabolism. Both methods, however, depend on key assumptions, which are outlined below. The first approach results in transgenic plants with increased expression of an enzyme involved in organic anion synthesis or decreased expression of an enzyme involved in its consumption, with the expectation that these changes would increase anion accumulation in the cytosol and greater efflux from the cell. If the target enzyme is rate-limiting in a pathway, changes in protein concentration will lead to changes in flux across that pathway; this strategy will have a better chance of succeeding. Doubling or tripling the amount of an enzyme that catalyses a near-equilibrium reaction is unlikely to have a phenotype-producing effect on metabolism. Similarly, enzymes whose function is tightly regulated by secondary modifications such as phosphorylation or other restrictive feedback mechanisms should be avoided. Unwanted regulation can be decreased by expressing the transgene in a separate subcellular compartment (for example, expressing a mitochondrial gene in the cytosol) or by using a gene from a different species (for example, expressing a bacterial gene in a plant; expression of a carrot gene in *Arabidopsis*). Finally, this technique suggests that the organic anions are transported through the plasma membrane and into the rhizosphere through an endogenous transport device. Plants with genes encoding proteins that promote organic anion movement through the plasma membrane are transformed in the second method to increase organic anion efflux. This method assumes that the transport protein's function is unaffected by cell activity and that efflux is not restricted by the plant's ability to synthesize organic anions.

4.2.8 Approach 1: Engineering Metabolic Pathways for Greater Organic Anion Efflux

De la Fuente et al. [41] published one of the first attempts to control organic anion release from plants. By transforming plants with a citrate synthase gene from *Pseudomonas aeruginosa* under the influence of a constitutive promoter, they increased citrate efflux from tobacco roots. The transgenic lines had 2- to 3-fold higher citrate synthase activity, which was linked to 3- to 10-fold higher citrate concentrations in root tissue and a 4-fold higher efflux of citrate from seedlings suspended in sterile water. Transgenic plants showed an increase in Al^{3+} resistance and an increased capacity to acquire phosphorus of poorly soluble forms present on an alkaline soil compared to control plants transformed with an empty vector [42].

Koyamah et al. [43] also transformed the suspension cells of carota (*Daucus carota*) with an *Arabidopsis* gene in mitochondrial citrate synthase and four-fold larger releases of enhanced cell growth citrate on media in addition to $AlPO_4$. *Arabidopsis* with the mitochondrial citrate synthase genes was subsequently transformed by Koyamah et al. [43] and although the increase in citrate release was relatively small by 60%, the increase in Al^{3+} resistance and the capacity to use sparingly soluble P was sufficient to achieve a marginal increase in transgenic lines compared to wild plants and zero segregants. Further studies of a similar approach reported a successful increase in the *Medicago sativa* [44] and *Brassica napus* organic anion efflux [45]. The *Arabidopsis* vacuolar H^+ pyrophosphatase gene AVP1 was recently used to transform rice and tomato plants. When treated with $AlPO_4$, the transgenic lines showed approximately 50% greater citrate and malate efflux than wild-type plants. The authors claimed that this, coupled with increased rhizosphere acidification, was sufficient to boost Al^{3+} stress resistance and insoluble phosphorus utilization [36]. Larger roots, longer root hairs, and more shoot biomass were also observed in the transgenic lines, which were thought to be caused by changes in auxin transport [38]. The statements of nutrient efficiency and Al^{3+} resistance must be carefully reconciled with the variations in plant size under control conditions, since the transgenic plants exhibited several strong phenotypes. Other studies that used this technique to manipulate organic anion efflux from roots were less effective [46]. For example, despite using identical constructs and the citrate synthase activity in transgenic tobacco lines to comparable, if not greater, levels than those reported originally [46], independent groups were unable to replicate the findings of de la Fuente et al. [41]. These findings show that increased enzyme activity does not always imply increased anion accumulation and efflux, and that metabolic or environmental factors can have an impact on the efficacy of this strategy.

4.2.9 Approach 2: Engineering Transport Proteins for Greater Organic Anion Efflux

Since the genes encoding the transport proteins had not been identified, there was no way to control organic anion efflux from roots by targeting the transport stage until recently. Fortunately, the discovery of candidate genes in two separate families – the Al^{3+}-activated malate transporter (ALMT) and the multidrug and toxic compound extrusion (MATE) gene families – has revolutionized the potential for engineering the rhizosphere in recent years. Membrane-bound proteins that promote malate and citrate efflux from plant cells are encoded by members of these families. New possibilities for controlling organic anion efflux from roots have arisen as a result of these recent developments.

4.2.9.1 ALMT Family

TaALMT1 from wheat (*Triticum aestivum*) was the first gene discovered to encode a transport protein that promotes organic anion efflux from plant cells [26]. TaALMT1 is the first member of a new membrane protein family that mediates malate efflux from roots by acting as an anion channel [47]. Tobacco suspension cells, rice (*Oryza sativa*) and *Xenopus* oocytes [26], barley (*Hordeum vulgare*) [48], and *Arabidopsis* (Peter Ryan, unpublished data) all produce an Al^{3+}-activated efflux of malate anions when TaALMT1 is expressed constitutively. Malate efflux also improved resistance to Al^{3+} stress in some cases (tobacco suspension cells, barley, and *Arabidopsis*). The fact that TaALMT1 induces the same phenotype in a wide range of plant and animal systems indicates that once a transport pathway across the membrane is formed, organic anions will be continuously effluxed from all of these cell types and tissues. TaALMT1 is a useful tool for monitoring malate release into the rhizosphere, but it has one significant drawback: it needs Al^{3+} to activate.

Malate efflux will not occur until Al^{3+} is applied to the bathing solution, even though TaALMT1 is constitutively expressed in a heterologous system. The most likely explanation is that TaALMT1 is a ligand-gated anion channel that needs direct interaction with Al^{3+} to activate. Although this property is ideal for conferring Al^{3+} resistance to transgenic plants, TaALMT1's functional application is restricted to acid soils with ample Al^{3+} to enable the protein.

As per reports of Delhaize et al. [49], among the members of protein family, TaALMT1 is widely distributed in various plant species of which three are involved in Al^{3+} resistance in *Arabidopsis* and *B. napus* with similar activation via Al^{3+} [50]. However, another protein, namely AtALMT9, binds to the tonoplast of shoot cells and transports malate from the cytoplasm to the vacuole without the need for Al^{3+} [51]. For controlling malate release from roots, AtALMT9 and other proteins that act similarly to it may be more useful members of the ALMT family.

4.2.9.2 MATE Family

In all the species of kingdom plantae, MATE genes are found where they act to export a variety of small organic compounds, such as secondary metabolites like flavanoids and alkaloids, as well as exogenous antibiotics [52]. Members of the MATE family have recently been discovered to aid citrate efflux from plant cells. MATE genes in sorghum (*Sorghum bicolor*) and barley (*H. vulgare*) regulate Al^{3+} resistance by promoting citrate efflux from the roots, according to recent research [53]. Citrate binds to the toxic Al^{3+} ions in the rhizosphere, protecting the root apices in the same way as malate efflux defends wheat against Al^{3+} stress. SbMATE1 heterologous expression in *Arabidopsis* and HvMATE heterologous expression in tobacco resulted in Al^{3+}-activated citrate efflux and improved Al^{3+} stress resistance [53]. Frd3 and OsFRDL1 from *Arabidopsis* and rice (*O. sativa*), respectively, are two more MATE genes that promote citrate efflux from cells, but not as part of an Al^{3+} resistance process. Instead, they regulate the flow of citrate from vascular cells to the xylem, where it binds to iron and travels to the shoot [54]. Since both Frd3 and OsFRDL1 facilitated citrate efflux in the absence of Al^{3+}, their expression in *Xenopus laevis* oocytes showed a radically different mode of action than MATEs involved in Al^{3+} resistance. Since the plants constitutively released citrate from their roots, overexpression of Frd3 in *Arabidopsis* under the influence of the CaMV 35S promoter conferred greater Al^{3+} resistance. FRD3 has a distinct advantage over the ALMT1 genes studied so far in that it can function without being exposed to Al^{3+}. FRD3 allows for increased citrate efflux into the rhizosphere through a broader variety of soil types. The metabolic and energy costs associated with both the synthesis and exudation of organic acid anions in the rhizosphere should be factored into this form of efflux engineering. The chemical properties of the rhizosphere can be influenced by biotechnology, as defined in this section. Manipulation of root exudates has the ability to enhance plant nutrition and protect plants from stress, as shown by the examples given. The composition and density of rhizosphere microbial populations capable of metabolizing these carbon sources are likely to be influenced by increased organic acid exudation by transgenic plants, but the extent and possible effects of increased nutrient availability are yet to be established.

4.2.10 Engineering Microbes

In addition to their direct influence on soil chemistry, root exudates support large microbial populations that afford a basal level of protection to roots against pathogens simply through their metabolic activity. In addition to their direct effect on soil chemistry, root exudates support large microbial populations that provide a basal level of protection to roots against pathogens simply through their metabolic activity.

Plant growth-promoting rhizobacteria (PGPR) – free-living strains that colonize roots and stimulate plant growth – are one form of isolate that has additional beneficial properties. Growth promotion can occur directly, such as increased nutrient acquisition or regulation of phytohormone synthesis, or indirectly, such as induction of the plant's own defensive responses or antagonism of soil-borne pathogens, and multiple mechanisms of promotion can operate simultaneously in a single strain.

Over the past 50 years, hundreds, if not thousands, of PGPR representing various genera have been represented. Despite their popularity as a 'natural' means of plant defence, only a few strains have been commercialized. This is partially due to the fact that uneconomically high doses are often used, and field output can be inconsistent. Genetically engineered strains have been useful in identifying some of the causes of variable strain output, and they provide a way to grow PGPR that function at low inoculum doses and in a variety of environments.

4.2.11 Strategic Issues for Strain Development

Introduced PGPR must create and maintain biologically active populations in competition with the resident microflora in order to be successful. Although a variety of cell surface molecules, as well as efficient utilization of the major carbon and nitrogen sources available in root exudates [55, 56], contribute to colonization, genetic engineering of individual fitness determinants has not generally resulted in improved strain efficiency. Genes involved in growth promotion, on the other hand, have proved to be successful targets for strain improvement, either by altering the timing or degree of their expression or by transferring and expressing them in different hosts with different desirable characteristics. The effects of the changed strain on non-target species in the environment, as well as whether the engineered trait is stably preserved and expressed, are all important considerations for field application of such strains.

4.2.12 PGPR Activity Is Enhanced in Engineered Strains

The chiA gene, which encodes a chitinase that targets chitin in fungal root pathogen cell walls, was one of the first to be transferred to heterologous bacteria with the goal of raising PGPR activity through pathogen suppression. *Sclerotium rolfsii* hyphal tips were rapidly and extensively bursting in *Escherichia coli* (which lacks the ability to control plant pathogens) expressing chiA, effectively reducing the pathogen's ability to cause disease on bean plants [57]. An engineered strain of Pseudomonas expressing a chiA gene from *Serratia marcescens* more effectively controlled *Fusarium oxysporum* f. sp. redolens and *Gaeumannomyces graminis* var. tritici and *P. fluorescens* expressing the introduced gene constitutively, either on a plasmid or integrated into the genome, provided improved control of fungal pathogens in the growth chamber or greenhouse [58]. While several rhizobacteria generate phytohormones that influence root development, attempts to engineer the rhizosphere through hormone manipulation have primarily focused on the degradation of so-called 'stress' ethylene, which is synthesized by plants in response to stresses such as flooding, drought, salt, and the presence of metals, organic pollutants, and pathogens. As we know, for the proper overall growth and development, ethylene plays an important role in plant defence mechanism to have a protection against pathogens. As per reports of Glick and coworkers [59], although ethylene is needed for proper plant development, growth, and survival, but if the concentration of ethylene will get increased it will affect the physiology that in turn will exacerbate environmental pressures and exaggerate disease symptoms. Rhizobacteria that produce 1-aminocyclopropane-1-carboxylate (ACC) deaminase, which catalyses the cleavage of ACC, the immediate precursor of ethylene, are a metabolic sink for ACC, diminishing ethylene concentrations to levels too low to initiate the physiological cascade leading to reduced growth or plant tissue damage. In seedlings of canola the root length have been enhanced as well for defence against *Pythium* in cucumber after incorporation of deaminase gene acdS from *Pseudomonas putida* UW4, which showed role of ethylene in plant pathogen defence [60]. According to reports of Glick et al. [59], on exposure to abiotic stresses in the greenhouse, canola, tomato, and tobacco plants grown in the presence of ACC deaminase-producing bacteria or transformed directly with the bacterial acdS gene grew better. For genetic engineers to target the rhizosphere is a versatile and appealing goal especially under abiotic stress conditions due to regulation of metabolic pathways importantly for overall development of plant growth as well as PGPR's ability to modulate stress ethylene levels. With pathogen-induced pressures, ethylene is required to initiate the systemic protection cascade, but stress ethylene levels may intensify pathogen-induced apoptosis and tissue damage, and a more complicated situation needing analytical assessment can prevail. Effective disease control by PGPR strains that manufacture antibiotics necessitates the production of active compounds in adequate quantities and within a time frame determined by the target pathogen's disease cycle. This is particularly difficult for *Pythium* and *Rhizoctonia* spp. caused pre- and post-emergence damping-off diseases, which can invade seeds and seedlings until introduced strains can synthesize inhibitory amounts of metabolites. *P. fluorescens* BL915, which synthesizes the antifungal compound pyrrolnitrin, was engineered by Ligon et al. [61] to solve this issue. The regulatory gene gacA, which regulates pyrrolnitrin synthesis in *Pseudomonas*, was expressed constitutively on a multicopy plasmid in BL915. In comparison to the parental strain, this regulatory derivative released 2.5 times more pyrrolnitrin. Another derivative, in which the entire 4-gene prnABCD operon was constitutively expressed from a plasmid, developed 4-fold more pyrrolnitrin than the parental strain, and when both plasmids were expressed in the same cells, development was increased 10-fold higher than the parental strain. The derivative strains were also as protective of cucumber and *Impatiens* in soil infested with *Rhizoctonia solani* in greenhouse trials as BL915 was added at 10-fold higher doses. *R. solani* and *Pythium ultimum* had greater management of cotton in the field than BL915 and were not substantially different from chemically treated and stable controls. *R. solani* influence on cotton was equal to the amount of pyrrolnitrin synthesized in vitro, according to other studies [61]. The studies have shown that constitutive production or overproduction of antibiotics can improve strain effectiveness and are supported by other works in which biosynthesis operons for the antibiotics phenazine-1-carboxylic acid (PCA) and 2,4-diacetylphloroglucinol (DAPG) were transferred to heterologous strains to enhance control of damping-off diseases [61]. Genetic stability and successful gene containment are critical in reducing the risk of horizontal gene transfer in the rhizosphere, and chromosomal insertion provides both. PCA-producing derivatives were found to be more efficient than wild-type SBW25 at reducing damping-off disease on pea seedlings, and the amount of PCA produced was directly linked to the bacteria's efficacy. Furthermore, pretreatment of the soil with the engineered strain decontaminated it successfully, lowering disease incidence [62]. Engineered derivatives of *P. fluorescens* 5-2/4 expressing an optimized cassette carrying the DAPG biosynthesis operon from *P. fluorescens* Q2-87 (B genotype strain) improved regulation of *P. ultimum* in another study [63]. *P. fluorescens* Q8r1-96, which develops DAPG, is highly successful against *G. graminis* var. tritici, but less successful against *R. solani*, which causes root rot. The phenazine biosynthesis operon is normally controlled by quorum sensing and, as predicted, a constitutively engineered derivative in Q8r1-96 developed more PCA in vitro and in the wheat rhizosphere [64]. However, it is surprising that derivatives containing PCAs have synthesized more DAPG than the parental strain, which indicated that an intermediate PCA biosynthesis may have a derepressed DAPG operon. The random PCA cassette insertion into the Q8r1-96 genome often varied between individual recombinants, possibly because of positions. As with transgenic lines, multiple individual recombinant clones must be compared with the control strains of parent and plasmid. Introduced genes may also be targeted to a neutral chromosomal site with vectors [65]. In the greenhouse, PCA-producing strains suppressed *Rhizoctonia* root rot at inoculum doses 10- to 100-fold lower than the dose of Q8r1-96 needed for comparable disease suppression [64], and in the field, wheat treated with the engineered strains consistently had yields 8–20% higher than those treated with Q8r1-96 alone.

4.2.13 Recombinant Strains and Rhizosphere Competence

PGPR's biological fitness is an important element in the assessment of possible threats linked to environmental release. De Leij et al. [66] found no impact of metabolic stress on bacteria in the rhizosphere of pea or wheat using isogenic derivatives of *P. fluorescens* SBW25 tagged with marker genes. In strain Q8r1-96 and its PCA-producing variants, the issue of fitness has also been raised. Since PCA helps *Pseudomonas* strains compete [67], it is possible that Q8r1-96 that produces PCA all the time will be more successful than the wild form. In addition, the PCA synthesis, combined with up-regulating DAPG synthesis in the recombinant strains, is energetically expensive and constitutive processing is likely to establish the metabolic burden that would limit the fitness of the recombinants. Because PCA contributes to the competitiveness of Pseudomonas strains, it is conceivable that Q8r1-96 constitutively producing PCA would be more competitive than the wild type. The persistence of Q8r1-96 wheat and its PCA-produced Z30-97 derivative was tracked under greenhouse conditions in a three-year field study to differentiate between these possibilities [68, 69]. When Q8r1-96 and Z30-97 colonized different rhizospheres, no consistent strain-specific discrepancies were found in rhizosphere competence. On the other hand, Q8r1-96 moved Z30-97 on a wheat rhizosphere in co-inoculated strains on the same plant to produce a greater competitive pressure [68]. Together, the data show that wheat is finally overwhelmed by its metabolic cost to the fabricated strain by some advantage of the PCA synthesis, but it seems that the cost is not sufficient to adversely affect the competitiveness of autochthonous microbes even during three field seasons. In the connected studies, wherein same strains were co-inoculated in controlled conditions on barley, pea, and marine beans, for the purposed assessment of the effect of plant host, the wild-variety Q8r1-96 showed that the Z30-97 surface was of pea [68, 69]. The findings show that plant species modulate the competitiveness of the strains and must be addressed when determining the possible fate and danger faced by the release into the ecosystem of recombinant strains.

4.2.14 Non-Target Effects of Wild-Type and Genetically Engineered PGPR

Pseudomonas putida WCS358r, modified to produce either PCA or DAPG, has been used in some of the best studies of population dynamics and non-target effects of recombinant rhizobacteria to date [70]. This study is noteworthy because it was performed in the field, PCA was demonstrated to be generated in the rhizosphere by the recombinant strain, and non-target effects were quantified using both cultivation-dependent and cultivation-independent methods. The findings showed that both the wild-type and recombinant strains have temporary effects on the structure of the wheat rhizosphere fungal and bacterial microflora, with the recombinant strains' effects remaining longer in certain cases. More specifically, the recombinant strains' effect varied from year to year and study to study. Given that WCS358r and other PGPR normally develop very high population densities immediately after inoculation, and then densities decline (sometimes precipitously) with time and distance from the inoculum, these findings, which mimic those of most other studies performed under managed and field conditions, are not unexpected. Furthermore, rhizobacteria inserted into the rhizosphere or among the roots of the same or separate plants do not grow uniformly. Non-target effects of wild-type and transgenic rhizobacteria have been studied extensively, and the results show that while the bacteria have definite effects on non-target bacterial, fungal, and protozoan species, the effects differ from study to study and are temporary.

4.3 Engineering Microbial Populations and Plant–Microbe Interactions

The number and structure of endogenous microorganisms, as well as the population dynamics of imported rhizobacteria, are all influenced by plant genotype. Genetic codes that are embedded in plants regulate the rhizodeposition, which in turn control the growth of rhizospheric microorganisms. In the earlier literature, there are several studies that demonstrate this idea [71]. Using chromosomal replacement lines of maize, Neal, Atkinson, and Larson showed wonderfully the effect of plant genotype on rhizospheric microflora [72]. Both the S-615 and Rescue lines were sensitive to common root-red; they contained greater rhizosphere-bacterial communities than the Apex root-resistant line. The chromosome pair 5B from Apex was substituted for its homologue in S-615, yielding the chromosomal replacement line S-A5B, which was as insensitive to root rot as Apex. In addition, the bacterial population in S-rhizosphere A5B's was close to that of Apex. Furthermore, immune lines had higher percentages of bacteria that were antagonistic to the widespread root rot pathogen than susceptible lines. Fluorescent pseudomonads were isolated from the rhizosphere, rhizoplane, or root interior of flax and tomato grown in a silty loam soil by Lemanceau et al. [73]. The enzyme activities and ability to use organic substrates were used to classify the isolates. The isolates were clustered using numerical analysis of these features, which revealed that the plant species had a selective effect on the populations. Nine clusters were formed by combining flax and tomato isolates. Six clusters of flax isolates were found, three of which were flax-specific. Six clusters of tomato isolates were produced, two of which were tomato-specific. Inositol, ribose, saccharose, trehalose, erythritol, m-hydroxybenzoate, and 5-cetogluconate may be assimilated by a much higher proportion of tomato isolates than flax isolates. More recently, studies targeting DAPG-producing *P. fluorescens* (*phlD*$^+$) strains known for their PGPR activity have demonstrated that plant species and varieties differentially enrich and support populations [74] and genotypes of this specific group of *P. fluorescens*, and that DAPG accumulation in the rhizosphere also differs among plant hosts [75]. In Dutch soil, the yield of the DAPG was found to be higher than lily while experimenting on maize, sugar beet, and pulp, where *PhlD*$^+$ isolates were more favourable [75]. Notz et al. [76] have shown that *P. fluorescens* CHA0 aggregation in DAPG has a correlation with expression of the phlA gene for the biosynthesis of DAPG, and expression in monocots (maize and wheat) has been substantially higher compared to dicots (bean and cucumber). Differences in gene expression were also found in six maize farmers. Wheat varieties also help *phlD*+ populations, *phlD*+ genotypes, and DAPG root aggregation. Mazzola et al. [74] observed that *phlD*+ isolates from wheat varieties grown in apple orchard soils had slightly different population densities, as the variety namely 'Lewjain' supported densities >105 colony-forming unit (CFU) g^{-1} root, but 'Eltan' had no *phlD*+ isolates. Okubara and Bonsall [77] measured DAPG on roots of wheat varieties 'Tara', 'Buchanan', and 'Finley' grown on moist filter paper and inoculated with *P. fluorescens* Q8r1-96 (D genotype) or Q2-87 (B genotype). While both strains colonized roots similarly, DAPG accumulation varied greatly between strains and wheat varieties.

4.4 Emerging Approaches in Rhizoremediation

Rhizoremediation is the reciprocal association of plant and microbial flora in the rhizosphere that leads to the degradation of pollutants. This is a novel remediation approach that is both efficient and low-cost [78]. The suitability of plants and associations with the rhizosphere microbiota, however, are critical to the establishment of rhizoremediation. For effective rhizoremediation of polycyclic aromatic hydrocarbons (PAH), Zhuang et al. [79] examined the use of various bacteria associated with various plants. It has been previously published how various plant growth stimulating bacteria extracted from the rhizosphere of *Jatropha curcas* are able to effectively degrade Dichlorobenzene [80]. Benzopyrene and pyrene (high molecular weight [HMW]) may be used as a single source of carbon by biofilm-forming plant growth promoters *Acinetobacter* sp. PDB4 and *P. aeruginosa* PDB1 [81]. PGPR are now being seen as a viable method for pollutant treatment, where the rhizobacteria help a plant flourish in polluted soil [82]. Exudation from the plant's root system assists microorganisms in surviving, growing, and penetrating the soil [83]. By fostering plant growth and reducing PAH tension in the rhizosphere, the microorganisms aid the host plants directly or indirectly. The degradation and decay of PAHs in the rhizosphere of plants benefits from these mutually beneficial relationships. Exudates from plants, such as flavonoids and fatty acids, can stimulate microbial activity and thus PAH degradation [84]. Microorganisms use root exudates as a carbon supply and mediate the initial ring hydroxylation in the rhizodegradation of PAH [85]. PAH dispersion has been shown to be slightly higher in plant rhizospheric soil than in unplanted soil, but no evidence of rhizosphere inhibitory effects on degradation has been found [86]. Releases of aromatic composites, PAH homologous, and plant root surfactant molecules create an influence of rhizosphere that can increase microbial activity in polluted soil and rhizodegrade PAH [87]. Cebron et al. [88] identified bacterial composition changes and degradation of PAH in the presence of the exudate root. In presence of root exudates the degradation of phenanthrene in soil was postponed and applied after 4 days of incubation followed by 68% deterioration that occurred after 12 days. In contrast to unplanted soil, Launen et al. [89] found a higher proportion of PAH-degrading bacteria in the rhizosphere of *Spartina* plants. As a result, plant–microbe interactions in the rhizosphere have a major impact on contaminant bioremediation. The variations in PAH concentrations between the rhizosphere and bulk soil suggest that the rhizosphere microbiome plays a role in PAH reduction [90]. A variety of studies have shown that hydrocarbon degradation in soil is improved when plants are present [91], owing to the symbiotic associations between the host plant and its rhizosphere-microbiome [92].

4.4.1 Impact of Rhizosphere Microbiome on Rhizoremediation

The function of rhizospheric bacterial populations in polluted soil and their significance in the deterioration of pollutants has been better understood thanks to advancements in 'omics' e-based technologies [93]. As a result, the effective remediation of toxins in the rhizosphere is found to be highly dependent on plant–microbiome interactions. As a result, the increased metabolic activity of the bacteria is the key cause for the change in rhizosphere degradation. Various experiments have recorded the degradation of PAH compounds using single species or consortiums, as well as their metabolic mechanisms, enzymatic modulation, and genetic controls [94]. However, the role of the complete rhizospheric microbiome in hydrocarbon rhizodegradation is understudied and requires further research. According to many studies, higher levels of PAHs' (both low molecular weight [LMW] and HMW) degradation occur near to the roots, where rhizospheric microbes play an important role [95]. Abiotic conditions in the rhizosphere, such as pH, moisture, oxygen, and usable nutrients, are more conducive for bacterial growth and survival, which has a beneficial impact on microbial processes for PAH degradation [96]. As a result of their 'rhizosphere impact', plants can help bioremediation in the root zone by supporting hydrocarbon-degrading rhizobacteria [97]. One of the root endophytic bacteria, namely *Bacillus* sp. SBER3, was recorded to degrade up to 83.4% and 75.1% of anthracene and naphthalene, respectively, under controlled conditions while in case of rhizosphere of *Populus deltoides*, Bisht and coworkers [98] recorded 45.6% degradation of anthracene. Moreover for vegetated soil, a decline of 74% of pyrene but only 40% or less in the unplanted soil after eight weeks of duration has been observed [99]. According to Yang et al. [100] pyrosequencing research showed a high level of diversity in the bacterial population in permafrost soil along the China–Russia crude oil pipeline, with 84834 reads per active taxonomic unit (OTU). The abundance of aliphatic and aromatic catabolic genes in the rhizosphere of plants [101] indicates natural abundance of microbial catabolic genes in the ecosystem, and thus the entire microbiome can play a significant role in rhizodegradation. While an isolated bacterial assemblage had a higher relative abundance of PAH-degrading genes, Bell et al. [102] found that a diverse soil microbiome could degrade more crude oil than advanced bacterial assemblages obtained in culture. Since the soil microbiome is made up of a wide variety of microorganisms such as microbes, fungi, algae, and viruses, a stable microbial community is critical for rhizoremediation.

As per reports of Raaijmakers and coworkers [103] and Fernández-Luqueño et al. [104], in the decomposition of organic matter including PAHs as well as in cycling of minerals in biogeochemical processes, fungi plays an immense role. Bacterial species including þ ryegrass (BR), þ mycorrhizae þ ryegrass (BMR), and mycorrhizae þ ryegrass (MR) showed elevated indulgence of phenanthrene and pyrene and bacteria. Yu et al. [105] reported the effects of arbuscular mycorrhizal fungi in multi-component bioremediation system, where higher dissipation of phenanthrene and pyrene were observed in treatments with bacteria þ ryegrass (BR) and mycorrhizae þ ryegrass (MR), and bacteria þ mycorrhizae þ ryegrass (BMR); than bacteria (B) and ryegrass (R) alone. The hyphae of arbuscular mycorrhizal fungi contribute to the uptake of PAHs by rye grass [106]. According to Siciliano et al. [107], the abundance of toxins in the rhizosphere plays a significant role in determining the microbiome's composition. The most abundant PAH degraders were found to be Alphaproteobacteria Sphingomonads [108]. Furthermore, PAH with different benzene rings may have different levels of toxicity in the soil microbiome, and thus the microbiome composition in the rhizosphere with PAH of different molecular weight may vary. If the number of benzene rings grows, toxicity rises, microbial activity to degrade HMW-PAH declines, and the microbiome population composition shifts [109]. The uptake and aggregation of toxins by roots in the rhizosphere is aided by rhizosphere microorganisms, which are then translocated to other sections of the plant and stored in biomass. However, a better understanding of the structure of the microbiome is needed to improve the efficiency of rhizoremediation. One factor is to learn all about how a host builds a protective microbiota and how it works under contaminant stress.

4.4.2 Current Approaches to Understand the Role of the Microbiome in Rhizoremediation

Most of the taxa in the microbial community are unknown and culture independent, which is the most common and frequent issue in studies of the function of the microbiome in PAH rhizoremediation. Novel methods and technology have arisen to further explain pollutant biodegradation, allowing us to research the anatomy, morphology, and metabolism of the microbiota along with pollutant degradation.

4.4.3 Metagenomics

The microbiome structure and composition can be studied using rhizosphere metagenome study. The microbial populations in bulk soil and the rhizosphere are distinct, according to a metagenome study of bacterial 16S rDNA [110]. In reality, metagenomics techniques have been used for a variety of purposes, including determining population composition by targeting conserved regions and determining the availability of usable genes in any context using traditional recombinant DNA techniques [111]. Scientists have been able to design methods and pipelines to gain deeper insights into metagenomes, thanks to advancements in next-generation technologies and associated bioinformatics approaches. Bell et al. [112] used amplicon sequencing to demonstrate that soil pollution concentration and plant phylogeny used for phytoremediation influence the microbial populations in the rhizosphere. As a result, metagenomics research helps in quantifying the microbial community's transition effectiveness. Several microbial strains with unknown functions may occupy the rhizosphere during phytoremediation and may possess novel unknown pathways for PAH degradation, according to a study of the rhizosphere microbiome.

Through defining functional targets and identifying various classes of dioxygenases from different genera of bacteria, studies using the metagenomics method have documented novel functional metabolic mechanisms for biodegradation of aromatic compounds [113]. Metagenomics of diverse microbial species gives researchers full access to the genetic material of a habitat's entire population [114]. The existence of the gene is determined by examining soil microbial diversity, which allows researchers to relate microbial phylogeny to the role of microbe diversity, as well as gain insight into the plant–microbe relationship for PAH removal in the presence of plants in the rhizosphere [115]. As a result, metagenomics analysis has allowed researchers to gain insight into uncultivable microorganisms in the rhizosphere microbiome, allowing them to tap into their capacity for PAH deprivation [116]. In most cases, the dominance of Alphaproteobacteria and Actinobacteria has been observed to increase in hydrocarbon-polluted soil, while Proteobacteria has been found to dominate the overall population in hydrocarbon-contaminated sites, according to recent research. According to Chemerys et al. [117], metagenomics sequence review on a contaminated site revealed that uncultured bacterial organisms were primarily responsible for pollutant elimination. Furthermore, current classifications underestimate the diversity of PAH-degrading bacterial enzymes in the soil, including RHDs [118]. In PAH-polluted soils and sediments, a number of gene sequences linked to RHDs from Gram-negative bacteria have been discovered [119]. The concentration of Gammaproteobacteria is greater than Alphaproteobacteria, according to a metagenomics study of anthracene-amended soil samples. Anthracene-spiked soil has the most Actinobacteria, while unspiked soil has the most Acidobacteria [120]. However, there is no clear explanation of how or whether the metagenome can affect the roles of individual microorganisms or meta-organisms, including the presence of biotic factors such as antagonism, quorum sensing, and horizontal gene transfer.

4.4.4 Metatranscriptomics

Transcriptomic or metatranscriptomics analysis is used to study mRNA expression profiles of genes in single particular microorganism or microbiome of a different ecosystem [121]. Therefore, metatranscriptomics refers to the analysis of transcription (mRNA) from microbial communities and its reduced sequencing cost allows to study microbial expression analysis for precise assessment of the taxonomic abundances to functionally active fraction of microflora for selected characteristic without polymerase chain reaction (PCR) bias [5]. The metatranscriptomics analysis is useful in the study of the genetic expression profile of microbiome in pollutant environments and the shift in microbiome expression following the introduction of contaminants. As a result, metatranscriptomics has been investigated for evaluating gene transcript through selectively enriching mRNA by subtractive hybridization of rRNA and has shown the abundance of several genes involved in PAH degradation and metabolism [122]. It enables quantification of microbial gene expression and control in natural and contaminated habitats using global transcriptional analysis and random sequencing of mRNA transcripts derived from the microbiota of the setting [123]. Such a culture-independent approach allows access for pollutant depletion to the incultivable microorganisms and the functions of the indigenous microbiome. Unlike metagenomics, this not only shows microbial genetic ability, but also correctly reflects behaviour and provides a clue to the taxonomic classes that contribute to pollutants' biodegradation. The sensitivity of soil microbial populations to phenanthrene resulted in a higher abundance of transcripts linked to dioxygenase, stress reaction, and detoxification [124]. In the transcripts relating to PAH depletion, willows rhizosphere grown in polluted soils were greatly enlarged [125] and subsequently members showed that they had a link with orders of Actinomycetales, Rhodospirillales, Burkholderiales, Alteromonadales, Solirubrobacterales, Caulobacterales, and Rhizobiales [126]. A recent research has shown that metatranscriptomes can specifically demonstrate the role of a certain microbiome where the relative abundance of mRNA and the atrazine-degrading activity are closely linked [127]. Many genes involved in hydrocarbon biodegradation were shown to be more abundantly expressed in polluted soils with plantations, implying that interspecies associations played a larger role in contaminant remediation. In general, metagenomics analysis reflects a small number of eukaryotes in the rhizosphere, while metatranscriptomic analysis reveals more than 25%

reads of eukaryotic representative, which may play a significant role in directing microbial activity in the polluted soil's rhizosphere [5]. However, a mixture of metagenomic and metatranscriptomic approaches may contribute to the relative efficacy of rhizobiomics in polluted sites' rhizoremediation.

4.4.5 Metaproteomics

The relative effectiveness of rhizobiomics in contaminated sites may therefore be contributing to the mixture of metagenomics–metatranscriptomic approaches. The measurement of the entire protein level in a given ecosystem involves protection or metaproteomics [128]. This technique is based on highly efficient separation methods by means of two-dimensional polyacrylamide gel electrophoresis (2-DE) and modern bioinformatics instruments combined with mass spectrometry (MS) [129]. The research of proteomics was used to examine the variation in the protein expression profile involved in the physiological responses to contaminants of microorganisms and to establish the particular proteins involved when exposed to some organic contaminant in a polluted setting [130]. Catalase-peroxidase, putative monooxygenases, dioxygenases, naphthalene-inducing dioxygenases, and aldehyde dehydrogenase have been detected by this study in the presence of HMW-PAH in *Mycobacterium vanbaalenii* PYR-1. This analysis was carried out on catabolic pathway [131]. Similarly, 80 distinct proteins including various dioxygenases, hydrolase, and thiolase were induced in presence of benzoates, *p*-hydroxybenzoates, and vanillin in *P. putida* KT2440. NidA is closely associated with the digestion of pyrene and nidA3 with fluoranthrene, meaning that the bacterium uses a separate initial RHO enzyme in reaction to HMW-PAH as a carbon source, according to proteomics analysis [132]. Protein expression profiles in *P. putida* KT2440, *Rhodococcus jostii* RHA1, *Burkholderia xenovorans* LB400, *Corynebacterium glutamicum*, and *Comamonas* sp. CNB-1 have revealed aromatic hydrocarbon metabolic pathways in response to many aromatic compounds [131]. The discovery of enzymes involved in aromatic compound degradation pathways was made using proteomics data from *Pseudomonas* sp. K82 and *Acinetobacter* strains. Biodegradation-related proteins were discovered after a proteomics study of *Novosphingobium pentaromativorans* US6-1, as well as the induction of LMW aromatic-hydrocarbon-degrading genes in plasmid pLA1 [133]. A proteomics analysis identified 25 proteins needed for the catabolic pathway of fluoranthene degradation in *Mycobacterium* sp. JS14, confirming a suggested fluoranthene degradation pathway [134]. The protein profiles of mycobacteria grown in the presence of HMW-PAHs were then recorded using a series of comparative proteomics analyses. The active group member for the degradation of PAH was identified as Burkholderiales by a metaproteomics analysis, which was validated by the existence of dioxygenase [135]. Proteomics offers biomarkers for measuring bioremediation processes and a greater knowledge of physiological functions than the microbial community's metatranscriptome [128]. To address the protein distribution and research in the hydrocarbon-degrading microbial population, as well as catabolic enzymes associated with the biodegradation pathways, metaproteomics studies used 2-DE, MS, and upgrades in protein sequence and structure databases [136]. The naphthalene degradation pathway for particular groups of bacteria from complex microbial communities was reconstructed using metaproteomics studies [137]. Proteome examination of Sphe3 cells grown on phenanthrene or glucose using two-dimensional gel electrophoresis in conjunction with MS detected PAH-degrading proteins, including 1-hydroxy-2-naphthoate dioxygenase and protocatechuate dioxygenase involved in the degradation of phenanthrene [138]. Proteomics methods have also been active in identifying 27 and 54 proteins that are needed for complete pyrene and fluoranthene degradation, respectively [139]. Also, 847 proteins, as active degraders from naphthalene and fluorene, have been identified by a practical metaproteemics strategy, and 70% of the proteins identified were from the Burkholderiales, Actinomycetales, and Rhizobiales. Naphthalene-degrading proteins known have helped classify bacteria that degrade naphthalene by 80% of their carbon. From the BACTRAP of the plume fringe, the catabolic enzymes salicylate-5-hydroxylase, fumaryl pyruvate hydrolase, and gentisate 1,2-dioxygenase, GST-like protein NagJ, and catechol 2,3-dioxygenase were identified, which are needed for the gentisate pathway [140]. *Rhodococcus*-degrading enzymes associated with fluorine degradation were also detected. Stable isotope probing of proteins may be a useful method for identifying metabolic primary enzymes and degradation pathways in real time [141]. Kim et al. [142] used two-dimensional gel electrophoresis to identify multiple proteins that were induced in *M. vanbaalenii* PYR-1 in the presence of PAHs, including catalase-peroxidase, putative monooxygenases, dioxygenase, naphthalene-inducible dioxygenase, and aldehyde dehydrogenase (2-DE). Kim et al. [132] investigated aromatic hydrocarbon catabolic pathways in *P. putida* KT 2440 and discovered 80 new proteins, including dioxygenases, hydrolases, and thiolases, that were induced in the presence of benzoate, *p*-hydroxybenzoate, and vanillin.

4.4.6 Genomics

Studies of a group of genes involved in complete hydrocarbon degradation benefit from genomic methods that study the complete genome or draught genome of species. Microbial genes involved in PAH degradation have been established by genomic analysis, and high-throughput molecular techniques are particularly useful for detecting microbial traits and operons that serve as a combined action of several genes and genes with identical functions that may be correlated with various other genes in different species. A large number of full and draught bacteria genome sequences with catabolic activities to degrade PAH are available [143]. *Pseudomonas*, *Shewanella*, *Deinococcus*, *Dehalococcoides*, *Serratia*, *Bacillus*, and *Alcaligenes* have complete and draught genome sequences that include structure and organization of various promoters, genomes, and degradative pathways relevant to PAH remediation [144, 145]. Cultural PAH-degrading bacteria can be isolated based on their ability to metabolize PAH as a single source of carbon and energy and genes and by isolating the opera required to fully degrade PAH from the genome. This Phn island was predicted to contain three clusters of genes encoding metabolism of phenanthrene to *ortho*-phthalate (phn genes), *ortho*-phthalate to protocatechuate (Oph genes) and protocatechuate to pyruvate (pmd genes). Complete genome sequence of Sphingobium yanoikuyae B1 stated about 5140 putative open reading frames in the genome and a total of 35 putative dioxygenase genes were found including catechol 1,2-dioxygenase, biphenyl 2,3-dioxygenase, biphenyl-2,3-diol 1,2-dioxygenase. Three clusters of genes encoding phenanthrene to ortho-phthalate (phn genes), ortho-phthalate to protocatechuate (Oph genes), and protocatechuate to pyruvate metabolism were expected to be present on this Phn Island (pmd genes). The genome sequencing of *Sphingobium yanoikuyae* B1 revealed 5140 putative open reading frames and 35 putative dioxygenase genes, including catechol 1,2-dioxygenase, biphenyl 2,3-dioxygenase, and biphenyl-2,3-diol 1,2-dioxygenase [146]. *P. putida* KT2440, *R. jostii* RHA1, *B. xenovorans* LB400, *C. glutamicum*, and *Comamonas* sp. CNB-1 all have a catabolic enzyme involved in PAH catabolic degradation [147]. Several genomes from various bacterial species have been sequenced and are available in public databases such as the NCBI. The genomics of bacteria capable of degrading polyaromatic hydrocarbons reveals the molecular structure and metabolic pathways of these bacteria. Table 4.1 lists some of these bacterial genomes and their importance in bioremediation.

Table 4.1 Bacterial strains having ability to degrade polycyclic aromatic hydrocarbons (PAH).

Species	Source	Significance	Reference
Alcanivorax borkumensis SK2	Oil-polluted marine water	Ability to degrade oil hydrocarbons	[148]
Celeribacter indicus P73T	Deep-sea sediment of the Indian Ocean	Ability to degrade a wide range of polyaromatic hydrocarbons	[149]

(*Continued*)

Table 4.1 (Continued)

Species	Source	Significance	Reference
Cycloclasticus sp. strain P1	Deep-sea sediment of Pacific Ocean	The strain has the ability to use polyaromatic hydrocarbons as the only carbon source for growth	[150]
Cycloclasticus sp. strain PY97M	Sea sediment of China	Degrade pyrene and fluoranthene	[151]
Pseudomonas putida strain KT2440	Contaminated soil	Metabolize a large number of aromatic compounds	[152]
Pseudomonas stutzeri A1501	Rhizosphere	Degradation of aromatic compounds	[153]
Pseudomonas stutzeri	Rhizosphere of a *Sorghum nutans* cultiva	Oxidize a large variety of organic compounds	[154]
Pseudomonas sp.	Plants growing at PAH-contaminated sites	Ability to degrade a wide range of polyaromatic hydrocarbons	[155]
Pseudomonas stutzeri	Oil spill sites	Ability to degrade a wide range of polyaromatic hydrocarbons	[156]
Mycobacterium vanbaalenii PYR-1	—	Ability to degrade a wide range of high-molecular-weight PAH	[157]
Mycobacterium rufum JS14T	PAH-contaminated soil	Capable of degrading polyaromatic hydrocarbons	[158]
Bacillus megaterium	Diep and Plankenburg rivers	Degrading polyaromatic hydrocarbons	[159]
Bacillus simplex	Oil spill sites	Degrading polyaromatic hydrocarbons	[156]
Bacillus pumilus	Oil spill sites	Degrading polyaromatic hydrocarbons	[156]
Cellulosimicrobium cellulans CWS2	PAH-contaminated soil	Degrading polyaromatic hydrocarbons	[160]
Stenotrophomonas sp.	Plants growing at PAH-contaminated sites	Degrading polyaromatic hydrocarbons	[155]
Raoultella ornithinolytica	Diep and Plankenburg rivers	Degrading polyaromatic hydrocarbons	[159]
Achromobacter xylosoxidans PY4	Arabian Gulf shoreline	Degrading polyaromatic hydrocarbons	[161]

4.5 Heritability of Rhizosphere Microbiome

The rhizosphere microbial population composition of 27 modern maize inbreeds planted in randomized and repeated plots in each of 5 field ecosystems at flowering time was investigated. To describe the bulk soil microbial community composition, the microbial communities of 15 bulk soil samples in random locations within each of the 5 field ecosystems were profiled [162]. The impact of maize host genotype on the rhizospheric microbial population through a variety of field environments was measured. This method has helped to determine how often these plant–microbe interactions are influenced by the field context in which they are measured [162]. Despite the limited effects, it was discovered that the maize genotype has a substantial impact on α- and β-diversity across field ecosystems. Three of our areas including Aurora, Ithaca, and Lansing were in the Northeastern state of New York, while the other two were in the Midwest (Columbia, MO; and Urbana, IL). The microbiota of the five fields was grouped significantly by regional proximity: the three New York soils had microbiota that was more comparable to the two Midwestern fields. The microbiota of the two Midwestern fields, on the other hand, were not very similar, with the Columbia, MO, soil microbiota showing greater similarities to the New York microbiota than the microbiota of Urbana, IL. We might assume significantly different microbiota to be present in the Northeastern and Midwestern soils if geographical distance results in isolation and diversification of microbiota, but this was not the case, implying that regional and climatic variations between the Midwestern and Northeastern soils are inadequate to explain the biogeographic trends observed herein. Fair enough, in relation to traditional management in the other areas, we observed that organically managed site in Ithaca, NY, did not lead to significantly different microbiota, which is in line with previous work that showed the longer-term impact of cultivation but not field management on soil microbial diversity [163]. Previous study on North and South American bacterial populations has shown that there has been no major effect on diversity on latitude or geographical distance [164]. However, pH on soil was later the most important component [165]. The findings suggest that the heterogeneity of the climate, such as pH and moisture, and geographical trends, is likely to interfere with the spatial scale of the microbial maize rhizosphere [166, 167] and the extent of such interactions could be environmentally different. The distinction between the bulk soil and the rhizosphere was the next major cause of heterogeneity in the makeup of the microbiota after interfield variations. In comparison to bulk soil microbiota in all five field ecosystems, we found a decrease in species abundance and higher diversity of the maize rhizosphere microbiota. It is well established that microbiota rhizosphere differs from bulk soil microbiota. The fact that a subset of rhizosphere microbiota, enriched by replicates of a certain plant genotype, may differ widely across ecosystems, contributes to our results. We emphasize that their relationship remains very significant even after taking into account genotypic variations and environmental differences. A few taxa have been found to be constantly enriched by the rhizosphere of maize, such as the Proteobacteria order Burkholderiales, Oceanospirillales, and Sphingobacteriales. The previously enriched maize rhizosphere was found to be Proteobacteria and the order of Burkholderiales in particular [168]. Recent studies have also reported the enrichment of Proteobacteria in Oak [169], *A. thaliana* [170], and rhizosphere (and *Arabidopsis* endophyte). In our research, while most of bacterial lower-grade taxa were not sequenced to a readable sufficiency to allow powerful comparisons, the *Sphingobium*, a well-known aerobic rhizosphere bacterium that has, for example, been isolated from the *P. deltoides* rhizosphere, has been enriched [32]. Taken together, these results underscore the fact that Proteobacteria are adapted to the plant rhizosphere generally and across diverse plant species. This discovery is not unexpected, as Proteobacteria are commonly known to react to labile carbon resources and are generally regarded as r-selected or weedy microbiota, whose populations fluctuate as an opportunity [171]. The bulk soil is usually assumed to be enriched with k-selected microbiota, or slower-growing microbiota with more robust population sizes, in comparison to rhizosphere. Our analysis consisted of bulk soil enriched phenol with Acidobacteria, Chloroflexi, Verrucomicrobia, and Planctomycetes, previously referred to as soil oligotrophs [171, 172]. We discovered that host genetics could account for a small but important portion of microbial diversity (both α- and β-diversity). Furthermore, we discovered major maize inbred by ecosystem interactions, which clarified a significant portion of the difference in rhizosphere diversity. In 21-day-old seedlings grown in a greenhouse, Bouffaud et al. found that maize genotype would affect rhizosphere microbiota [173]. Similarly, we recently demonstrated that genotype describes a similar fraction of heterogeneity in the endophyte microbiota in a pyrosequencing-based sample of eight *A. thaliana* inbred genotypes grown in two soil types under regulated conditions [174]. Greenhouse or well-controlled growth-chamber studies may more closely monitor variables like temperature and soil heterogeneity than field trials, and therefore can produce higher heritability. Field tests carried out here allow us to monitor variables, like soil and climate heterogeneity by blocking and replicating statistically, and to estimate environmental dependence. However, given the high proportion of microbial variation in field-speaking maize inbred genotypes, several associations in the greenhouse are likely not to be replicable in the field. Both field and greenhouse research are necessary and would benefit from a wider sequence depth, which may partially account for the low-heritability estimates. While the heritability of microbiota in rhizosphere was statistically important, we could not link the matrix for the corn inbreeds to the diversity profiles of the microbiota in this trial. With such a small heritage and minimal efficiency of population, it is most definitely due to the lack of statistical strength that we cannot justify a significant part of the legacy variation by family. Nonetheless, a lack of interaction between the maize kindergarden matrix and microbial diversity may mean a few main alloys rather than a lot of small-impact alloys around the genome dominate the microbial communities. It is also conceivable that the microbiota impacts are just indirect. One such scenario may be that the diversity of microbiota represents the allocation of resources and the carbon supply to reduce the connections in the factory. The carbohydrate transfer from the kernels will, for example, help to reduce carbon exudation from the roots. We also included covariates for bloom time and height in our analyses in order to start addressing this option. Neither trait, however, had a substantial impact on diversity. Related interactions with microbial variation can be investigated in the future through a wider range of phenotypes. Future research may benefit from a focus on functional classes of microbes rather than taxonomic relatedness of the microbial population. Furthermore, surveying maize landraces and a pool of diversity capturing allelic variation that existed prior to breeding for adaptation to the heavily fertilized field environments of modern industrial agriculture may expose functional alleles and symbiotic relationships that were not captured in this study.

4.6 Future Course of Orientations

The rhizosphere may be engineered by suitable crop species and hybrids, the incorporation of microorganisms or soil amendments, and genetic alteration of plant and microbial biological activities, according to research so far. The development of molecular techniques now allows for the direct modification of genes that affect rhizosphere functions, and biotechnology advancements can ensure that advancement continues in the future. Screening and analysing massive and diverse microbial populations in the soil is now achievable, thanks to high throughput and 'omics' techniques. The phenomenal advancement of mass sequencing procedures has given rise to metagenomics, a method that will enable us to investigate the microbial diversity of the rhizosphere more quickly that too in greater detail due to advancement in genomics. Although development has been promising in many ways, the ability to develop rhizosphere reliably and predictably remains a problem. One significant science hindrance prevents further progress: an in-depth understanding of the diverse chemical and biological interactions in this area. The uncertainty of chemistry and biology of rhizosphere continues to give our interpretation a host of black boxes. Basic problems related to microbial abundance and soil diversity remain unresolved. The function of predation is poorly understood, including the phenomenon of resilience. The dynamic interactions between microbial community structure and function make attempts to forecast and control their ecology extremely challenging, and rhizospheric ecology will undoubtedly remain a touchstone for many more years. Failure to recognize emerging technology in society is often an obstacle. While comparatively limited in Canada, China, or the United States, in Europe and parts of South America, even among scientists, the controversy is raging over genetically modified organisms.

In reality, in Italy, Germany, and France, in the short term, the testing of genetically modified species, including plants that may support the ecosystem, is quite unlikely. If more stringent rules are enforced on the use of agrochemicals, the situation will change. There is, however, a drive to ban agrochemicals, which will promote the advancement of more environmentally sustainable solutions, including non-polluting modifications, new bio-control agents, and potentially genetically modified options. These requirements, along with the need for an ever-growing global population, are only rising where there is a chance of arable surface reduction (for example, in fertile deltas or lowlands). Scientists must continue their study so that a more responsive and environmentally friendly farming approach will benefit their future public.

References

1 Rosenberg, E. and Zilber-Rosenberg, I. (2016). Microbes drive evolution of animals and plants: the hologenome concept. *MBio* 7: e01395-15. https://doi.org/10.1128/mBio.01395-15.

2 Brundrett, M.C. (2002). Coevolution of roots and mycorrhizas of land plants. *New Phytol.* 154: 275–304. https://doi.org/10.1046/j.1469-8137.2002.00397.x.

3 Masson-Boivin, C. and Sachs, J.L. (2017). Symbiotic nitrogen fixation by rhizobiathe roots of a success story. *Curr. Opin. Plant Biol.* 44: 7–15. https://doi.org/10.1016/j.pbi.2017.12.001.

4 Leach, J.E., Triplett, L.R., Argueso, C.T., and Trivedi, P. (2017). Communication in the phytobiome. *Cell* 169: 587–596. https://doi.org/10.1016/j.cell.2017.04.025.

5 Turner, T.R., James, E.K., and Poole, P.S. (2013). The plant microbiome. *Genome Biol.* 14: 209. https://doi.org/10.1186/gb-2013-14-6-209.

6 Mendes, R., Garbeva, P., and Raaijmakers, J.M. (2013). The rhizosphere microbiome: significance of plant beneficial, plant pathogenic, and human pathogenic microorganisms. *FEMS Microbiol. Rev.* 37: 634–663. https://doi.org/10.1111/1574-6976.12028.

7 Badri, D.V., Weir, T.L., van der Lelie, D., and Vivanco, J.M. (2009). Rhizosphere chemical dialogues: plant-microbe interactions. *Curr. Opin. Biotechnol.* 20: 642–650. https://doi.org/10.1016/j.copbio.2009.09.014.

8 Venturi, V. and Keel, C. (2016). Signaling in the rhizosphere. *Trends Plant Sci.* 21: 187–198. https://doi.org/10.1016/j.tplants.2016.01.005.

9 Jessup, C.M., Forde, S.E., and Bohannan, B. (2005). Microbial experimental systems in ecology. *Adv. Ecol. Res.* 37: 273–307. https://doi.org/10.1016/S0065-2504(04) 37009-37001.

10 Swenson, W., Wilson, D.S., and Elias, R. (2000). Artificial ecosystem selection. *Proc. Natl. Acad. Sci. U. S. A.* 97: 9110–9114. https://doi.org/10.1073/pnas.150237597.

11 Panke-Buisse, K., Poole, A.C., Goodrich, J.K. et al. (2015). Selection on soil microbiomes reveals reproducible impacts on plant function. *ISME J.* 9: 980–989. https://doi.org/10.1038/ismej.2014.196.

12 Faust, K. and Raes, J. (2002). Microbial interactions: from networks to models. *Nat. Rev. Microbiol.* 10: 538–550. https://doi.org/10.1038/nrmicro2832.

13 Leff, J.W., Bardgett, R.D., Wilkinson, A. et al. (2018). Predicting the structure of soil communities from plant community taxonomy, phylogeny, and traits. *ISME J.* 12: 1794–1805. https://doi.org/10.1038/s41396-018-0089-x.

14 Grobkopf, T. and Soyer, O.S. (2014). Synthetic microbial communities. *Curr. Opin. Microbiol.* 18: 72–77.

15 Schenk, P.M., Carvalhais, L.C., and Kazan, K. (2012). Unraveling plant-microbe interactions: can multi-species transcriptomics help? *Trends Biotechnol.* 30: 177–184. https://doi.org/10.1016/j.tibtech.2011.11.002.

16 van Loon, L.C. (2007). Plant responses to plant growth-promoting rhizobacteria. *Eur. J. Plant Pathol.* 119: 243–254. https://doi.org/10.1007/s10658-007-9165-1.

17 Wolf, J.B.W. (2013). Principles of transcriptome analysis and gene expression quantification: an RNA-seq tutorial. *Mol. Ecol. Resour.* 13: 559–572. https://doi.org/10.1111/1755-0998.12109.

18 Swenson, T.L., Karaoz, U., Swenson, J.M. et al. (2018). Linking soil biology and chemistry in biological soil crust using isolate exometabolomics. *Nat. Commun.* 9: 19. https://doi.org/10.1038/s41467-017-02356-9.

19 Baran, R., Brodie, E.L., Mayberry-Lewis, J. et al. (2015). Exometabolite niche partitioning among sympatric soil bacteria. *Nat. Commun.* 6: 8289. https://doi.org/10.1038/ncomms 9289.

20 White, R.A., Rivas-Ubach, A., Borkum, M.I. et al. (2017). The state of rhizospheric science in the era of multi-omics: a practical guide to omics technologies. *Rhizosphere* 3: 212–221.

21 Bortesia, L. and Fischerab, R. (2015). The CRISPR/Cas9 system for plant genome editing and beyond. *Biotechol. Adv.* 33: 41–52.

22 Nemudryi, A.A., Valetdinova, K.R., Medvedev, S.P., and Zakian, S.M. (2014). TALEN and CRISPR/Cas genome editing systems: tools of discovery. *Acta Nat.* 6: 19–40.

23 Pinto, R., Varanini, Z., and Nannipieri, P. (2007). *The Rhizosphere: Biochemistry and Organic Substances at the Soil-Plant Interface*. Boca Raton: CRC.

24 Bowen, G.D. and Rovira, A.D. (1999). The rhizosphere and its management to improve plant growth. *Adv. Agron.* 66: 1–102.

25 Ryan, P.R., Delhaize, E., and Randall, P.J. (1995). Characterisation of Alstimulated efflux of malate from the apices of Al-tolerant wheat roots. *Planta* 196: 103–111.

26 Sasaki, T., Yamamoto, Y., Ezaki, B. et al. (2004). A wheat gene encoding an aluminum-activated malate transporter. *Plant J.* 37: 645–653.

27 de Sousa, C.N.A. (1998). Classification of Brazilian wheat cultivars for aluminium toxicity in acid soil. *Plant Breed.* 117: 217–221.

28 Hinsinger, P., Plassard, C., Tang, C.X., and Jaillard, B. (2003). Origins of root-mediated pH changes in the rhizosphere and their responses to environmental constraints: a review. *Plant Soil* 248: 43–59.

29 Neumann, G. and Römheld, V. (2007). The release of root exudates as affected by the root physiological status. In: *The Rhizosphere: Biochemistry and Organic Substances at the Soil-Plant Interface* (ed. R. Pinton, Z. Varanini and P. Nannipieri), 23–72. Boca Raton: CRC.

30 Vance, C., Uhde-Stone, C., and Allan, D.L. (2003). Phosphorus acquisition and use: critical adaptations by plants for securing a nonrenewable resource. *New Phytol.* 157: 423–447.

31 Ma, J.F., Ryan, P.R., and Delhaize, E. (2001). Aluminium tolerance in plants and the complexing role of organic acids. *Trends Plant Sci.* 6: 273–278.

32 Clough, S.J. and Bent, A.F. (1998). Floral dip: a simplified method for Agrobacterium-mediated transformation of Arabidopsis thaliana. *Plant J.* 16: 735–743.

33 Gaxiola, R.A., Palmgren, M.G., and Schumacher, K. (2007). Plant proton pumps. *FEBS Lett.* 581: 2204–2214.

34 Gévaudant, F., Duby, G., von Stedingk, E. et al. (2007). Expression of a constitutively activated plasma membrane H+-ATPase alters plant development and increases salt tolerance. *Plant Physiol.* 144: 1763–1776.

35 Palmgren, M.G. (1991). Regulation of plant plasma-membrane H+-ATPase activity. *Physiol. Plant.* 83: 314–323.

36 Yang, H., Knapp, J., Koirala, P. et al. (2007). Enhanced phosphorus nutrition in monocots and dicots over-expressing a phosphorus-responsive type IH+ -pyrophosphatase. *Plant Biotechnol. J.* 5: 735–745.

37 Young, J.C., DeWitt, N.D., and Sussman, M.R. (1998). A transgene encoding a plasma membrane H+-ATPase that confers acid resistance in Arabidopsis thaliana seedlings. *Genetics* 149: 501–507.

38 Li, J.S., Yang, H.B., Peer, W.A. et al. (2005). Arabidopsis H+ -PPase AVP1 regulates auxin-mediated organ development. *Science* 310: 121–125.

39 López-Bucio, J., Nieto-Jacobo, M.F., Ramirez-Rodriguez, V., and Herrera-Estrella, L. (2000). Organic acid metabolism in plants: from adaptive physiology to transgenic varieties for cultivation in extreme soils. *Plant Sci.* 160: 1–13.

40 Ryan, P.R., Delhaize, E., and Jones, D.L. (2001). Function and mechanism of organic anion exudation from plant roots. *Annu. Rev. Plant Physiol. Plant Mol. Biol.* 52: 527–560.

41 de la Fuente, J.M., Ramírez-Rodríguez, V., Cabrera-Ponce, J.L., and Herrera-Estrella, L. (1997). Aluminum tolerance in transgenic plants by alteration of citrate synthesis. *Science* 276: 1566–1568.

42 López-Bucio, J., Martínez de la Vega, O., Guevara-García, A., and Herrera-Estrella, L. (2000). Enhanced phosphorus uptake in transgenic tobacco plants that overproduce citrate. *Nat. Biotech.* 18: 450–453.

43 Koyama, H., Takita, E., Kawamura, A. et al. (1999). Over expression of mitochondrial citrate synthase gene improves the growth of carrot cells in Al-phosphate medium. *Plant Cell Physiol.* 40: 482–488.

44 Tesfaye, M., Temple, S.J., Allan, D.L. et al. (2001). Overexpression of malate dehydrogenase in transgenic alfalfa enhances organic acid synthesis and confers tolerance to aluminum. *Plant Physiol.* 127: 1836–1844.

45 Anoop, V.M., Basu, U., McCammon, M.T. et al. (2003). Modulation of citrate metabolism alters aluminum tolerance in yeast and transgenic canola overexpressing a mitochondrial citrate synthase. *Plant Physiol.* 132: 2205–2221.

46 Delhaize, E., Ryan, P.R., Hocking, P.J., and Richardson, A.E. (2003). Effects of altered citrate synthase and isocitrate dehydrogenase expression on internal citrate concentrations of tobacco (Nicotiana tabacum L.). *Plant Soil* 248V: 137–144.

47 Zhang, W.H., Ryan, P.R., Sasaki, T. et al. (2008). Electrophysiological characterisation of the TaALMT1 protein in transfected tobacco (*Nicotiana tabacum* L.) cells. *Plant Cell Physiol.* 49: 1316–1330.

48 Delhaize, E., Ryan, P.R., Hebb, D.M. et al. (2004). Engineering high-level aluminum tolerance in barley with the ALMT1 gene. *Proc. Natl. Acad. Sci. U. S. A.* 101: 15249–15254.

49 Delhaize, E., Gruber, B.D., and Ryan, P.R. (2007). The roles of organic anion permeases in aluminium tolerance and mineral nutrition. *FEBS Lett.* 581: 2255–2262.

50 Hoekenga, O.A., Maron, L.G., Cançado, G.M.A. et al. (2006). AtALMT1, which encodes a malate transporter, is identified as one of several genes critical for aluminium tolerance in Arabidopsis. *Proc. Natl. Acad. Sci. U. S. A.* 103: 9738–9743.

51 Kovermann, P., Meyer, S., Hortensteiner, S. et al. (2007). The arabidopsis vacuolar malate channel is a member of the ALMT family. *Plant J.* 52: 1169–1180.

52 Omote, H., Hiasa, M., Matsumoto, T. et al. (2006). The MATE proteins as fundamental transporters of metabolic and xenobiotic organic cations. *Trends Pharmacol. Sci.* 27: 587–593.

53 Magalhaes, J.V., Liu, J., Guimaraes, C.T. et al. (2007). A gene in the multidrug and toxic compound extrusion (MATE) family confers aluminum tolerance in sorghum. *Nat. Genet.* 39: 1156–1161.

54 Yokosho, K., Yamaji, N., Ueno, D. et al. (2009). OsFRDL1 is a citrate transporter required for efficient translocation of iron in rice. *Plant Physiol.* 149: 297–305.

55 Kamilova, F., Kravchenko, L.V., Shaposhnikov, A.I. et al. (2006). Organic acids, sugars, and L-tryptophane in exudates of vegetables growing on stonewool and their effects on activities of rhizosphere bacteria. *Mol. Plant-Microbe Interact.* 19: 250–256.

56 Kamilova, F., Validov, S., Azarova, T. et al. (2005). Enrichment for enhanced competitive root tip colonizers selects for a new class of biocontrol bacteria. *Environ. Microbiol.* 7: 1809–1817.

57 Shapira, R., Ordentlich, A., Chet, I., and Oppenheim, A.B. (1989). Control of plant diseases by chitinase expressed from cloned DNA in Escherichia coli. *Phytopathology* 79: 1246–1249.

58 Downing, K. and Thomson, J.A. (2000). Introduction of the Serratia marcescens chiA gene into an endophytic Pseudomonas fluorescens for the biocontrol of phytopathogenic fungi. *Can. J. Microbiol.* 46: 363–369.

59 Glick, B.R., Cheng, Z., Czarny, J., and Duan, J. (2007). Promotion of plant growth by ACC deaminase-producing soil bacteria. *Eur. J. Plant Pathol.* 119: 329–339.

60 Wang, C.X., Knill, E., Glick, B.R., and Defago, G. (2000). Effect of transferring 1-aminocyclopropane-1-carboxylic acid (ACC) deaminase genes into Pseudomonas fluorescens strain CHA0 and its gacA derivative CHA96 on their growth-promoting and disease-suppressive capacities. *Can. J. Microbiol.* 46: 898–907.

61 Ligon, J.M., Hill, D.S., Hammer, P.E. et al. (2000). Natural products with antifungal activity from Pseudomonas biocontrol bacteria. *Pest Manag. Sci.* 56: 688–695.

62 Timms-Wilson, T.M., Ellis, R.J., Renwick, A. et al. (2000). Chromosomal insertion of phenazine-1-carboxylic acid biosynthetic pathway enhances efficacy of damping-off disease control by Pseudomonas fluorescens. *Mol. Plant Microbe Interact.* 13: 1293–1300.

63 Alsanius, B.W., Hultberg, M., and Englund, J.E. (2002). Effect of lacZYmarking of the 2, 4-diaceyl-phloroglucinol producing Pseudomonas fluorescens strain 5–2/4 on its physiological performance and root colonization ability. *Microbiol. Res.* 157: 39–45.

64 Huang, Z.Y., Bonsall, R.F., Mavrodi, D.V. et al. (2004). Transformation of Pseudomonas fluorescens with genes for biosynthesis of phenazine-1-carboxylic acid improves biocontrol of rhizoctonia root rot and in situ antibiotic production. *FEMS Microbiol. Ecol.* 49: 243–251.

65 Craig, N. (2002). Tn7. In: *Mobile DNA II* (ed. N. Craig, R. Craigie, M. Gellert and A. Lambowitz), 423–456. Washington, DC: ASM.

66 De Leij, F.A.A.M., Thomas, C.E., Bailey, M.J. et al. (1998). Effect of insertion site and metabolic load on the environmental fitness of a genetically modified Pseudomonas fluorescens isolate. *Appl. Environ. Microbiol.* 64: 2634–2638.

67 Mazzola, M., Cook, R.J., Pierson, L.S. et al. (1992). Contribution of phenazine antibiotic biosynthesis to the ecological competence of fluorescent pseudomonads in soil habitats. *Appl. Environ. Microbiol.* 58: 2616–2624.

68 Bankhead, S.B., Landa, B.B., Lutton, E. et al. (2004). Minimal changes in rhizobacterial population structure following root colonization by wild type and transgenic biocontrol strains. *FEMS Microbiol. Ecol.* 49: 307–318.

69 Bankhead, S.B., Schroeder, K., Son, M.Y. et al. (2004). Population dynamics and movement of transgenic Pseudomonas fluorescens in the rhizosphere of field-grown wheat. In: *Rhizosphere 2004: Perspectives and Challenges*. A Tribute to Lorenz Hiltner (ed. A. Hartmann, M. Schmid, W. Wenzel and P. Hinsinger), 165. GSF Forschungszentrum.

70 Viebahn, M., Glandorf, D.C.M., Ouwens, T.W.M. et al. (2003). Repeated introduction of genetically modified Pseudomonas putida WCS358r without intensified effects on the indigenous microflora of field-grown wheat. *Appl. Environ. Microbiol.* 69: 3110–3118.

71 Miller, H.J., Henken, G., and van Veen, J.A. (1989). Variation and composition of bacterial populations in the rhizosphere of maize, wheat, and grass cultivars. *Can. J. Microbiol.* 35: 656–660.

72 Atkinson, T.G., Neal, J.L. Jr., and Larson, R.I. (1975). Genetic control of the rhizosphere microflora of wheat. In: *Biology and Control of Soil-Borne Plant Pathogens* (ed. G.W. Bruehl), 116–122. Saint-Paul: American Phytopathological Society.

73 Lemanceau, P., Corberand, T., Gardan, L. et al. (1995). Effect of two plant species, flax (Linum usitatissinum L.) and tomato (Lycopersicon esculentum Mill.), on the diversity of soilborne populations of fluorescent pseudomonads. *Appl. Environ. Microbiol.* 61: 1004–1012.

74 Mazzola, M., Funnell, D.L., and Raaijmakers, J.M. (2004). Wheat cultivar-specific selection of 2,4-diacetylphloroglucinolproducing fluorescent Pseudomonas species from resident soil populations. *Microb. Ecol.* 48: 338–348.

75 Bergsma-Vlami, M., Prins, M.E., and Raaijmakers, J.M. (2005). Influence of plant species on population dynamics, genotypic diversity and antibiotic production in the rhizosphere by indigenous Pseudomonas spp. *FEMS Microbiol. Ecol.* 52: 59–69.

76 Notz, R., Maurhofer, M., Schnider-Keel, U. et al. (2001). Biotic factors affecting expression of the 2,4-diacetylphloroglucinol biosynthesis gene phlA in

Pseudomonas fluorescens biocontrol strain CHA0 in the rhizosphere. *Phytopathology* 91: 873–881.

77 Okubara, P.A. and Bonsall, R.F. (2008). Accumulation of Pseudomonasderived 2,4-diacetylphloroglucinol on wheat seedling roots is influenced by host cultivar. *Biol. Control* 46: 322–333.

78 Kuiper, E.L., Lagendijk, G.V., and Lugtenberg, B. (2004). Rhizoremediation: a beneficial plant microbe interaction. *Mol. Plant Microbe Interact.* 17: 6–15.

79 Zhang, K., Martiny, A.C., Reppas, N.B. et al. (2006). Sequencing genome from single cells by polymerase cloning. *Nat. Biotechnol.* 24: 680–686.

80 Pant, R., Pandey, P., and Kotoky, R. (2016). Rhizosphere mediated biodegradation of 1,4-dichlorobenzene by plant growth promoting rhizobacteria of Jatropha curcas. *Ecol. Eng.* 94: 50–56.

81 Kotoky, R., Das, S., Singha, L.P. et al. (2017). Biodegradation of Benzo(a)pyrene by biofilm-forming and plant growth promoting Acinetobacter sp. strain PDB4. *Environ. Technol. Innov.* 8 (2017): 256–268.

82 Zhuang, P., Yang, Q.W., Wang, H.B., and Shu, W.S. (2007). Phytoextraction of heavy metals by eight plant species in the field. *Water Air Soil Pollut.* 184: 235–242.

83 Gentry, T.J., Rensing, C., and Pepper, L.I. (2004). New approaches for bioaugmentation as a remediation technology. *Crit. Rev. Environ. Sci. Technol.* 34: 447–494.

84 Toussaint, J.P., Pham, T., Barriault, D., and Sylvestre, M. (2012). Plant exudates promote PCB degradation by a rhodococcal rhizobacteria. *Appl. Microbiol. Biotechnol.* 95: 1589–1603. https://doi.org/10.1007/s00253-011-3824-z.

85 Rentz, J.A., Alvarez, P.J.J., and Schnoor, J.L. (2005). Benzo[a]pyrene co-metabolism in the presence of plant root extracts and exudates: implications for phytoremediation. *Environ. Pollut.* 136 (3): 477–484.

86 Kamath, R., Schnoor, J.L., and Alvarez, P.J.J. (2004). Effect of root-derived substrates on the expression of nah-lux genes in Pseudomonas fluorescens HK44: implications for PAH biodegradation in the rhiziosphere. *Environ. Sci. Technol.* 38 (6): 1740–1745.

87 Goel, G., Pandey, P., Sood, A. et al. (2012). Transformation of pWWO in Rhizobium leguminosarum DPT to engineer toluene degrading ability for rhizoremediation. *Indian J. Microbiol.* 52 (2): 197–202.

88 Cebron, A., Norini, M.P., Beguiristain, T., and Leyval, C. (2008). Real-time PCR quantification of PAH-ring hydroxylating dioxygenase (PAH-RHDa) genes from Gram positive and Gram negative bacteria in soil and sediment samples. *J. Microbiol. Methods* 73: 148–159. https://doi.org/10.1016/j.mimet.2008.01.009.

89 Launen, L.A., Dutta, J., Turpeinen, R. et al. (2008). Characterization of the indigenous PAH-degrading bacteria of Spartina dominated salt marshes in the New York/New Jersey Harbor. *Biodegradation* 19 (3): 347–363.

90 Parrish, Z.D., White, J.C., Isleyen, M. et al. (2006). Accumulation of weathered polycyclic aromatic hydrocarbons (PAH) by plant and earthworm species. *Chemosphere* 64 (4): 609–618.

91 Muratova, A., Hubner, T., Tischer, S. et al. (2003). Plant-rhizosphere-microflora association during phytoremediation of PAHcontaminatedsoil. *Int. J. Phytorem.* 5 (2): 137–151.

92 Venkata, M.S., Takuro, K., Takeru, O. et al. (2006). Bioremediation technologies for treatment of PAH-contaminated soil and strategies to enhance process efficiency. *Rev. Environ. Sci. Biotechnol.* 5: 347–374.

93 Reinhold-Hurek, B., Bünger, W., Burbano, C.S. et al. (2015). Roots shaping their microbiome: global hotspots for microbial activity. *Annu. Rev. Phytopathol.* 53: 403–424.

94 Bujang, M., Ibrahim, N.A., and Rak, A. (2013). Biodegradation of oily wastewater by pure culture of Bacillus cereus. *ARPN J. Agric. Biol. Sci.* 8: 2.

95 Liu, S.L., Cao, Z.H., and Liu, H.E. (2013). Effect of ryegrass (Lolium multiflorum L.) growth on degradation of phenanthrene and enzyme activity in soil. *Plant Soil Environ.* 59 (6): 247–253.

96 Haritash, A.K. and Kaushik, C.P. (2009). Biodegradation aspects of polycyclic aromatic hydrocarbons (PAHs): a review. *J. Hazard. Mater.* 169: 1–15.

97 Eskandary, S., Tahmourespour, A., Hoodaji, M., and Abdollahi, A. (2017). The synergistic use of plant and isolated bacteria to clean up polycyclic aromatic hydrocarbons from contaminated soil. *J. Environ. Health Sci. Eng.* 15: 12.

98 Bisht, S., Pandey, P., Sood, A. et al. (2010). Biodegradation of naphthalene and anthracene by chemo-tactically active rhizobacteria of populus deltoides. *Braz. J. Microbiol.* 41: 922–930.

99 Liste, H.H.M. (2000). Plant-promoted pyrene degradation in soil. *Chemosphere* 40: 7–10.

100 Yang, S., Wen, X., Jin, H., and Wu, Q. (2012). Pyrosequencing investigation into the bacterial COMMUNITY in permafrost soils along the China-Russia crude oil pipeline (CRCOP). *PLoS One* 7 (12): e52730. https://doi.org/10.1371/journal.pone.0052730.

101 Mitter, B., Petric, A., Shin, M.W. et al. (2013). Comparative genome analysis of Burkholderia phytofirmans PsJN reveals a wide spectrum of endophytic lifestyles based on interaction strategies with host plants. *Front. Plant Sci.* 4: 120. https://doi.org/10.3389/fpls.2013.00120.

102 Bell, T.H., Stefani, F.O.P., Abram, K. et al. (2016). A diverse soil microbiome degrades more crude oil than specialized bacterial assemblages obtained in culture. *Appl. Environ. Microbiol.* https://doi.org/10.1128/AEM.01327-16.

103 Raaijmakers, J.M., Paulitz, C.T., Steinberg, C. et al. (2009). The rhizosphere: a playground and battlefield for soilborne pathogens and beneficial microorganisms. *Plant Soil* 321: 341–361.

104 Fernández-Luqueño, F., Valenzuela-Encinas, C., Marsch, R. et al. (2011). Microbial communities to mitigate contamination of PAHs in soil – possibilities and challenges: a review. *Environ. Sci. Pollut. Res. Int.* 18: 12–30. https://doi.org/10.1007/s11356-010-0371-6.

105 Yu, X.Z., Wu, S.C., Wu, F.Y., and Wong, M.H. (2011). Enhanced dissipation of PAHs from soil using mycorrhizal ryegrass and PAH-degrading bacteria. *J. Hazard. Mater.* 186: 1206–1217.

106 Gao, Y., Cheng, Z., Ling, W., and Huang, J. (2010). Arbuscular mycorrhizal fungal hyphae contribute to the uptake of polycyclic aromatic hydrocarbons by plant roots. *Bioresour. Technol.* 101: 6895–6901.

107 Siciliano, S.D., Fortin, N., Mihoc, A. et al. (2001). Selection of specific endophytic bacterial genotypes by plants in response to soil contamination. *Appl. Environ. Microbiol.* 67: 2469–2475. https://doi.org/10.1128/AEM.67.6.2469-2475.2001.

108 Aislabie, J., Chhour, K.-L., Saul, D. et al. (2006). Dominant bacteria in soils of marble point and wright valley, Victoria Land, Antarctica. *Soil Biol. Biochem.* 38: 3041–3056.

109 Sawulski, P., Clipson, N., and Doyle, E. (2014). Effects of polycyclic aromatic hydrocarbons on microbial community structure and PAH ring hydroxylating dioxygenase gene abundance in soil. *Biodegradation* 25: 835–847.

110 Bulgarelli, D., Rott, M., Schlaeppi, K. et al. (2012). Revealing structure and assembly cues for Arabidopsis root-inhabiting bacterial microbiota. *Nature* 488: 91–95.

111 El Amrani, A., Dumas, A., Wick, L.Y. et al. (2015). Omics insights into PAH degradation toward improved Green remediation biotechnologies. *Environ. Sci. Technol.* 49: 11281–11291.

112 Bell, T.H., Hassan, S.E., Lauron-Moreau, A. et al. (2014). Linkage between bacterial and fungal rhizosphere communities in hydrocarbon-contaminated soils is related to plant phylogeny. *ISME J.* 8: 331–343.

113 Nazir, A. (2016). Review on metagenomics and its applications. *Imp. J. Interdiscip. Res.* 2 (3): 277–286.

114 Siddhapura, P.K., Vanparia, S., Purohit, M.K., and Singh, S.P. (2010). Comparative studies on the extraction of metagenomic DNA from the saline habitats of Coastal Gujarat and Sambhar Lake, Rajasthan (India) in prospect of molecular diversity and search for novel biocatalysts. *Int. J. Biol. Macromol.* 47 (3): 375–379. https://doi.org/10.1016/j.ijbiomac.2010.06.004.

115 Wren, B.W. (2000). Microbial genome analysis: insights into virulence, host adaptation and evolution. *Nat. Genet.* 1: 30–39.

116 Kotoky, R., Rajkumari, J., and Pandey, P. (2018). The rhizosphere microbiome: significance in rhizoremediation of polyaromatic hydrocarbon contaminated soil. *J. Environ. Manag.* 217: 858–870. https://doi.org/10.1016/j.jenvman.2018.04.022.

117 Chemerys, A., Pelletier, E., Cruaud, C. et al. (2014). Characterization of novel PAH dioxygenases from the bacterial metagenomic DNA of a contaminated soil. *Appl. Environ. Microbiol.* 80 (21): 6591–6660.

118 Iwai, S., Chai, B.L., Sul, W.J. et al. (2010). Gene-targeted metagenomics reveals extensive diversity of aromatic dioxygenase genes in the environment. *ISME J.* 4: 279–285.

119 Zhou, H.W., Guo, C.L., Wong, Y.S., and Tam, N.F.Y. (2006). Genetic diversity of dioxygenase genes in polycyclic aromatic hydrocarbon-degrading bacteria isolated from mangrove sediments. *FEMS Microbiol. Lett.* 262: 148–157.

120 Castro-silva, C., Ruiz-valdiviezo, V.M., Valenzuela-encinas, C. et al. (2013). The bacterial community structure in an alkaline saline soil spiked with anthracene. *Electron. J. Biotechnol.* 16 (5).

121 Jennings, L.K., Chartrand, M.M., Lacrampe-Couloume, G. et al. (2009). Proteomics and Transcriptomics analysis reveals genes upregulated by cis-dichloroethane in Polaromonas sp. strain JS666. *Appl. Environ. Microbiol.* 11: 3733–3744.

122 Poretsky, R.S., Bano, N., Buchan, A. et al. (2005). Analysis of microbial gene transcripts in environmental samples. *Appl. Environ. Microbiol.* 71 (7): 4121–4126. https://doi.org/10.1128/AEM.71.7.4121-4126.2005.

123 Moran, M.A. (2009). Metatranscriptomics: eavesdropping on complex microbial communities. *Microbe* 4 (7): 329–335.

124 deMenezes, A., Clipson, N., and Doyle, E. (2012). Comparative metatranscriptomics reveals widespread community responses during phenanthrene degradation in soil. *Environ. Microbiol.* 14: 2577–2588.

125 Yergeau, E., Sanschagrin, S., Maynard, C. et al. (2014). Microbial expression profiles in the rhizosphere of willows depend on soil contamination. *ISME J.* 8: 344–358. https://doi.org/10.1038/ismej.2013.163.

126 Page, A.P., Yergeau, E., and Greer, C.W. (2015). Salix purpurea stimulates the expression of specific bacterial xenobiotic degradation genes in a soil contaminated with hydrocarbons. *PLoS One* 10 (7): e0132062. https://doi.org/10.1371/journal.pone.0132062.

127 Helbling, D.E., Ackermann, M., Fenner, K. et al. (2012). The activity level of a microbial community function can be predicted from its metatranscriptome. *ISME J.* 6: 902–904.

128 Wilmes, P. and Bond, P.L. (2006). Metaproteomics: studying functional gene expression in microbial ecosystems. *Trends Microbiol.* 14: 92–97.

129 Singh, O. and Nagaraj, N.S. (2006). Transcriptomics, proteomics and interactomics: unique approaches to track the insights of bioremediation. *Brief. Funct. Genomic. Proteomic.* 4 (4): 355–362.

130 Keller, M. and Hettich, R. (2009). Environmental proteomics: a paradigm shift in characterizing microbial activities at the molecular level. *Microbiol. Mol. Biol. Rev.* 73: 62–70.

131 Kim, S.J., Kweon, O., and Cerniglia, C.E. (2009). Proteomic applications to elucidate bacterial aromatic hydrocarbon metabolic pathways. *Curr. Opin. Microbiol.* 12: 301–309.

132 Kim, Y.H., Cho, K., Yun, S.H. et al. (2006). Analysis of aromatic catabolic pathways in Pseudomonas putida KT 2440 using a combined proteomic approach: 2-DE/MS and cleavable isotope-coded affinity tag analysis. *Proteomics* 6: 1301–1318.

133 Yun, S.H., Choi, C.W., Lee, S.Y. et al. (2014). Proteomic characterization of plasmid pLA1 for biodegradation of polycyclic aromatic hydrocarbons in the marine bacterium, Novosphingobium pentaromativorans US6-1. *PLoS One* 9 (3): e90812.

134 Lee, S.E., Seo, J.S., Keum, Y.S. et al. (2007). Fluoranthene metabolism and associated proteins in Mycobacterium sp. JS14. *Proteomics* 7: 2059–2069.

135 Rabus, R. (2013). Environmental microbial proteomics: new avenues for a molecular understanding of the functional role of microorganisms in the natural environment. *Proteomics* 13: 2697–2699.

136 Wilmes, P., Andersson, A.F., Lefsrud, M.G. et al. (2008). Community proteogenomics highlights microbial strain- variant protein expression within activated sludge performing enhanced biological phosphorus removal. *ISME J.* 2: 853–864.

137 Guazzaroni, M.E., Herbst, F.A., Lores, I. et al. (2013). Metaproteogenomic insights beyond bacterial response to naphthalene exposure and bio-stimulation. *ISME J.* 7: 122–136.

138 Koukkou, A.I. and Drainas, C. (2008). Addressing PAH biodegradation in Greece: biochemical and molecular approaches. *IUBMB Life* 60 (5): 275–280.

139 Kim, S.I., Choi, J.S., and Kahng, H.Y. (2007). A proteomics strategy for the analysis of bacterial biodegradation pathways. *OMICS* 11: 280–294.

140 Herbst, A.F., Bahr, A., Duarte, M. et al. (2013). Elucidation of in situ polycyclic aromatic hydrocarbon degradation by functional metaproteomics (protein-SIP). *Proteomics* 13: 2910–2920.

141 Jones, M.D., Crandell, D.W., Singleton, D.R., and Aitken, M.D. (2011). Stable-isotope probing of the polycyclic aromatic hydrocarbon-degrading bacterial guild in a contaminated soil. *Environ. Microbiol.* 13 (10): 2623–2632.

142 Kim, S.I., Kweon, S.M., Kim, E.A. et al. (2004). Characterization of RNase-like major storage protein from the ginseng root by proteomic approach. *J. Plant Physiol.* 161: 837–845. https://doi.org/10.1016/j.jplph.2004.01.001.

143 Buermans, H.P. and den Dunnen, J.T. (2014). Next generation sequencing technology: advances and applications. *Biochim. Biophys. Acta* 1842 (10): 1932–1941. https://doi.org/10.1016/j.bbadis.2014.06.015. Epub 2014 Jul 1.

144 Kotoky, R., Singha, L.P., and Pandey, P. (2017). Draft genome sequence of heavy metal resistant soil bacterium Serratia marcescens S2I7, which has the ability to degrade polyaromatic hydrocarbons. *Genome Announc.* 5: e01338e17.

145 Kotoky, R., Singha, L.P., and Pandey, P. (2017). Draft genome sequence of polyaromatic hydrocarbon-degrading bacterium Bacillus subtilis SR1, which has plant growthpromoting attributes. *Genome Announc.* 5: e01339e17.

146 Zhao, Q., Hu, H., Wang, W. et al. (2015). Genome sequence of Sphingobium yanoikuyae B1, a polycyclic aromatic hydrocarbon-degrading strain. *Genome Announc.* 4 (1): 1–2.

147 McLeod, M.P., Warren, R.L., Hsiao, W.W.L. et al. (2006). The complete genome of Rhodococcus sp. RHA1 provides insights into a catabolic powerhouse. *Proc. Natl. Acad. Sci. U. S. A.* 103: 15582–15587.

148 Schneiker, S., dos Santos, V.A.P.M., Bartels, D. et al. (2006). Genome sequence of the ubiquitous hydrocarbon degrading marine bacterium Alcanivorax borkumensis. *Nat. Biotechnol.* 24 (8): 997–1004.

149 Cao, J., Wei, C., Liu, D. et al. (2015). DNA methylation landscape of body size variation in sheep. *Sci. Rep.* 5: 13950. https://doi.org/10.1038/srep13950 pmid:26472088.

150 Lai, Q., Li, W., Wang, B. et al. (2012). Complete genome sequence of the pyrene-degrading bacterium Cycloclasticus sp. strain P1. *J. Bacteriol.* 194 (23): 6677.

151 Cui, Z., Xu, G., Li, Q. et al. (2013). Genome sequence of the pyrene- and fluoranthene-degrading bacterium Cycloclasticus sp. strain PY97M. *Genome Announc.* 1 (4). https://doi.org/10.1128/genomeA.00536-13: e00536e13.

152 Nelson, K.E., Weinel, C., Paulsen, I.T. et al. (2002). Complete genome sequence and comparative analysis of the metabolically versatile Pseudomonas putida KT2440. *Environ. Microbiol.* 4: 799–808.

153 Yan, Y., Yang, J., Dou, Y. et al. (2008). Nitrogen fixation island and rhizosphere competence traits in the genome of root-associated Pseudomonas stutzeri A1501. *Proc. Natl. Acad. Sci. U. S. A.* 105: 7564–7569.

154 Gupta, A., Gopal, M., Thomas, G.V. et al. (2014). Whole genome sequencing and analysis of plant growth promoting bacteria isolated from the rhizosphere of plantation crops coconut, cocoa and arecanut. *PLoS One* 9 (8): e104259. https://doi.org/10.1371/journal.pone.0104259.

155 Zhu X, Ni X, Waigi MG, Liu J, Sun K Gao Y (2016) Biodegradation of mixed PAHs by PAH-degrading endophytic bacteria. *Int. J. Environ. Res. Public Health* 13: 805.

156 Rabodonirina, S., Rasolomampianina, R., Krier, F. et al. (2019). Degradation of fuorene and phenanthrene in PAHs-contaminated soil using Pseudomonas and Bacillus strains isolated from oil spill sites. *J. Environ. Manag.* 232: 1–7.

157 Kim, S.J., Kweon, O., Jones, R.C. et al. (2008). Genomic analysis of polycyclic aromatic hydrocarbon degradation in Mycobacterium vanbaalenii PYR-1. *Biodegradation* 19 (6): 859–881. https://doi.org/10.1007/s10532-008-9189-z.

158 Kwak, Y., Li, Q.X., and Shin, J.H. (2016). Draft genome sequence of Mycobacterium rufum JS14T, a polycyclic-aromatic hydrocarbon-degrading bacterium from the petroleum-contaminated soil in Hawaii. *Stand. Genomic Sci.* 11 (47): 1–9.

159 Alegbeleye, O.O., Opeolu, B.O., and Jackson, V. (2017). Bioremediation of polycyclic aromatic hydrocarbon (PAH) compounds: (acenaphthene and fuorene) in water using indigenous bacterial species isolated from the Diep and Plankenburg rivers, Western Cape, South Africa. *Braz. J. Microbiol.* 48 (2): 314–325.

160 Qin, W., Fan, F., Zhu, Y. et al. (2018). Anaerobic biodegradation of benzo(a) pyrene by a novel Cellulosimicrobium cellulans CWS2 isolated from polycyclic aromatic hydrocarbon-contaminated soil. *Braz. J. Microbiol.* 49 (2): 258–268.

161 Nzila, A., Ramirez, C.O., Musa, M.M. et al. (2018). Pyrene biodegradation and proteomic analysis in Achromobacter xylosoxidans, PY4 strain. *Int. Biodeterior. Biodegradation.* https://doi.org/10.1016/j.ibiod.2018.03.014.

162 Peiffer, J., Spor, A., Koren, O. et al. (2013). Diversity and heritability of the maize rhizosphere microbiome under field conditions. *Proc. Natl. Acad. Sci. U. S. A.* 110: 6548–6553. https://doi.org/10.1073/pnas.1302837110.

163 Buckley, D.H. and Schmidt, T.M. (2001). The structure of microbial communities in soil and the lasting impact of cultivation. *Microb. Ecol.* 42 (1): 11–21.

164 Fierer, N. and Jackson, R.B. (2006). The diversity and biogeography of soil bacterial communities. *Proc. Natl. Acad. Sci. U. S. A.* 103 (3): 626–631.

165 Lauber, C.L., Hamady, M., Knight, R., and Fierer, N. (2009). Pyrosequencing-based assessment of soil pH as a predictor of soil bacterial community structure at the continental scale. *Appl. Environ. Microbiol.* 75 (15): 5111–5120.

166 Berg, G. and Smalla, K. (2009). Plant species and soil type cooperatively shape the structure and function of microbial communities in the rhizosphere. *FEMS Microbiol. Ecol.* 68 (1): 1–13.

167 Green, J. and Bohannan, B.J. (2006). Spatial scaling of microbial biodiversity. *Trends Ecol. Evol.* 21 (9): 501–507.

168 Salles, J.F., van Veen, J.A., and van Elsas, J.D. (2004). Multivariate analyses of Burkholderia species in soil: effect of crop and land use history. *Appl. Environ. Microbiol.* 70 (7): 4012–4020.

169 Uroz, S., Buee, M., Murat, C. et al. (2010). Pyrosequencing reveals a contrasted bacterial diversity between oak rhizosphere and surrounding soil. *Environ. Microbiol. Rep.* 2 (2): 281–288.

170 Bulgarelli, D., Rott, M., Schlaeppi, K. et al. (2012). Revealing structure and assembly cues for Arabidopsis rootinhabiting bacterial microbiota. *Nature* 488 (7409): 91–95.

171 Fierer, N., Bradford, M.A., and Jackson, R.B. (2007). Toward an ecological classification of soil bacteria. *Ecology* 88 (6): 1354–1364.

172 Fuerst, J.A. (1995). The planctomycetes: emerging models for microbial ecology, evolution and cell biology. *Microbiology* 141 (Pt 7): 1493–1506.

173 Bouffaud, M.-L., Kyselková, M., Gouesnard, B. et al. (2012). Is diversification history of maize influencing selection of soil bacteria by roots? *Mol. Ecol.* 21 (1): 195–206.

174 Lundberg, D.S., Lebeis, S.L., Paredes, S.H. et al. (2012). Defining the core Arabidopsis thaliana root microbiome. *Nature* 488 (7409): 86–90.

5

Microbial Communities in Phyllosphere

CONTENTS

5.1 Introduction

The phyllosphere is a habitat for microorganisms found in the aerial parts of living plants, such as shoots and roots, with dominance of bacteria, which are present on surface of plant parts, which are called epiphytes, and endophytes as well; archaea, filamentous fungi, and yeasts also comprise the microbial community [1–4]. Phyllosphere nearly covers around 4.108 km^2 of the global surface area

Microbiomics and Sustainable Crop Production, First Edition. Mohammad Yaseen Mir and Saima Hamid.

supporting bacterial populations in the region of 1026 cells. Latest estimates of phyllosphere bacteria diversity in the Brazilian Atlantic forests of 20 000 vascular plants mean that 2–13 million phyllosphere bacterial species alone can exist in that habitat [5]. The phyllosphere is an agricultural and environmental niche with a lot of potential. There is mounting evidence of significant interactions between phyllosphere microbial inhabitants that may influence the fitness of natural plant populations as well as the quality and productivity of agricultural crops. Plant pathogens may colonize and invade tissues by phyllosphere bacteria, which can both inhibit and encourage their colonization [6, 7]. In the same way, fungal endophytes on leaves can discourage herbivores, protect against pathogens, and improve drought tolerance [1, 8]. Furthermore, associations in the phyllosphere region determine the degree to which human pathogens can colonize and thrive on plant tissues, an environment that is becoming increasingly important as the number of cases of human disease linked to fresh lettuce, fruit, and vegetable intake rises [9]. The confirmation about the adaptive functions in the phyllosphere microbial population, considering the habitat's complexity, may have global implications in which nitrogen fixation is the most researched among these processes. Phyllosphere bacterial nitrogen-fixation concentrations differ greatly, but rates of over 60 kg N per ha have been recorded in the phyllosphere of trees in some tropical environments, notwithstanding the fact that quantities fixed in the phyllosphere of temperate trees are usually far smaller [9]. As per reports of Murty [10] and Miyamoto et al. [11], the phyllosphere of various crop plants possess nitrogen-fixing bacteria, while for the degradation of methanol and nitrification, which involves mechanism, it is unclear, but these processes hold significant environmental potential [9, 12–14]. Culture-dependent approaches have been used to develop the majority of information about the function and behaviours of phyllosphere microbial communities. However, it is widely acknowledged that these estimates greatly underestimate diversity, with just 1–3% of environmental bacteria deemed culturable [15]. As a result, the knowledge collected using these techniques is limited to culturable members of the population and does not include the overwhelming number of microbes found in samples. The latest introduction of culture-independent approaches focused on the characterization of small subunit ribosomal RNA (rRNA) gene sequences for microbial community analysis is offering new insights into the dynamics of phyllosphere microbial communities and their relationships with plants and the broader environment, much as it has in other areas of environmental microbiology. Some reviewers examined how the use of culture-independent methods has altered our perception of the structure and composition of phyllosphere cultures in the current article. The establishment of microbial communities in the phyllosphere is regulated by a number of plant, microbial, and environmental factors, but plant genotype has recently been recognized as a factor in phyllosphere community selection.

Proof of the various factors regulating the organization of the phyllosphere populations is examined with particular regard to the function of the plant genotype. Finally, for future studies on phyllosphere microbiology, we propose methods and focus areas. While much of the literature on phyllosphere concerns the relationships between herbal and vegetal pathogenics (bacterial and fungal pathogens causing disease in plants), focus in the present analysis is focused on studies of microorganisms living in the phyllosphere with no clear effects on the plant, as lack of disease is common in nature.

5.2 Diversity of Microbes in Phyllospheric Environment

The phyllosphere's microbial populations are complex, including bacteria, filamentous fungi, yeasts, algae, and, in some cases, protozoans and nematodes coexisting [3, 16]. Bacteria colonize leaves in the greatest number and diversity, with cultural counts ranging from 10^2 to 10^{12} cells g^{-1} leaf [2, 17]. More than 78 bacterial species covering 37 recognized bacterial genera were discovered during culture-based studies of sugar beet over the entire growing season [17]. Also, 88 bacterial species comprising 37 recognized bacterial genera were discovered in wheat in similar experiments [18]. Current research has shown that using culture-dependent approaches to profile phyllosphere cultures is likely to be unreliable and underestimate diversity [19]. The use of culture-independent methods in the phyllosphere has shown that, although hypotheses about the dominant inhabitants are generally right, the complexity of phyllosphere cultures is much greater than previously understood. Proteobacteria are the most common group present on leaves, according to 16S recombinant DNA (rDNA) cloned directly from leaf samples, supporting data obtained using culture-dependent methods [17]. The phyllosphere is dominated by a- and c-Proteobacteria, with bacteroids being particularly significant. In certain cases, b-Proteobacteria and Firmicutes will make up a considerable portion of the bacterial population, with Acidobacteria, Actinobacteria, and Cyanobacteria appearing just infrequently [5, 19–21]. In a sample of phyllosphere bacterial populations in a tropical Brazilian jungle, 97% of the bacterial sequences were from previously unknown organisms, with phyllospheres of various plant species supporting between 95 and 671 bacterial species [5]. It is unknown how much variation exists in other plant species; 15.2% of the sequences collected from *Thlapsi geosingense* had 95% or less homology to recognized bacterial species database entries [21], and 7.9, 2.3, 3.5, and 1.2% of those sequences obtained from Crocus, potato, pepper, and maize, respectively [19, 20, 22]. Furthermore, 5 of 17 bands cut from 16S rRNA denaturant gradient gel electrophoresis gels had less than 90% similarity to database entries in a sample of a variety of temperate agricultural crop species, implying that phyllosphere of crop plants may sustain large numbers of novel bacteria in some circumstances [23]. Since the number of sequences examined in culture-independent experiments to date has been reduced, only dominant members of the group are likely to have been found leaving the true extent of bacterial diversity in the phyllosphere unknown. Excluding on older plants, yeasts are the most common epiphytic fungal community in the phyllosphere, with filamentous fungi often occurring as dormant spores rather than active mycelia [24, 25]. The populations of culturable yeast will range from 10 to 10^{10} colony-forming unit (CFU) per g^{-1} leaf [2, 17]. Mostly, yeasts belonging to *Cryptococcus*, *Sporobolomyces*, and *Rhodotorula* with total maximum species number of 40 comprise the phyllospheric part while coexistence for multiple microbial populations occurs at lower diversity forms [2, 17, 26]. As per reports of Thompson et al. [17] and Inacio et al. [2], *Aspergillus*, *Acremonium*, *Mucor*, and *Penicillium* are among the least colonial microflora while with dominating ones are *Alternaria* and *Cladosporium*; however, the range between 10^2 and 10^8 CFU g^{-1} leaf is for filamentous fungus with overall presence as endophytes, with long span on tropical leaves with extensive diversity. According to Arnold and coworkers [27], over 340 genetically distinct taxa were retrieved from individuals of two tropical forest understory plant species at two sites using culture-dependent approaches. Furthermore, there was evidence of endophyte population host choice. To classify fungal diversity in the phyllosphere, culture-independent methods are yet to be used. There are a number of emerging technologies that have the potential to greatly increase research throughput and provide a better resolution of understanding about the diversity and composition of phyllosphere populations, as well as to relate diversity to work. The use of phylogenetic-based primers and culture-independent analysis to investigate the dynamics and distribution of specific bacterial groups of interest is an effective tool [28]. Additionally, multiplex terminal restriction fragment length polymorphism (TRFLP), which allows multiple phylogenetic classes or functional genes to be analysed at the same time, offers a cost-effective way to increase sample throughput [29]. The phylogenetic microarrays provide a direct path for the simultaneous identification of the presence and number of thousands of microorganisms and for the detection of new phylogenetic community members. Functional gene arrays often help characterize the behaviour of the phyllosphere population, and relate community structure to work when used with phylogenetic microarray; nevertheless, the technique takes lot of time and further implementation depends on high-throughput technologies [30]. Understanding the biological and environmental factors that influence the establishment and dynamics of microbial communities on the leaf surface is important in order to understand and forecast the diversity and structure of phyllopsheric colonies.

5.2.1 Sources of Microbes Colonizing the Phyllosphere

On the phyllosphere, microbes may come from a variety of places as the sources for epiphytic filamentous fungi, yeasts, and bacteria to the surface of the leaf includes insects, seeds, or even animals. Since they are a significant reservoir of bacteria already adapted to the phyllosphere, tree buds, seeds of annual plants, and crop litter are likely to be the most important sites for invasion of new plants and leaves [31]. Transient epiphytes are microorganisms that do not or just do minimal replication in the phyllosphere, while residual epiphytes have the ability to multiply even though there are no wounds [32]. Microbial communities can vary in size among and within plant species over short periods of time [33] as well as over the course of the growing season [2, 17, 18], with few epiphytic bacteria present on leaves shortly after emergence from buds or seeds, but rising in quantity later [33]. Over the course of the growing season, bacterial colonies predominate on leaves, followed by yeasts, and eventually filamentous fungi [34]. The composition and concentration of the atmospheric microflora will change during the day and season, as well as in response to environmental events such as rainfall and strong winds [34, 35], affecting microorganism migration to the phyllosphere. Local flora, as well as agricultural activities like harvesting and planting in areas where crops are grown, have an effect on atmospheric microbiology and plant colonization [36–38]. Microorganisms can immigrate to leaves from the atmosphere through impaction onto the leaf surface, sedimentation, or rain splash, as well as contamination with soil [37, 39]. Microorganisms on seeds or roots may become endophytic in the roots, join the vascular system, and be transported internally to the aerial sections of plants, where they establish as phyllosphere endophytes, [40, 41]. Bacterial colonization is most widespread in epidermal cell wall junctions, particularly in protected areas such as grooves along stomata veins and at the base of trichomes; or often found under the cuticle; in depressions in the cuticle; near hydrathodes; or in rare locations that only occur on specific plants such as stomata pits in oleander and pectate hairs in olive [42–49]. Bacteria are present in larger numbers on lower leaf surfaces than on upper leaf surfaces, likely due to a higher density of stomata or trichomes on the lower leaf surface, or a thinner cuticle coating on the lower leaf surface [48–54]. It is clearly not possible to colonize and evolve all microorganisms arriving at the phyllosphere. This illustrates emigration processes by dispersal mechanisms including rain splash, wash-off, bounce-off, water flow, or insect elimination to some degree [34]. The plant's ability to thrive and expand is determined by its environmental, physicochemical, and genetic characteristics, as well as the phyllosphere microorganisms' unique properties, which together define the microbial community's composition and diversity. The evidence for such an analysis is provided by the discovery of Miyamoto et al. [11], which showed the occurrence of diverse Clostridia populations in the

Miscanthus sinensis in the 16S rRNA-TRFLP with Clostridia-specific primers that were distinct from those inhabitants of Clostridia in the plant environment. In addition, because many of the bacteria inhabiting the phyllosphere seem new to this habitat, it was suggested that some of them might be singular and experts in this habitat [5, 23]. The colonization of the phyllosphere involves a variety of fields that need more detailed understanding. The transmission of microorganisms from roots to aerial plant parts seems to be a neglected research field and clarity is needed about the importance of the route towards colonization of this environmentally protected phyllosphere. For soils polluted with human pathogens, this may be especially significant.

5.2.2 Leaf Characteristics and Environmental Factors Controlling Phyllosphere Microbiology

A number of factors decide when microbial cells or propagules are able to colonize the leaf and where they become clustered when they arrive on the leaf surface. The interaction of leaf and environmental characteristics that interact to regulate conditions in the phyllosphere habitat determines establishment. The cuticle is the first point of contact for microbial cells migrating to the phyllosphere. The waxy surface, which is also microcrystalline in nature, has various functionalities like diffusion barrier, water reduction, solution loss, and aqueous pollution; as a reflector for minimizing the variations in temperature; providing repellence to water; and protecting against pathogens [55]. Water repellence is especially crucial for preventing microorganism migration to the leaf surface. This is particularly true on young leaves with intact cuticles compared to older leaves, since wettability improves as the cuticle erodes [55]. Furthermore, the cuticle's function in limiting nutrient loss from the leaf is critical for the survival of epiphytic microbial species. Sucrose, fructose, and glucose are only found in isolated localized locations on the leaf, according to whole cell bacterial biosensors [56, 57]. This, combined with recent microscopical evidence [58], indicates that the vast majority of microbial immigrants to the phyllosphere are introduced to nutrient-poor environments, with only a few cells landing in areas with relatively abundant nutrients that sustain development. Such discrete sites of nutrient depletion, such as wounds or glandular trichomes, as well as sites of nutrient enrichment, such as pollen [59] or honeydew, provide microbial growth opportunities [58, 60]. Limited access to carbon sources limits the growth of phyllospheric microbes but not of nutrients like N, Fe, or P [3]. When on the resurrection fern *Polypodium polypodioides* is subjected to rainfall after a time of desiccation, the diverse phyllosphere group changes in overall composition and function, indicating the use of labile organic substrates in the form of an enrichment culture [61]. It is unclear if this happens in other species. Volatile organic compounds (VOCs) are being released around peripheries of plant leaves, which include carbon dioxide, acetone, terpenoids, aldehydes, alcohols, sesquiterpenoids, sulphides, and nitrogen-containing compounds. It is not certain if these can directly be food sources, but some of these can be inhibitory or harmful to some fungi [62]. Similarly, certain glandular trichoma secreted proteins can inhibit certain pathogens [63]. It is also possible for plants to release a variety of compounds in response to damage that not only promotes microbial production but also selectively hampers the growth of microbials [64]. The microbial ability of the leaf can also affect the characteristics of the plant species themselves. Broad-leaved plants like cucumber, lettuce, and bean yield higher number of culturable bacteria than narrow-structured leaves like grasses [65–67]. Group structure on leaves from individuals of the same species is identical, but differs considerably between species, according to culture-independent approaches [5]. According to Lambais et al. [5], only 5% of the bacteria found in tropical tree canopies were endemic to all tree species. Furthermore, the scale of bacterial and fungal populations on leaves has been linked to leaf location, plant architecture, and canopy height [68–70]. Characterizing the arrangement and dynamics of societies in time and space, particularly at the landscape scale, requires data. It is especially important to consider the relative importance of environmental conditions, location, and plant species in deciding the structure and dynamics of phyllosphere populations. Biogeographical approaches to microbial community research have the ability to reveal these underlying relationships [71].

5.3 Microbial Adaptation to the Phyllosphere

The degree to which microbial colonists may develop themselves on the leaf surface is determined by properties of the colonists themselves, in addition to plant and environmental factors. This represents certain microorganisms' innate capacity to live in their current habitat, while others are capable of altering the atmosphere to reduce the amount of tension they are subjected to. Tolerance to ultraviolet (UV) radiation is likely to be an effective selection pressure for survival and development in this habitat, according to culture-independent studies [20, 72], and most isolated phyllosphere microorganisms can tolerate high UV radiation levels on the leaf surface [72]. Black melanin-type pigments, as well as UV-B-induced hyphal wall thickening, are believed to play a vital role as defensive pigments in fungi, with the latter defending lower levels of the fungal colony [73, 74]. Surprisingly, the bacteria strains from the peanut phyllosphere that developed pink or orange pigments were the most UV-B resistant [75], implying that phyllosphere microbes have numerous UV-B protectant mechanisms. Since water and nutrients are main limiting factors for microbial growth in the phyllosphere, epiphytes have developed a number of mechanisms to resolve them. Some epiphytic *Pseudomonas* spp., for example, can release surfactants that improve the wettability of leaf surfaces, making it simpler for microorganisms to use water and increase nutrient solubilization and diffusion, increasing epiphytic bacteria's substrate supply [76]. Phyllosphere bacteria have been found to improve cuticle permeability, increasing water and nutrient supply in the phyllosphere [77]. Another factor for increasing nutrient supply may be the ability to create toxins that disrupt ion transfer through plant cell plasma membranes [78]. The toxin syringomycin is secreted by the plant pathogenic *Pseudomonas syringae* pv. *syringae*, which causes cell lysis. Non-pathogenic epiphytic strains of *P. syringae* pv. *syringae*, on the other hand, develop low levels, preventing necrosis and disease while also stimulating plant nutrient release [79]. Syringomycin also functions as a surfactant, offering two potential pathways for the nutrient supply in the phyllosphere. The synthesis and release of plant growth regulators is another, perhaps more widely used mechanism. Glickmann et al. [80] and Brandl et al. [81] found that bacterial epiphytes produce indole-3-acetic acid (IAA), which is linked to increased nutrient leakage and microbial health [3, 82]. The existence of a functional type III secretion pathway in *Pseudomonas fluorescens* and *Pseudomonas putida* [83], which provides the ability to change the local habitat, may be needed for growth and survival in the phyllosphere [3]. Bacterial attachment and invasion of the phyllosphere can also be aided by pili and flagellae development [84]. Molecular methods are now being used to identify a wide variety of genes and gene products that are essential for phyllosphere colonization [85, 86], which may lead to new insights into epiphytic growth mechanisms as, previously stated, bacterial distribution on the leaf surface is not standardized, and cell aggregates are common [87]. The epiphytes' ability to colonize and thrive in the phyllosphere, as well as change the local climate, can be aided by the inclusion of these aggregates. Bacteria producing extracellular polysaccharides (EPS) can protect bacteria from water stress and aid in cell attachment to the leaf surface [52, 85]. In the same way that biofilms protect against UV radiation, predation, and bacteriocides, these aggregates can also moderate pH and gas exchange, enhance genetic exchange, especially through plasmid transfer, and enable cell-density-dependent behaviour. The latter, which is also mediated by quorum sensing (QS) and the aggregation of diffusible molecules like *N*-acylhomoserine lactones (AHLs), has a variety of effects on microbial behaviour, including EPS and antibiotic activity, as well as pathogenicity traits [74, 87, 88]. If molecular signals and receptors important for aggregate behaviour can be established, it may be possible to influence microbial communities on the phyllosphere in the future.

5.3.1 Plant Genotype and Phyllosphere Microbiology

Microbial communities in the phyllosphere can vary significantly in size and composition both spatially and temporally on the same plant, differ between different plants and sections of plants in the same location, and even differ on the same plant species in different locations [34, 89, 90]. Many of the variation must be due to the environmental conditions at the time and place of sampling, which influence microbial immigration, emigration, development, and death processes. However, the microbial species that does emerge must be closely related to the plant's phenotypic characteristics, which are largely controlled by its genetic makeup. Certainly, there are 'hotspots' of microbial development on the leaf associated with particular areas, and these can be assumed to be influenced by plant genetic characteristics as well. We propose that genotype plays an important role in the colonization and formation of microbial populations in the phyllosphere within plant species. Despite its possible relevance, few studies have addressed the interaction between plant genetic regulation of phenotypic characteristics and their concomitant effects on microbial communities in the phyllosphere. Several experiments have used cultivation-based methods to examine the effect of plant genotype on microbiology of phyllosphere. Endophytic bacteria population sizes and composition varied among 9 cotton cultivars, and endophytic bacteria were present in 5 of 11 pea cultivars, with one having a higher colonization level than the others [91, 92]. In an external shooting system, one of four cultivars supported less *Pseudomonas* sp. after applying the bacterium to the seed in a gnotobiotic system [93]. There were also differences between various cultivars of snap bean in the ability to support *P. syringae* pv. *syringae* [94, 95]. However, no variations were found between the 3 and 10 major coffee cultivars or between clones of the same genus of natural, epiphytic mycoparasites [96]. As no variations were found between epiphytes on three apple cultivars [97] or endophytes on three wheat cultivars [19, 97] for elucidating interactions between plant genotype and phyllosphere microbial community structure, culture-independent community profiling approaches have been especially useful. Phyllosphere microbial communities vary between cultivars of the same plant genus, according to several reports. Bacterial species in the phyllosphere varied between sweet pepper and tomato cultivars [4, 15], and endophytes and epiphytes differed between potato varieties [4, 15]. Differences in plant genotype can have a greater impact on certain microbial communities than others. The composition of the phyllosphere bacterial population, for example, was observed to differ between wheat cultivars, despite no differences in archaeal populations [98]. Klerks et al. [98] found that lettuce cultivar influences *Salmonella enterica* serovar colonization of leaves, with major serovar–cultivar interactions. Furthermore, the diversity of endophyte bacterial communities differed between the three lettuce cultivars studied, suggesting that competitive encounters with the normal endophyte bacterial culture influenced the ability of *S. enterica* serovars to colonize plants endophytically. While magainin-expressing potato tubers had lower total numbers of bacteria than unmodified plants, culture-dependent research revealed that genetic modification of potato with the antibacterial peptide magainin had little effect on the number or composition of phyllosphere bacterial or fungal populations [99]. In comparison to differences across three cultivars, genetic alteration of potato with a gene generating antibacterial T4-lysozyme or attacin cecropin caused a greater disparity in phyllosphere microbial community composition [99]. Field location and plant growth point, on the other hand, had a greater impact on bacterial population structure than cultivar or genetic alteration. Furthermore, according to Rasche et al. [99] various genotypes choose different microbial communities, which may have different responses to environmental variables. Chilling sweet pepper plants altered endophyte bacterial population composition, with the degree of the effect varying between cultivars and depending on cultivar chilling resistance. In the same way, the reaction of phyllosphere bacterial communities to UV-B radiation was found to be dependent on host genotype in a study of wheat cultivars. However, it was unclear if these variations were due to direct effects on the bacterial population or indirect effects related to plant responses to UV-B [4]. Microbial inoculants' colonization and persistence in the phyllosphere may also be influenced by plant genotype. Correa et al. [97] showed that the survival in the phyllosphere of the contrasting genotypes, for tomatoes, of plant growth that promoted *Azospirillum* inoculant differs and that the bacterial community response to inoculation of phyllosphere is different among genotypes.

5.4 Relationship between Phyllosphere Microbial Communities and Functional Traits of Plants

Different bacterial species exist in tropical trees. On Barro Colorado Island, over 450 tree species have been recorded [100]. There are about as many bacterial taxa as tree species on the whole 16 km^2 Island, but only 50–100 g of a leaf tissue from a tree. There are so many unrevealed bacterial taxa living on tropical tree leaves and at a sampling depth of hundreds of thousands of sequences. In this forest, the majority of bacterial taxa found on leaves are uncommon. In comparison to a previous analysis of temperate phyllosphere bacterial diversity [101], we discovered a central microbiome of typical bacterial taxa on nearly all leaves. Many of the most common bacterial taxa in the phyllosphere are diazotrophic (nitrogen-fixing) and methylotrophic (methanol- and other one-carbon compound-consuming) taxa like *Beijerinckia* and *Methylobacterium*, which are considered to be closely associated with plant hosts. Many bacterial taxa from groups that are usually found in other habitats, such as soil and water, were also discovered. It is difficult to classify the origins of bacterial colonies on tropical tree leaves due to a lack of thorough information regarding the ecological niches of bacteria [102]. The populations we studied are likely to be a combination of resident taxa that live on leaves in biofilms and on leaf surface substrates all of the time [103, 104], as well as seasonal taxa that come from the atmosphere, rainwater, and interaction with animal and plant dispersal vectors. To classify the meta-colonies [105] contributing to microbial diversity on leaf surfaces, future studies comparing bacterial population composition across various possible source populations, such as soil, air, water, and animals, will be needed. Plant traits related to the leaf economics continuum [106, 107] and the wood density–growth/mortality trade-off were among those that clarified heterogeneity in bacterial population structure on leaves [37, 107]. The leaf economics continuum is a set of associated traits that describe plants' resource uptake and retention strategies. This range distinguishes taxa with an 'acquisitive' resource strategy and traits such as relatively high resource uptake rates, short-lived leaves, and high concentrations of nutrients, such as nitrogen and phosphorus, from taxa with a 'retentive' resource strategy and traits such as relatively low resource uptake rates, long-lived leaves, and low concentrations of nutrients, such as nitrogen and phosphorus [106, 107]. The profound effect of leaf resource uptake strategies on leaf morphology and physiology is possibly the basis of evidence that bacterial population composition is linked to the leaf economics continuum.

The abundance of leaf nutrients and carbon compounds, the rate of development of VOCs, such as methanol, and the amount of structural and antimicrobial compounds released by leaves are all factors that could affect the resources available to bacteria living on the leaf surface [106]. The wood density–growth/mortality trade-off continuum was another major axis of plant trait and functional covariation linked to bacterial diversity on leaves [3, 36]. Fast-growing species with high mortality rates and low wood density (low structural investment) suited to rapid growth in favourable conditions, such as canopy openings, are separated from species with the opposite set of traits by this continuum. While wood density and growth/mortality rate are not supposed to have a direct impact on bacterial communities on leaves, they are linked to a plant's overall growth and life cycle strategy. Our findings indicate that plant ecological strategies and functional traits have a significant impact on phyllosphere bacterial populations and that integrating data on plant-associated bacteria could help us better understand the processes influencing plant trait associations at biogeographic scales [108]. We found evidence of non-neutral population assembly on virtually all sampled leaves, in the form of widespread phylogenetic clustering, using ecological null models to determine the degree of phylogenetic clustering in bacterial populations. The correlation between bacterial phylogenetic clustering and host features such as mortality and growth rate suggests that leaf conditions on fast-growing trees are an ecologically stronger filter for potential colonists in the phyllosphere leading to more phylogenetically clustered communities. Since certain tropical leaves have a long life space, future succession research on leaves of various ages [3] would demand that they be completely distinguished from each other. Plants and phyllosphere bacteria have evolved in tandem, according to evolutionary processes. Alphaproteobacteria, for example, were abundant on Ochnaceae and Vochysiaceae plants but much less numerous on Violaceae plants, which were commonly colonized by Sphingobacteria. Given the relationship between microbial population variances explained by host traits and taxonomy, these evolutionary correlations are likely to be linked to phylogenetic heterogeneity in host traits [109]. Several traits not evaluated in this analysis, such as the production of VOCs, cuticle chemistry, and antibiotic production, are also likely to play a part. To completely comprehend the mechanisms responsible for these evolutionary interactions, additional functional traits must be studied in the future, as well as thorough investigations of the impact of host traits on microbial population development. The findings of correlations between bacterial phyllosphere population composition and host characteristics pose the issue of who has more influence over phyllosphere interactions. Is the connection between the relative abundance of certain bacterial taxa and host plant growth rates due to bacteria directly affecting their hosts' growth, or do the traits of fast-growing host species filter for a specific bacterial population composition by promoting or inhibiting the growth of different bacterial clades. The fact that bacterial taxa respond to changes in leaf chemistry and microenvironment means that leaf chemistry has a significant impact on bacterial growth and survival on leaves [1, 3], which is reinforced by our observation that leaf chemistry is closely associated with bacterial population composition. Microbial communities, on the other hand, have been found to have direct effects on plant growth through resource exchange and pathogen defence [110, 111]. Experiments on plant–microbe interactions would be needed to fully disentangle the relative role of plants versus microbes in explaining the associations we found [7, 107]. Any trait with potential to affect fitness by associations with development, reproduction, or survival is referred to as a functional trait [112]. It has been revealed that biodiversity in tree leaves of bacteria correlates with host growth and mortality, meaning that the composition of the plant microbiome may be considered a functional feature of the plant. Many well-measured plants are easy measurements of the possible impact of microbial species on the work of plants and ecosystems, which includes combination of nitrogen-fixing bacteria or mycorrhizal fungi. The ability to quantify plant-associated microbial biodiversity in depth opens up the possibility of moving beyond binary measures of plant–microbial associations (e.g. mycorrhizal status) to a more detailed understanding of how microbial biodiversity is related to host function, both directly through considering microbial biodiversity as a potential functional trait and indirectly by understanding samples of plant–microbial associations (e.g. mycorrhizal status). A research goal for biogeography and macroecology is to broaden our understanding of plant functional biogeography beyond phenotypic traits to include biotic associations and species interaction networks [113–115]. Our understanding of plant–microbe interactions at biogeographic scales is currently limited [116]; our research takes a step towards integrating this information to better understand plant functional biogeography and predict species and ecosystem responses to global change [104, 117].

5.5 Metabolic Dynamics of Phyllosphere Microbiota

Arabidopsis thaliana holds Flavobacteria as the dominant genera in terms of leaf microbiota, rhizosphere, and phyllosphere [118]. According to a new genomic comparison of Flavobacteria isolated from marine ecosystems and plants, it may be strongly adapted to plant carbohydrate metabolism. Genes encoding glycoside hydrolase families GH78 and GH106, which are responsible for the utilization of rhamnogalacturonan, which is exclusively associated with terrestrial plant hemicelluloses, were found only in the genomes of Flavobacteria from terrestrial plant communities and not in the genomes of Flavobacteria from aquatic plant communities [119]. Microorganisms associated with the phyllosphere reside in a sunlight-exposed setting. The photochemical conversion of this light resource into carbon and oil, which could supplement carbon supplies from the host plant, could be a significant advantage for growth in nutrient-limited environments. In phyllospheric cultures, metagenomic data showed the existence of bacterial rhodopsin genes [120]. It appears that certain epiphytic microorganisms have retinal-dependent rhodopsin proton pumps that can be light-activated by radiations that are different from the absorption spectrum of chlorophylls and carotenoids, which drive plant photosynthetic processes and thus the production of the plant carbon resources that are eventually available to epiphytic microorganisms [120]. According to Tolonen et al. [121], and Moran and Miller [122], it has been observed that the N and C metabolic pathways of host plant affects the nutrient availability of epiphytic bacteria by involving regulations of light-dependent processes and photoassimilate production in the host plant [123, 124]. In maize, it has been discovered that global relations between plant nitrogen balance and leaf epiphytic bacterial species richness are correlated. Ren et al. [125] found significant differences in phyllosphere bacterial species in rice exposed to varying combinations of elevated CO_2 and multiple levels of nitrogen fertilization in a free air CO_2 enrichment trial. Phyllospheric microorganisms' enzymatic processes, on the other hand, tend to function on essential plant metabolites [126], increasing the likelihood of complex metabolic feedback loops between plant tissues and phyllospheric microorganisms [127]. As per findings of Atamna-Ismaeel et al. [119], in phototrophic bacteria the potential presence of gene markers regulates metabolic pathways for aerobic anoxygenic conditions. Five separate metagenomes from phyllosphere microbiota have homologs of bchY, which encodes the Y subunit of chlorophyllide reductase, and pufM, which encodes the M subunit of the photosynthetic reaction core (rice, soybean, tobacco, *Tamarix*, clover). Epifluorescence microscopy was used to detect the presence of special pigments associated with aerobic anoxygenic phototrophic bacteria. With the inclusion of the genus *Methylobacterium*, as well as a mysterious group of bacteria that seem to be peculiar to the phyllosphere, it was discovered that the genus *Methylobacterium* accounts for 1–7% of the total population of epiphytic bacteria [119]. The phyllosphere's bacterial diversity is close in many plant organisms, according to rarefaction curves using ribosomal genes, which will be comparable to the diversity of the human microbiome. However, bacterial diversity in the phyllosphere seems to be much smaller than in the rhizosphere, soil, or aquatic environments [128, 129]. The identification of phyllosphere-specific bacterial communities is limited by sequencing depth, particularly when low-abundance species are present. Thus, indicators of aerobic anoxygenic phototrophic bacteria populations, which have a low abundance (0.4%), were not found in metagenomic evidence of *Tamarix* leaf microbiota [119], despite their inclusion in a number of plant species [119]. As a result, complementary approaches to detecting low-abundance bacterial species in the phyllosphere should be established, especially through microscope observation methods such as Fluorescence in situ hybridization (FISH) or fluidic force microscope approaches (FluidFM) [120]. To understand metabolic and enzymatic pathways of phyllospheric microbes including the effects of abiotic factors that regulate their roles, one must advent with newer technologies like proteomics, transcriptomics, and metabolomics especially to understand auxotrophic relationships and their interdependence on available abiotic factors like sunlight, etc. [128, 129].

5.6 Impact of Phyllospheric Microorganisms on Plant–Plant, Plant–Insect, and Plant Atmosphere Chemical Exchanges

Microbial species composition is being regulated by different types of chemicals or compounds that are being released by plants, like VOCs, as they allow or inhibit microbial growth for allowing precise species to grow or restrict. According to Shiojiri et al. [130], plants emit scent like chemicals that interfere with microbial population thus ultimately affecting plant signalling process with other plants or animals. Plant surface microbiota knowledge may show the mechanisms that

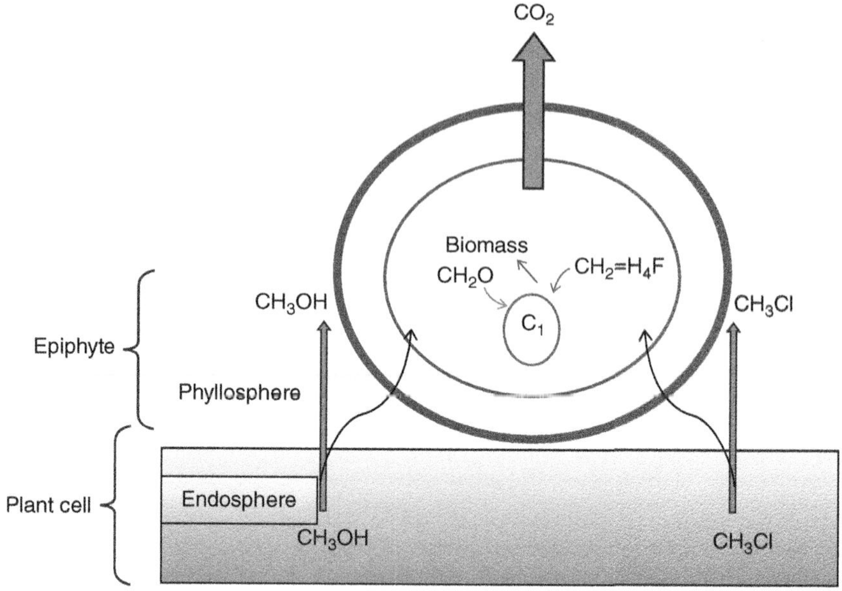

Figure 5.1 Biocapture of plant C1 VOCs by methylotrophs of the phyllosphere.

regulate processes at the interface between plants, microorganisms, and plant-interacting organisms, or between plants, microorganisms, and the atmosphere (Figure 5.1), in both natural and anthropogenic environs. Many herbivores induce the gaseous hormone ethylene, the oxylipin hormone jasmonate and its volatile derivative methyl jasmonate, and phenolic hormone salicylate and its volatile derivative methyl salicylate, which are induced by many bacterial pathogens. A correlation between the population composition of the phyllosphere microbiota and plant ethylene signalling has been discovered by genetic analysis of *A. thaliana* mutants [131]. A genome-wide association analysis of the associated leaf microbial population using 196 recombinant inbred lines of field-grown *A. thaliana* showed that plant loci involved in protection, such as virus replication and cell wall integrity, trichome branching, and morphogenesis, form the microbial species richness of the leaf microbiota [132]. Deficiency in the phytohormone jasmonic acid biosynthesis has little impact on the structuring of bacterial populations, as revealed in another research focused on tobacco plants [133]. A variety of plant defensive and signalling compounds, both volatile and non-volatile, are used in plant biotic interactions in addition to phytohormones [134]. Plant protective chemicals have been found to be degraded by foliage-associated bacteria, resulting in decreased defence against insect defoliators [134]. When secreted on plant surfaces by Colorado potato beetle larvae, bacterial symbionts of the genera *Stenotrophomonas*, *Pseudomonas*, and

Enterobacter suppress tomato anti-herbivore defences by enhancing the microbial defence response, favouring larval development [135]. Experimental bacterial infections of the phyllosphere revealed that individual *Pseudomonas* spp. strains encouraged herbivore host preference and that bacterial strains demonstrated heterogeneity in how they ecologically impacted insect herbivores in a new analysis of interactions between a specialist chewing insect herbivore and its single plant host, *Cardamine cordifolia* [136]. As previously stated, pesticides have a significant impact on phyllosphere group composition, implying that pesticide treatments can disrupt natural interactions between phyllosphere microbiota and plant defences [137–139]. As a result of a better understanding of defence mechanisms involving multiple biotic interactions and phyllospheric bacteria, novel pesticides in the sense of sustainable agriculture may be shaped. Epiphytic microorganisms have the microbial capability to degrade substances that are harmful to plants, humans, or the ecosystem. Depollution processes dependent on phyllospheres may benefit from such detoxification ability. List of organic compounds including nicotine, phenol [140, 141], which are known to be deteriorated epiphytic microbes, as well as acenaphthylene, acenaphthene, fluorene, and phenanthrene comes under the category of polycyclic aromatic hydrocarbons (PAH) and also chloromethane and isoprene, which are being produced by plants [142–144]. According to Scheublin et al. [145], genus *Arthrobacter* is known to flourish, persist in the phyllosphere, and degrade various organic compounds like pesticides and herbicides including phenylurea, glyphosate, and malathion; besides, degradation of aromatic hydrocarbons such as phenol, chlorophenol, benzene, toluene, ethylbenzene and xylene (BTEX), phenanthrene, triazines, atrazine, and cyanazine has also been reported. As per several reports, few *cph* genes are involved in 4-chlorophenol degradation in the group of Gram-positive *Arthrobacter* belonging to epiphytic bacterial colonies and also based on phyllosphere-induced expression as the comparative transcriptome profiling revealed presence of bacterial species in leaves of *Phaseolus vulgaris*, which is a common bean and possesses some natural plant-excreted phenolic compounds [146, 147]. Plants with sufficient microbial communities that degrade a given collection of organic compounds may be used for processes of atmospheric depollution in urban or industrial settings, as well as for the depollution of phytosanitary substance atmospheric drifts in agricultural environs. Lastly, epiphytic microorganisms with plant-friendly properties may be used as probiotic agents [148].

5.7 Quorum Sensing in Phyllosphere

QS is a form of microbial chemical communication that uses complex diffusible signal molecules [149]. QS bacteria control a variety of functions, including virulence factor generation, bacterial motility, and symbiosis, by using a variety of small, diffusible signal molecules [150]. One of the most well-studied signal molecules is AHLs [151]. Many Gram-negative plant-associated bacteria pathogens have been confirmed to use AHL-based QS to control their virulence [152]. *Pseudomonas, Pectobacterium, Ralstonia,* and *Pantoea* genus all have plant pathogenic bacteria [153–155]. Interfering with these plant pathogens' contact systems is a more modern technique for suppressing virulence expression and eliminating bacterial contamination. Quorum quenching (QQ) [156] is a concept used to describe processes that conflict with QS. Biocontrol techniques dependent on QQ pathways seem to be more sustainable than biocide therapies because they have a lower selective pressure for microbial survival [157]. Enzymatic depletion of AHL signals can be used to obtain QQ. Transgenic plants expressing the AHL-degradation enzyme demonstrated increased defence against the plant pathogen *Erwinia carotovora* in the lab [158]. However, large-scale field deployment of the genetically modified plant poses possible ecological threats, and negative public perceptions stymie transgenic crop adoption [159]. It is thought that using bacteria that are naturally capable of destroying AHL to combat plant pathogens is a better way [160]. Bacteria that degrade AHL have been found in a variety of soil samples. The phyllosphere, an essential microorganism ecosystem, should be studied as a source of quenching bacteria [3]. They have the unique potential to coexist with their pathogenic relatives in the same ecological niche, successfully protecting the plant [161]. The native diversity and abundance of AHL-degrading phyllosphere bacterial species, on the other hand, are much less well known, particularly in terms of understanding their ecological functions. Nonetheless, a thorough understanding of leaf-dwelling QQ bacteria should help to emphasize the relevance of phyllosphere bacteria community interactions. This expertise would also help to advance biotechnological studies in the field of plant disease biocontrol using AHL-degrading phyllosphere bacterial materials. As a result, the current research centred on determining the identity and prevalence of cultivable bacteria that may interfere with AHL-mediated QS systems and are found on leaves. The possible AHL-degrading ability of leaf-associated strains was determined using a bioassay after they were harvested from the tobacco phyllosphere using a culture-based approach. The phylogenetic profiles of AHL-degrading phyllosphere bacteria were then investigated using the 16S rRNA gene in detail. A QS mechanism based on AHL is used by many Gram-negative plant pathogenic bacteria to control their virulence traits. The use of QQ bacteria to interact with QS and protect plants from pathogens has been introduced as a sustainable biocontrol technique. The frequency and variety of QQ strains found on tobacco leaf surfaces were investigated in this study. Using the biosensor *Chromobacterium violaceum* CV026, 1177 leaf-associated isolates were screened for their ability to inhibit AHL-mediated QS. A total of 168 strains (14%) will inhibit AHL activity. Among these, 106 (63%) of the culturable quenchers may degrade AHL molecules enzymatically, while the remaining strains can use other QS inhibitors to disrupt chemical contact. Furthermore, lactonase activity is found in nearly 79% of QQ strains capable of enzymatically inactivating AHLs. *Bacillus* sp., *Acinetobacter* sp., *Lysinibacillus* sp., *Serratia* sp., *Pseudomonas* sp., and *Myroides* sp. were identified as leaf-associated QQ bacteria in a phylogenetic study based on 16S rDNA. The bacterial quenchers can be used as efficient biocontrol reagents to kill in situ pathogens of plants because of their natural diversity.

5.8 Applications for Phyllosphere Microbiology

The plant-based beneficial microbes are valuable agricultural bio-sources and can boost plant growth by solvability and mobilization (of P, K, and Zn), nitrogen attachment, and siderophore generation (microbes-mediated bio-fortification of Fe in different crops). Advantageous microbes can play a major role in raising crop yields, removing pollutants, inhibiting pathogenic substances, and producing new substances. Beneficial microbes may stimulate plant growth by producing plant growth regulators such as IAA, gibberellic acids, and cytokines, as well as biocontrol of phytopathogens by producing antibiotic, antifungal, or antibacterial compounds; Fe-chelating compounds; inducing acquired host tolerance; and increasing mineral bioavailability [162–164]. The phyllosphere microbes can have a positive impact on the host plant's growth and create an antagonistic compound against pathogens in this competition. There have been reports of phyllosphere endophytes with nitrogen fixation [9, 165, 166], bioremediation of toxic chemicals/pollutants, and as biocontrol agents (BCAs) against essential foliar plant pathogens [50, 167, 168]. Furthermore, the phyllosphere microbiome is a reflection of environmental factors, and it can play a major role in global food chains and nutrient linkages. Many beneficial microbes have been found in the phyllosphere of various crop plants, including *Achromobacter*, *Bacillus*, *Beijerinckia*, *Burkholderia*, *Flexibacterium*, *Methylobacterium*, *Micrococcus*, *Micromonospora*, *Nocardioides*, *Pantoea*, *Penicillium*, *Planomonospora*, *Pseudomonas*, *Streptomyces*, and *Xanthomonas* [169–171]. The study of phyllosphere bacteria, however, is restricted in comparison to most other microbial environments. The following are some of its major biotechnological potentials.

5.8.1 Biocontrol Agents

Biocontrol refers to using other beneficial bacteria or harmless living materials to manage viruses and disease-causing pests such as nematodes, weeds, flies, and mites. Plant diseases are caused by bacterial pathogens in nature, and they result in a significant reduction in agricultural product production. Instead of using poisonous hazardous chemicals as biological control, scientific methods for sustainable agriculture use the antagonistic properties of beneficial microbes against harmful pathogens [172, 173]. Biological therapy is a desirable method for managing plant pathogens [174], and a growing number of BCAs are being commercialized for different crops, including *Bacillus* spp., *Pseudomonas* spp., *Trichoderma* spp., and others [175, 176]. Most of them live in the phyllosphere or in the dirt, and they can help to reduce the amount of plant pathogens on the surface of the leaves by using the competitive theory. Plant fitness, crop production, and the protection of horticultural goods for human use are all affected by pathogenic microbial interactions in the phyllosphere. Plant pathogens have been shown to be inhibited by phyllosphere Actinomycetes' growth and colonization [3]. For example, the endophytic isolate *Gordonia* sp. has been confirmed to generate antifungal imidazole-2-yl amino acids and an acidic polysaccharide named Gordon as the key component in biofilms, both of which are considered important for pathogenicity against plant disease [177]. Diverse *Streptomyces* sp., like *Streptomyces griseus*, developed a number of antifungal substances such as 1-H-pyrrol2-carboxylic acid (PCA), cycloheximide, and streptomycin, used effectively to manage fungal and bacterial diseases in plants [178, 179]. Various researchers analysed the possible BCAs against fungal leaf blast disease for Actinomycetes in rice phyllosphere. The exogenous Actinomycete XN-1 was reported to work as an antagonistic agent to monitor the occurrence and growth of a Cucumber Leaf Spot in greenhouse [180]. This also confirms that microorganisms of the phyllosphere play an important role in the fight against infections and have a promising future in biocontrol products' growth. The bacteria colonize on plant surface by means of microbes with compounds such as the IAA and *N*-acyl homoserin (AHL) [3]. Sartori et al. [181] investigated the biocontrol ability of maize phyllosphere microorganisms against the causal agent of leaf blight, *Exserohi lumturcicum*. Shrestha et al. [182] explored the potential of rice-associated *Bacillus* for biological regulation of sheath blight and panicle blight induced by *Rhizoctonia solani* and *Burkholderia glumae*, respectively. *Bacillus* isolates were found to suppress the fungus's sclerotial germination, which may be due to the bacteria's antimicrobial secondary metabolites. Plant defence is also shown by Gram-negative bacteria. *Pseudomonas graminis* isolated from the apple phyllosphere, for example, demonstrated control against *Erwinia amylovora*-caused fire blight [183], and *Pseudomonas protegens* CS1 isolated from the lemon phyllosphere is used as a biocontrol against citrus canker [183, 184].

In addition, microbial siderophores quench phytopathogens and shield the host plant from contamination [185–187] as a BCA. *P. syringae* pv. *syringae* 22d/93, for example, produces a siderophore with biocontrol activity against *Pseudomonas syringae* pv. *glycinea* 1a/96, a plant pathogen [188]. *Pyricularia oryzae* is responsible for the siderophore pyochelin released by the endophyte rice blast [187].

Table 5.1 Endophytic fungi as biocontrol agent.

Endophyte	Host	Pathogen	Reference
Acremonium zeae	Maize	*Aspergillus flavus* and *Fusarium verticillioides*	[191]
Clonostachys rosea	*Theobroma cacao*	*Phytophthora palmivora*	[192]
Trichoderma martiale	*Theobroma cacao*	*Phytophthora palmivora*	[193]
Penicillium spp.	*Vitis vinifera* L.	*Botrytis cinerea*	[194]
Ramularia endophylla	Plant	*Mycosphaerella labyrinth*	[195]
Clonostachys rosea	Wheat	*Gibberella zeae*	[196]
Cladosporium and *Colletotrichum*	*Smallanthus sonchifolius*	*Lecythophora* sp.	[197]
Botryosphaeria dothidea	*Cornus officinalis*	*Alternaria alternata, A. arborescens, Botryosphaeria dothidea,* and *Colletotrichum gloeosporioides*	[198]
Alternaria sp., *Botryosphaeria ribis, Phoma medicaginis, Bionectria ochroleuca, Aureobasidium pullulans*, and *Chaetomium spirochaete*	*Vitis vinifera*	*Botrytis cinerea*	[199]
Fusarium solani sensu lato	*Vitis labrusca*	*Botrytis* sp.	[200]

Commercial BCAs based on plant-associated *Pseudomonas* spp. have proven to be effective [189, 190]. In the presence of the heavy chelator 8-Hydroxyquinoline, cyanogenic fluorescent, *Pseudomonas* produces siderophores that inhibit pathogens including *R. solani* and *Sclerotium rolfsii* [190]. Endophytes as BCAs are mentioned in Table 5.1. For compound processing, most BCAs use the non-ribosomal peptide synthetase (NRPS) gene and/or type 1 polyketide synthase gene.

5.8.2 Plant Growth-Promoting Compounds

Growth hormones, sufficient nutrients, healthy environmental conditions, and beneficial microbial interactions all play a role in plant growth. Plant-associated or phyllosphere microbial communities develop IAA, gibberellic acids, and cytokines, and they can fix nitrogen and mobilize nutrients [201]. Many bacteria and fungi, including trees, develop IAA [202–204]. Plant tryptophan is used by microbes to generate IAA, which can boost plant growth and overall health [205–210].

The genus *Methylobacterium* is among the most commonly observed leaf epiphytes and represents an abundant and stable member of the phyllosphere microbial community of a wide range of crop plants such as sugarcane (*Saccharum officinarum* L.), pigeon pea (*Cajanus cajan* L.), mustard (*Brassica campestris* L.), potato (*Solanum tuberosum* L.), and radish (*Raphanus sativus* L.) [169], and has produced variety of growth-promoting phytohormones. Plant growth is greatly aided by the association of plant growth-promoting bacteria (PGPB), especially *Methylobacterium* sp., with plant hosts due to the production of phytohormones such as auxins and cytokinins, as well as increased activity of enzymes such as urease and 1-aminocyclopropane-1-carboxylate deaminase (ACCD), which promotes growth and enhances the production of siderophores; thereby, the various inoculation approaches, such as soil, foliar, and a mixture of both soil and foliar inoculations, affect the benefits associated with plant–microbe interactions [170]. The effects of *Erwinia herbicola* inoculation on plant growth by IAA cultivation were investigated in a report. According to the findings, about 65% of the *E. herbicola* strain retrieved from the leaves had higher ipdC gene expression than in culture. The study indicated that physical or chemical microclimates directly influence the differential expression of ipdC [171].

5.8.3 Biopharmaceutical Importance

Medicinal plant biological behaviour and uses of different healing products have been well established. Microbes associated with plants, especially endophytes, have shown to have high therapeutic utility in recent years. Endophytic microbes are well known for their positive effects on the host, including phytohormones, metabolites, and stress-resistant physiology, as well as their biotechnological ability [211–213]. Endophytes have been shown to contain bioactive metabolites, and have been used in the medicinal and cosmetic industry as a powerful medicine [213]. Alkaloids, flavonoids, phenolic acids, quinones, steroids, saponins, terpenoids, tetralones, and xanthones are examples of commercially useful secondary metabolites developed by endophytic bacteria, Actinomycetes, and fungi [214–216]. Endophytic bacteria, for example, are well-known sources of taxol, a diterpene alkaloid, and lignin, which are used in cancer treatments such as cathartics, emetics, and cholagogue [217–219]. Many antibacterial, antifungal, antiviral, anticancer, and antihelminthic metabolites have been isolated from plant-associated microbes.

5.8.4 Other Applications

Some bacteria do play role in toxic elimination from plant parts, thus the phyllospheric microbes act as BCAs and improve growth. Moreover, as per Grobelak and Hiller [220], in *A. thaliana*, few bacteria that are hydroxamate siderophores add the property in host plant to eliminate heavy metal toxicity, and hence act as phytoremediators. Even they provide some other defence properties like tolerance to drought, which has been seen in *Brachypodium distachyon* grass where endophytic *Bacillus subtilis* strain B26 enhances this ability to this grass. Gagné-Bourque et al. [221] found that the drought tolerance process is due to a complex carbohydrate metabolism in which the endophytic bacteria enhance stress-responsive raffinose-related family carbohydrates in the host. In another case, the endophytic interaction boosts the host plant's osmotic responses. Endophytic strains in pepper plants, such as *Arthrobacter* sp. and *Bacillus* spp., improve proline aggregation, resulting in osmotic resistance [222]. Furthermore, the extracellular enzymes of endophytic bacteria inoculants provide the host with an abiotic stress resistance system. Under elevated sodium concentrations, the endophytic interaction of various *Bacillus* spp. enhances the production of superoxide dismutase, phenylalanine lyase, catalase, and peroxidase enzymes in gladiolus plants [223]. Endophytic bacteria isolation and enzyme production potential vary when they colonize plant tissues, hence in this regard few studies have been published. Furthermore, Jalgaonwala et al. [224] discovered that *Lactobacillus fermentum* isolated from *Vinca rosea* leaves has the highest proteolytic activity, which is considered superior to non-endophytic isolates. Endophytic fungi isolated from *Ocimum sanctum* and *Aloe vera*, on the other hand, have higher enzymatic activity [225, 226]. Plant-associated microorganisms also aid nutrient acquisition by supplying minerals and other micro/macronutrients from the soil [227]. Overall, the merits offer new insights into the field of the phyllosphere microbiome, including the importance of interactions to host plant growth and defence, as well as the ecosystem's significance.

5.8.5 Conclusion and Future Prospects

The phyllosphere is a unique ecosystem inhabited by a vast range of microorganisms, including epiphytes and endophytes, beneficial and harmful bacteria, fungi, viruses, and other microbes. Understanding the structure, networking, and physiology of the phyllosphere community is a huge problem. However, substantial study on phyllosphere microbiota shows that it has a lot of potential for uses in agriculture and forestry, as well as ecosystem cleanup and wellness. To date, both in vitro and in vivo investigations have been needed to better understand microbial aggregations in the phyllosphere and their dynamic role in the ecosystem. Furthermore, phyllosphere microbiome research ensures that present concerns, such as terrestrial ecosystem change and the impact of global warming, are better understood, particularly the dominance of pathogenicity.

References

1 Arnold, A.E., Mejia, L.C., Kyllo, D. et al. (2003). Fungal endophytes limit pathogen damage in a tropical tree. *PNAS* 100: 15649–15654.
2 Inacio, J., Pereira, P., de Carvalho, M. et al. (2002). Estimation and diversity of phylloplane mycobiota on selected plants in a Mediterranean-type ecosystem in Portugal. *Microb. Ecol.* 44: 344–353.
3 Lindow, S.E. and Brandl, M.T. (2003). Microbiology of the phyllosphere. *Appl. Environ. Microbiol.* 69: 1875–1883.
4 Stapleton, A.E. and Simmons, S.J. (2006). Plant control of phyllosphere diversity: genotype interactions with ultraviolet-B radiation. In: *Microbial Ecology of the Aerial Plant Surface* (ed. M.J. Bailey, A.K. Lilley, P.T.N. Timms-Wilson and P.T.N. Spencer-Phillips), 223–238. Wallingford, UK: CABI International.
5 Lambais, M.R., Crowley, D.E., Cury, J.C. et al. (2006). Bacterial diversity in tree canopies of the Atlantic forest. *Science* 312: 1917.
6 Schweitzer, J.A., Bailey, J.K., Bangert, R.K. et al. (2006). The role of plant genetics in determining above- and below-ground microbial communities. In: *Microbial Ecology of the Aerial Plant Surface* (ed. M.J. Bailey, A.K. Lilley, P.T.N. Timms-Wilson and P.T.N. Spencer-Phillips), 107–119. Wallingford, UK: CABI International.
7 Rasche, F., Marco-Noales, E., Velvis, H. et al. (2006). Structural characteristics and plant-beneficial effects of bacteria colonizing the shoots of field grown conventional and genetically modified T4-lysozyme producing potatoes. *Plant Soil* 289: 123–140.
8 Whipps, J.M., Hand, P., Pink, D.A.C., and Bending, G.D. (2008). Human pathogens and the phyllosphere. *Adv. Appl. Microbiol.* 64: 183–221.
9 Freiberg, E. (1998). Microclimatic parameters influencing nitrogen fixation in the phyllosphere in a Costa Rican premontane rain forest. *Oecologia* 117: 9–18.
10 Murty, M.G. (1983). Nitrogen fixation (acetylene-reduction) in the phyllosphere of some economically important plants. *Plant Soil* 73: 151–153.
11 Miyamoto, T., Kawahara, M., and Minamisawa, K. (2004). Novel endophytic nitrogen-fixing clostridia from the grass Miscanthus sinensis as revealed by terminal restriction fragment length polymorphism analysis. *Appl. Environ. Microbiol.* 70: 6580–6586.
12 Wagner, M., Amann, R., Lemmer, H., and Schleifer, K.H. (1993). Probing activated-sludge with oligonucleotides specific for proteobacteria – inadequacy of culture-dependent methods for describing microbial community structure. *Appl. Environ. Microbiol.* 59: 1520–1525.
13 Van Aken, B., Peres, C.M., Doty, S.L. et al. (2004). Methylobacterium populi sp nov., a novel aerobic, pink-pigmented, facultatively methylotrophic, methane-utilizing bacterium isolated from poplar trees (Populus deltoides x nigra DN34). *Int. J. Syst. Evol. Microbiol.* 54: 1191–1196.

14 Papen, H., Gessler, A., Zumbusch, E., and Rennenberg, H. (2002). Chemolithoautotrophic nitrifiers in the phyllosphere of a spruce ecosystem receiving high atmospheric nitrogen input. *Curr. Microbiol.* 44: 56–60.

15 Corpe, W.A. and Rheem, S. (1989). Ecology of the methylotrophic bacteria on living leaf surfaces. *FEMS Microbiol. Ecol.* 62: 243–249.

16 Morris, C.E., Barnes, M.B., and McLean, R.C.J. (2002). Biofilms on leaf surfaces: implications for the biology, ecology and management of populations of epiphytic bacteria. In: *Phyllosphere Microbiology* (ed. S.E. Lindow, E.I. Hecht-Poinar and V.J. Elliott), 139–155. St Paul, USA: APS Press.

17 Thompson, I.P., Bailey, M.J., Fenlon, J.S. et al. (1993). Quantitative and qualitative seasonal changes in the microbial community from the phyllosphere of sugar beet (Beta vulgaris). *Plant Soil* 150: 177–191.

18 Legard, D.E., McQuilken, M.P., Whipps, J.M. et al. (1994). Studies of seasonal changes in the microbial populations on the phyllosphere of spring wheat as a prelude to the release of a genetically modified microorganism. *Agric. Ecosyst. Environ.* 50: 87–101.

19 Rasche, F., Trondl, R., Naglreiter, C. et al. (2006). Chilling and cultivar type affect the diversity of bacterial endophytes colonizing sweet pepper (Capsicum anuum L.). *Can. J. Microbiol.* 52: 1036–1045.

20 Kadivar, H. and Stapleton, A.E. (2006). Ultraviolet radiation alters maize phyllosphere bacterial diversity. *Microb. Ecol.* 45: 353–361.

21 Idris, R., Trifonova, R., Puschenreiter, M. et al. (2004). Bacterial communities associated with flowering plants of the Ni hyperaccumulator Thlaspi goesingense. *Appl. Environ. Microbiol.* 70: 2667–2677.

22 Reiter, B. and Sessitsch, A. (2006). Bacterial endophytes of the wildflower Crocus albiflorus analysed by characterisation of isolates and by a cultivation-independent approach. *Can. J. Microbiol.* 52: 140–149.

23 Yang, C.H., Crowley, D.E., Borneman, J., and Keen, N.T. (2001). Microbial phyllosphere populations are more complex than previously realized. *PNAS* 98: 3889–3894.

24 Andrews, J.H. and Harris, R.F. (2000). The ecology and biogeography of microorganisms of plant surfaces. *Annu. Rev. Phytopathol.* 38: 145–180.

25 de Jager, E.S., Wehner, F.C., and Korsten, L. (2001). Microbial ecology of the mango phylloplane. *Microb. Ecol.* 42: 201–207.

26 Glushakova, A.M. and Chernov, I.Y. (2004). Seasonal dynamics in a yeast population on leaves of the common wood sorrel Oxalis acetosella L. *Microbiology* 73: 184–188.

27 Arnold, A.E., Maynard, Z., Gilbert, G.S. et al. (2000). Are tropical fungal endophytes hyperdiverse? *Ecol. Lett.* 3: 267–274.

28 Sessitsch, A., Reiter, B., Pfeifer, U., and Wilhelm, E. (2002). Cultivation-independent population analysis of bacterial endophytes in three potato varieties

based on eubacterial and Actinomycetes-specific PCR of 16S rRNA genes. *FEMS Microbiol. Ecol.* 39: 23–32.

29 Singh, B.K., Nazaries, L., Munro, S. et al. (2006). Use of multiplex terminal restriction fragment length polymorphism for rapid and simultaneous analysis of different components of the soil microbial community. *Appl. Environ. Microbiol.* 72: 7278–7285.

30 Sessitsch, A., Hackl, E., Wenzl, P. et al. (2006). Diagnostic microbial microarrays in soil ecology. *New Phytol.* 171: 719–736.

31 Manceau, C.R. and Kasempour, M.N. (2002). In Endophytic versus epiphytic colonization of plants: what comes first? In: *Phyllosphere Microbiology* (ed. S.E. Lindow, E.I. Hecht-Poinar and V.J. Elliott), 115–123. St Paul, USA: APS Press.

32 Suslow, T.V. (2002). Production practices affecting the potential for persistent contamination of plants by microbial foodborne pathogens. In: *Phyllosphere Microbiology* (ed. S.E. Lindow, E.I. Hecht-Poinar and V.J. Elliott), 241–256. St Paul, USA: APS Press.

33 Hirano, S.S. and Upper, C.D. (1989). Diel variation in population size and ice nucleation activity of Pseudomonas syringae on snap bean leaflets. *Appl. Environ. Microbiol.* 55: 623–630.

34 Kinkel, L.L. (1997). Microbial population dynamics on leaves. *Annu. Rev. Phytopathol.* 35: 327–347.

35 Zak, J.C. (2002). Implications of a leaf surface habitat for fungal community structure and function. In: *Phyllosphere Microbiology* (ed. S.E. Lindow, E.I. Hecht-Poinar and V.J. Elliott), 299–315. St Paul, USA: APS Press.

36 Lindemann, J., Constantinidou, H.A., Barchet, W.R., and Upper, C.D. (1982). Plants as sources of airborne bacteria, including ice nucleation-active bacteria. *Appl. Environ. Microbiol.* 44: 1059–1063.

37 Lacey, J. (1996). Spore dispersal – its role in ecology and disease: the British contribution to fungal aerobiology. *Mycol. Res.* 100: 641–660.

38 Lighthart, B. (1997). The ecology of bacteria in the alfresco atmosphere. *FEMS Microbiol. Ecol.* 23: 263–274.

39 Venette, J.R. and Kennedy, B.W. (1975). Naturally produced aerosols of Pseudomonas glycinea. *Phytopathology* 65: 737–738.

40 Lamb, T.G., Tonkyn, D.W., and Kluepfel, D.A. (1996). Movement of Pseudomonas aureofaciens from the rhizosphere to aerial plant tissue. *Can. J. Microbiol.* 42: 1112–1120.

41 Wulff, E.G., van Vuurde, J.W.L., and Hockenhull, J. (2003). The ability of the biological control agent Bacillus subtilis, strain BB, to colonise vegetable brassicas endophytically following seed inoculation. *Plant Soil* 255: 463–474.

42 Blakeman, J.P. (1985). Ecological succession of leaf surface microorganisms in relation to biological control. In: *Biological Control on the Phylloplane* (ed.

C.E. Windels and S.E. Lindow), 6–30. St. Paul, USA: American Phytopathological Society.

43 Davis, C.L. and Brlansky, R.H. (1991). Use of immunogold labeling with scanning electron microscopy to identifphytopathogenic bacteria on leaf surfaces. *Appl. Environ. Microbiol.* 57: 3052–3055.

44 Mansvelt, E.L. and Hattingh, M.J. (1987). Scanning electron microscopy of colonization of pear leaves by Pseudomonas syringae pv syringae. *Can. J. Bot.* 65: 2517–2522.

45 Mew, T.W. and Vera Cruz, C.M. (1986). Epiphytic colonization of host and non-host plants by phytopathogenic bacteria. In: *Microbiology of the Phyllosphere* (ed. N.J. Fokkema and J. van den Heuvel), 269–282. Cambridge, UK: Cambridge University Press.

46 Mew, T.W., Mew, I.P.C., and Huang, J.S. (1984). Scanning electron microscopy of virulent and avirulent strains of Xanthomonas campestris pv oryzae on rice leaves. *Phytopathology* 74: 635–641.

47 Mariano, R.L.R. and McCarter, S.M. (1993). Epiphytic survival of Pseudomonas viridiflava on tomato and selected weed species. *Microb. Ecol.* 26: 47–58.

48 Leben, C. (1988). Relative humidity and the survival of epiphytic bacteria with buds and leaves of cucumber plants. *Phytopathology* 78: 179–185.

49 Surico, G. (1993). Scanning electron microscopy of olive and oleander leaves colonized by Pseudomonas syringae subsp savastanoi. *J. Phytopathol.* 138: 31–40.

50 Beattie, G.A. and Lindow, S.E. (1999). Bacterial colonization of leaves: a spectrum of strategies. *Phytopathology* 89: 353–359.

51 Last, F.T. (1955). Seasonal incidence of Sporobolomyces on cereal leaves. *Trans. Br. Mycol. Soc.* 38: 221–239.

52 Jacques, M.A., Josi, K., Darrasse, A., and Samson, R. (2005). Xanthomonas axonopodis pv. phaseoli var. fuscans is aggregated in stable biofilm population sizes in the phyllosphere of field-grown beans. *Appl. Environ. Microbiol.* 71: 2008–2015.

53 Morris, C.E., Monier, J.M., and Jacques, M.A. (1997). Methods for observing microbial biofilms directly on leaf surfaces and recovering them for isolation of culturable microorganisms. *Appl. Environ. Microbiol.* 63: 1570–1576.

54 Morris, C.E., Monier, J.M., and Jacques, M.A. (1998). A technique to quantify the population size and composition of the biofilm component in communities of bacteria in the phyllosphere. *Appl. Environ. Microbiol.* 64: 4789–4795.

55 Beattie, G.A. (2002). Leaf surface waxes and the process of leaf colonization by microorganisms. In: *Phyllosphere Microbiology* (ed. S.E. Lindow, E.I. Hecht-Poinar and V.J. Elliott), 3–26. St. Paul, USA: APS Press.

56 Leveau, J.H.J. and Lindow, S.E. (2001). Appetite of an epiphyte: quantitative monitoring of bacterial sugar consumption in the phyllosphere. *PNAS* 98: 3446–3453.

57 Miller, W.G., Brandl, M.T., Quinones, B., and Lindow, S.E. (2001). Biological sensor for sucrose availability: relative sensitivities of various reporter genes. *Appl. Environ. Microbiol.* 67: 1308–1317.

58 Monier, J.M. and Lindow, S.E. (2005). Aggregates of resident bacteria facilitate survival of immigrant bacteria on leaf surfaces. *Microb. Ecol.* 49: 343–352.

59 Diem, H.G. (1974). Microorganisms of the leaf surface: estimation of mycoflora of barley phyllosphere. *J. Gen. Microbiol.* 80: 77–83.

60 Dik, A.J. and Vanpelt, J.A. (1992). Interaction between phyllosphere yeasts, aphid honeydew and fungicide effectiveness in wheat under field conditions. *Plant Pathol.* 41: 661–675.

61 Jackson, E.F., Echlin, H.L., and Jackson, C.R. (2006). Changes in the phyllosphere community of the resurrection fern, Polypodium polypodioides, associated with rainfall and wetting. *FEMS Microbiol. Ecol.* 58: 236–246.

62 Mechaber, W. (2002). Mapping uncharted territory: nanoscale leaf surface topology. In: *Phyllosphere Microbiology* (ed. S.E. Lindow, E.I. Hecht-Poinar and V.J. Elliott), 43–50. St. Paul, USA: APS Press.

63 Shepherd, R.W., Bass, W.T., Houtz, R.L., and Wagner, G.J. (2005). Phylloplanins of tobacco are defensive proteins deployed on aerial surfaces by short glandular trichomes. *Plant Cell* 17: 1851–1861.

64 Dingman, D.W. (2000). Growth of Escherichia coli O157:H7 in bruised apple (Malus domestica) tissue as influenced by cultivar, date of harvest, and source. *Appl. Environ. Microbiol.* 66: 1077–1083.

65 O'Brien, R.D. and Lindow, S.E. (1989). Effect of plant species and environmental conditions on epiphytic population sizes of Pseudomonas syringae and other bacteria. *Phytopathology* 79: 619–627.

66 Lindow, S.E. and Andersen, G.L. (1996). Influence of immigration on epiphytic bacterial populations on navel orange leaves. *Appl. Environ. Microbiol.* 62: 2978–2987.

67 Kinkel, L.L., Wilson, M., and Lindow, S.E. (2000). Plant species and plant incubation conditions influence variability in epiphytic bacterial population size. *Microb. Ecol.* 39: 1–11.

68 Wildman, H.G. and Parkinson, D. (1979). Micro-fungal succession on living leaves of Populus tremuloides. *Can. J. Bot.* 57: 2800–2811.

69 Oliveira, J.R., Romeiro, R.S., and Muchovej, J.J. (1991). Population tendencies of Pseudomonas cichorii and P. syringae pv garcae in young and mature coffee leaves. *J. Phytopathol.* 131: 210–214.

70 Jacques, M.A., Kinkel, L.L., and Morris, C.E. (1995). Population sizes, immigration, and growth of epiphytic bacteria on leaves of different ages and positions of field-grown endive (Cichorium endivia var latifolia). *Appl. Environ. Microbiol.* 61: 899–906.

71 Ramette, A. and Tiedje, J.M. (2007). Biogeography: an emerging cornerstone for understanding prokaryotic diversity, ecology, and evolution. *Microb. Ecol.* 53: 197–207.

72 Sundin, G.W. (2002). Ultraviolet radiation on leaves: its influence on microbial communities and their adaptations. In: *Phyllosphere Microbiology* (ed. S.E. Lindow, E.I. Hecht-Poinar and V.J. Elliott), 27–41. St Paul, USA: APS Press.

73 Fourtouni, A., Manetas, Y., and Christias, C. (1998). Effects of UV-B radiation on growth, pigmentation, and spore production in the phytopathogenic fungus Alternaria solani. *Can. J. Bot.* 76: 2093–2099.

74 Swift, S., Bainton, N.J., and Winson, M.K. (1994). Gram-negative bacterial communication by N-acyl homoserine lactones: a universal language? *Trends Microbiol.* 2: 193–119.

75 Sundin, G.W. and Jacobs, J.L. (1999). Ultraviolet radiation (UVR) sensitivity analysis and UVR survival strategies of a bacterial community from the phyllosphere of fieldgrown peanut (Arachis hypogeae L.). *Microb. Ecol.* 38: 27–38.

76 Bunster, L., Fokkema, N.J., and Schippers, B. (1989). Effect of surface-active Pseudomonas spp. on leaf wettability. *Appl. Environ. Microbiol.* 55: 1340–1345.

77 Schreiber, L., Krimm, U., Knoll, D. et al. (2005). Plant–microbe interactions: identification of epiphytic bacteria and their ability to alter leaf surface permeability. *New Phytol.* 166: 589–594.

78 Quigley, N.B. and Gross, D.C. (1994). Syringomycin production among strains of Pseudomonas syringae pv. syringae: conservation of the syrb and syrd genes and activation of phytotoxin production by plant signal molecules. *Mol. Plant-Microbe Interact.* 7: 78–90.

79 Hutchison, M.L., Tester, M.A., and Gross, D.C. (1995). Role of biosurfactant and ion channel-forming activities of syringomycin in transmembrane ion flux: a model for the mechanism of action in the plant pathogen interaction. *Mol. Plant–Microbe Interact.* 8: 610–620.

80 Glickmann, E., Gardan, L., Jacquet, S. et al. (1998). Auxin production is a common feature of most pathovars of Pseudomonas syringae. *Mol. Plant–Microbe Interact.* 11: 156–162.

81 Brandl, M.T. and Lindow, S.E. (1998). Contribution of indole3-acetic acid production to the epiphytic fitness of Erwinia herbicola. *Appl. Environ. Microbiol.* 64: 3256–3263.

82 Manulis, S., Haviv-Chesner, A., Brandl, M.T. et al. (1998). Differential involvement of indole3-acetic acid biosynthetic pathways in pathogenicity and epiphytic fitness of Erwinia herbicola pv. gypsophilae. *Mol. Plant-Microbe Interact.* 11: 634–642.

83 Preston, G.M., Bertrand, N., and Rainey, P.B. (2001). Type III secretion in plant growth-promoting Pseudomonas fluorescens SBW25. *Mol. Microbiol.* 41: 999–1014.

84 Romantschuck, M., Boureau, T., Roine, E. et al. (2002). The role of pili and flagella in leaf colonization by Pseudomonas syringae. In: *Phyllosphere Microbiology* (ed. S.E. Lindow, E.I. Hecht-Poinar and V.J. Elliott), 101–113. St. Paul, USA: APS Press.

85 Gal, M., Preston, G.M., Massey, R.C. et al. (2003). Genes encoding a cellulosic polymer contribute toward the ecological success of Pseudomonas fluorescens SBW25 on plant surfaces. *Mol. Ecol.* 12: 3109–3121.

86 Gourion, B., Rossignol, M., and Vorholt, J.A. (2006). A proteomic study of Methylobacterium extorquens reveals a response regulator essential for epiphytic growth. *PNAS* 103: 13186–13191.

87 Morris, C.E. and Monier, J.M. (2003). The ecological significance of biofilm formation by plant-associated bacteria. *Annu. Rev. Phytopathol.* 41: 429–453.

88 Greenberg, E.P. (1997). Quorum sensing in gram-negative bacteria. *Am. Soc. Microbiol. News* 63: 371–377.

89 Lindemann, J., Arny, D.C., and Upper, C.D. (1984). Epiphytic populations of Pseudomonas syringae pv syringae on snap bean and nonhost plants and the incidence of bacterial brown spot disease in relation to cropping patterns. *Phytopathology* 74: 1329–1333.

90 Morris, C.E. and Lucotte, T. (1993). Dynamics and variability of bacterial population density on leaves of field-grown endive destined for ready-to-use processing. *Int. J. Food Sci. Technol.* 28: 201–209.

91 Elvira-Recuenco, M. and van Vuurde, J.W.L. (2000). Natural incidence of endophytic bacteria in pea cultivars under field conditions. *Can. J. Microbiol.* 46: 1036–1041.

92 Adams, P.D. and Kloepper, J.W. (2002). Effect of host genotype on indigenous bacterial endophytes of cotton (Gossypium hirsutum L.). *Plant Soil* 240: 181–189.

93 Pillay, V.K. and Nowak, J. (1997). Inoculum density, temperature, and genotype effects on in vitro growth promotion and epiphytic and endophytic colonization of tomato (Lycopersicon esculentum L) seedlings inoculated with a pseudomonad bacterium. *Can. J. Microbiol.* 43: 354–361.

94 Hirano, S.S., Baker, L.S., and Upper, C.D. (1996). Raindrop momentum triggers growth of leaf-associated populations of Pseudomonas syringae on field-grown snap bean plants. *Appl. Environ. Microbiol.* 62: 2560–2566.

95 Hoopen, G.M., Rees, R., Aisa, P. et al. (2003). Population dynamics of epiphytic mycoparasites of the genera Clonostachys and Fusarium for the biocontrol of black pod (Phytophthora palmivora) and moniliasis (Moniliophthora roreri) on cocoa (Theobroma cacao). *Mycol. Res.* 107: 587–596.

96 Becker, C.M. and Manning, W.J. (1983). Phylloplane microflora of apple cultivars immune and susceptible to Venturia inaequalis. *Phytopathology* 73: 1342.

97 Correa, O.S., Romero, A.M., Montecchia, M.S., and Soria, M.A. (2007). Tomato genotype and *Azospirillum* inoculation modulate the changes in bacterial communities associated with roots and leaves. *J. Appl. Microbiol.* 102: 781–786.

98 Klerks, M.M., Franz, E., van Gent-Pelzer, M. et al. (2007). Differential interaction of Salmonella enterica serovars with lettuce cultivars and plant–microbe factors influencing the colonization efficiency. *ISME J.* 1: 620–631.

99 Rasche, F., Velvis, H., Zachow, C. et al. (2006). Impact of transgenic potatoes expressing anti-bacterial agents on bacterial endophytes is comparable with the effects of plant genotype, soil type and pathogen infection. *J. Appl. Ecol.* 43: 555–566.

100 Croat, T. (1978). *Flora of Barro Colorado Island*. Redwood City, CA: Stanford University Press.

101 Redford, A.J., Bowers, R.M., Knight, R. et al. (2010). The ecology of the phyllosphere: geographic and phylogenetic variability in the distribution of bacteria on tree leaves. *Environ. Microbiol.* 12 (11): 2885–2893.

102 Philippot, L., Andersson, S.G.E., Battin, T.J. et al. (2010). The ecological coherence of high bacterial taxonomic ranks. *Nat. Rev. Microbiol.* 8 (7): 523–529.

103 Wilson, M. and Lindow, S.E. (1994). Coexistence among epiphytic bacterial populations mediatedthrough nutritional resource partitioning. *Appl. Environ. Microbiol.* 60 (12): 4468–4477.

104 Sarmento, H., Montoya, J.M., Vázquez-Domínguez, E. et al. (2010). Warming effects on marine microbial food web processes: how far can we go when it comes to predictions? *Philos. Trans. R. Soc. Lond. Ser. B Biol. Sci.* 365 (1549): 2137–2149.

105 Martiny, J.B.H., Bohannan, B.J.M., Brown, J.H. et al. (2006). Microbial biogeography: putting microorganisms on themap. *Nat. Rev. Microbiol.* 4 (2): 102–112.

106 Westoby, M., Falster, D.S., Moles, A.T. et al. (2002). Plant ecological strategies: some leading dimensions of variation between species. *Annu. Rev. Ecol. Syst.* 33: 125–159.

107 Wright, S.J., Kitajima, K., Kraft, N.J.B. et al. (2010). Functional traits and the growth-mortality trade-off in tropical trees. *Ecology* 91 (12): 3664–3674.

108 Westoby, M. and Wright, I.J. (2006). Land-plant ecology on the basis of functional traits. *Trends Ecol. Evol.* 21 (5): 261–268.

109 Kraft, N.J.B. and Ackerly, D.D. (2010). Functional trait and phylogenetic tests of community assembly across spatial scales in an Amazonian forest. *Ecol. Monogr.* 80 (3): 401–422.

110 Arnold, A.E., Mejía, L.C., Kyllo, D. et al. (2003). Fungal endophytes limit pathogen damage in a tropical tree. *Proc. Natl. Acad. Sci. U. S. A.* 100 (26): 15649–15654.

111 Innerebner, G., Knief, C., and Vorholt, J.A. (2011). Protection of Arabidopsis thaliana against leaf-pathogenic Pseudomonas syringae by Sphingomonas strains in a controlled model system. *Appl. Environ. Microbiol.* 77 (10): 3202–3210.

112 Lundberg, D.S., Lebeis, S.L., Paredes, S.H. et al. (2012). Defining the core Arabidopsis thaliana root microbiome. *Nature* 488 (7409): 86–90.

113 Violle, C., Navas, M.-L., Vile, D. et al. (2007). Let the concept of trait be functional! *Oikos* 116: 882–892.

114 Thébault, E. and Loreau, M. (2003). Food-web constraints on biodiversity-ecosystem functioningrelationships. *Proc. Natl. Acad. Sci. U. S. A.* 100 (25): 14949–14954.

115 Poulin, R., Krasnov, B.R., and Mouillot, D. (2011). Host specificity in phylogenetic and geographicspace. *Trends Parasitol.* 27 (8): 355–361.

116 Beck, J., Ballesteros-Mejia, L., Buchmann, C.M. et al. (2012). What's on the horizon for macroecology? *Ecography* 35: 673–683.

117 Andrews, J.H. and Harris, R.F. (2000). The ecology and biogeography of microorganisms on plant surfaces. *Annu. Rev. Phytopathol.* 38: 145–180.

118 Kolton, M., Sela, N., Elad, Y., and Cytryn, E. (2013). Comparative genomic analysis indicates that niche adaptation of terrestrial Flavobacteria is strongly linked to plant glycan metabolism. *PLoS One* 8: e76704. https://doi.org/10.1371/journal.pone.0076704.

119 Atamna-Ismaeel, N., Finkel, O., Glaser, F. et al. (2012). Bacterial anoxygenic photosynthesis on plant leaf surfaces. *Environ. Microbiol. Rep.* 4: 209–216. https://doi.org/10.1111/j.1758-2229.2011.00323.

120 Stiefel, P., Zambelli, T., and Vorholt, J.A. (2013). Isolation of optically targeted single bacteria using FluidFM applied to aerobic anoxygenic phototrophs from the phyllosphere. *Appl. Environ. Microbiol.* 79: 4895–4905. https://doi.org/10.1128/AEM.01087-13.

121 Tolonen, A.C., Aach, J., Lindell, D. et al. (2006). Global gene expression of Prochlorococcus ecotypes in response to changes in nitrogen availability. *Mol. Syst. Biol.* 2: 53. https://doi.org/10.1038/msb4100087.

122 Moran, M.A. and Miller, W.L. (2007). Resourceful heterotrophs make the most of light in the coastal ocean. *Nat. Rev. Microbiol.* 5: 792–800. https://doi.org/10.1038/nrmicro1746.

123 Athanasiou, K., Dyson, B.C., Webster, R.E., and Johnson, G.N. (2010). Dynamic acclimation of photosynthesis increases plant fitness in changing environments. *Plant Physiol.* 152: 366–373. https://doi.org/10.1104/pp.109.149351.

124 Sulmon, C., Gouesbet, G., Ramel, F. et al. (2011). Carbon dynamics, development and stress responses in Arabidopsis: involvement of the APL4 subunit of ADP-Glucose pyrophosphorylase (starch synthesis). *PLoS One* 6: e26855. https://doi.org/10.1371/journal.pone.0026855.

125 Ren, G., Zhang, H., Lin, X. et al. (2014). Response of phyllosphere bacterial communities to elevated CO2 during rice growing season. *Appl. Microbiol. Biotechnol.* 98: 9459–9471. https://doi.org/10.1007/s00253-014-5915-0.

126 Huang, S.H., Zhang, J.Y., Tao, Z. et al. (2014). Enzymatic conversion from pyridoxal to pyridoxine caused by microorganisms within tobacco phyllosphere. *Plant Physiol. Biochem.* 85: 9–13. https://doi.org/10.1016/j.plaphy.2014.10.006.

127 Manching, H.C., Balint-Kurti, P.J., and Stapleton, A.E. (2014). Southern leaf blight disease is correlated with decreased maize leaf epiphytic bacterial species richness and the phyllosphere bacterial diversity decline is enhanced by nitrogen fertilisation. *Front. Plant Sci.* 5: 403. https://doi.org/10.3389/fpls.2014.00403.

128 Delmotte, N., Knief, C., Chaffron, S. et al. (2009). Community proteogenomics reveals insights into the physiology of phyllosphere bacteria. *Proc. Natl. Acad. Sci. U. S. A.* 106: 16428–16433. https://doi.org/10.1073/pnas.0905240106.

129 Knief, C., Delmotte, N., Chaffron, S. et al. (2012). Metaproteogenomic analysis of microbial communities in the phyllosphere and rhizosphere of rice. *ISME J.* 6: 1378–1390. https://doi.org/10.1038/ismej.2011.192.

130 Shiojiri, K., Kishimoto, K., Ozawa, R. et al. (2006). Changing green leaf volatile biosynthesis in plants: an approach for improving plant resistance against both herbivores and pathogens. *Proc. Natl. Acad. Sci. U. S. A.* 103: 16672–16676. https://doi.org/10.1073/pnas.0607780103.

131 Bringel, F. and Couée, I. (2015). Pivotal roles of phyllosphere microorganisms at the interface between plant functioning and atmospheric trace gas dynamics. *Front. Microbiol.* 6: 486. https://doi.org/10.3389/fmicb.2015.00486.

132 Horton, M.W., Bodenhausen, N., Beilsmith, K. et al. (2014). Genome-wide association study of Arabidopsis thaliana leaf microbial community. *Nat. Commun.* 5: 5320. https://doi.org/10.1038/ncomms6320.

133 Santhanam, R., Groten, K., Meldau, D.G., and Baldwin, I.T. (2014). Analysis of plant-bacteria interactions in their native habitat: bacterial communities associated with wild tobacco are independent of endogenous jasmonic acid levels and developmental stages. *PLoS One* 9: e94710. https://doi.org/10.1371/journal.pone.0094710.

134 Mason, C.J., Couture, J.J., and Raffa, K.F. (2014). Plant-associated bacteria degrade defense chemicals and reduce their adverse effects on an insect defoliator. *Oecologia* 175: 901–910. https://doi.org/10.1007/s00442-014-2950-6.

135 Chung, S.H., Rosa, C., Scully, E.D. et al. (2013). Herbivore exploits orally secreted bacteria to suppress plant defenses. *Proc. Natl. Acad. Sci. U. S. A.* 110: 15728–15733. https://doi.org/10.1073/pnas.1308867110.

136 Humphrey, P.T., Nguyen, T.T., Villalobos, M.M., and Whiteman, N.K. (2014). Diversity and abundance of phyllosphere bacteria are linked to insect herbivory. *Mol. Ecol.* 23: 1497–1515. https://doi.org/10.1111/mec.12657.

137 Karlsson, I., Friberg, H., Steinberg, C., and Persson, P. (2014). Fungicide effects on fungal community composition in the wheat phyllosphere. *PLoS One* 9: e111786. https://doi.org/10.1371/journal.pone.0111786.

138 Ottesen, A.R., Gorham, S., Pettengill, J.B. et al. (2014). The impact of systemic and copper pesticide applications on the phyllosphere microflora of tomatoes. *J. Sci. Food Agric.* 95: 1116–1125. https://doi.org/10.1002/jsfa.7010.

139 Glenn, D.M., Bassett, C., and Dowd, S.E. (2015). Effect of pest management system on 'Empire' apple leaf phyllosphere populations. *Sci. Hortic.* 183: 58–65. https://doi.org/10.1016/j.scienta.2014.12.009.

140 Sguros, P.L. (1955). Microbial transformations of the tobacco alkaloids. I. Cultural and morphological characteristics of a nicotinophile. *J. Bacteriol.* 69: 28–37.

141 Sandhu, A., Halverson, L.J., and Beattie, G.A. (2007). Bacterial degradation of airborne phenol in the phyllosphere. *Environ. Microbiol.* 9: 383–392. https://doi.org/10.1111/j.1462-2920.2006.01149.

142 Yutthammo, C., Thongthammachat, N., Pinphanichakarn, P., and Luepromchai, E. (2010). Diversity and avtivity of PAH-degrading bacteria in the phyllosphere of ornamental plants. *Microb. Ecol.* 59: 357–368. https://doi.org/10.1007/s00248-009-9631-8.

143 Nakamiya, K., Nakayama, T., Ito, H. et al. (2009). Isolation and properties of a 2-chlorovinylarsonic acid-degrading microorganism. *J. Hazard. Mater.* 165: 388–393. https://doi.org/10.1016/j.jhazmat.2008.10.030.

144 Nadalig, T., Greule, M., Bringel, F. et al. (2014). Probing the diversity of chloromethane-degrading bacteria by comparative genomics and isotopic fractionation. *Front. Microbiol.* 5: 523. https://doi.org/10.3389/fmicb.2014.00523.

145 Scheublin, T.R., Deusch, S., Moreno-Forero, S.K. et al. (2014). Transcriptional profiling of Gram-positive Arthrobacter in the phyllosphere: induction of pollutant degradation genes by natural plant phenolic compounds. *Environ. Microbiol.* 16: 2212–2225. https://doi.org/10.1111/1462-2920.12375.

146 Rastogi, G., Sbodio, A., Tech, J.J. et al. (2012). Leaf microbiota in an agroecosystem: spatiotemporal variation in bacterial community composition on field-grown lettuce. *ISME J.* 6: 1812–1822. https://doi.org/10.1038/ismej.2012.32.

147 Berlec, A. (2012). Novel techniques and findings in the study of plant microbiota: search for plant probiotics. *Plant Sci.* 193–194: 96–102. https://doi.org/10.1016/j.plantsci.2012.05.010.

148 Waters, C.M. and Bassler, B.L. (2005). Quorum sensing: cell-to-cell communication in bacteria. *Annu. Rev. Cell Dev. Biol.* 21: 319–346.

149 Pacheco, A.R. and Sperandio, V. (2009). Inter-kingdom signaling: chemical language between bacteria and host. *Curr. Opin. Microbiol.* 12: 192–198.

150 Manefield, M. and Whiteley, A.S. (2007). Acylated homoserine lactones in the environment: chameleons of bioactivity. *Philos. Trans. R. Soc. B* 362: 1235–1240.

151 Loh, J., Pierson, E.A., Pierson, L.S. III et al. (2002). Quorum sensing in plant-associated bacteria. *Curr. Opin. Plant Biol.* 5: 285–290.

152 Cha, C., Gao, P., Chen, Y.C. et al. (1998). Production of acyl-homoserine lactone quorum-sensing signals by gram-negative plant-associated bacteria. *Mol. Plant-Microbe Interact.* 11: 1119–1129.

153 Quinones, B., Dulla, G., and Lindow, S.E. (2005). Quorum sensing regulates exopolysaccharide production, motility, and virulence in *Pseudomonas syringae*. *Mol. Plant-Microbe Interact.* 18: 682–693.

154 Morohoshi, T., Nakamura, Y., Yamazaki, G. et al. (2007). The plant pathogen *Pantoea ananatis* produces *N*-acylhomoserine lactone and causes center rot disease of onion by quorum sensing. *J. Bacteriol.* 189: 8333–8338.

155 Vonbodman, S.B. and Farrand, S.K. (1995). Capsular polysaccharide biosynthesis and pathogenicity in *Erwinia stewartii* require induction by an *N*-acylhomoserine lactone autoinducer. *J. Bacteriol.* 177: 5000–5008.

156 Dong, Y.H., Wang, L.H., and Zhang, L.H. (2007). Quorum-quenching microbial infections: mechanisms and implications. *Philos. Trans. R. Soc. B* 362: 1201–1211.

157 Uroz, S., Dessaux, Y., and Oger, P. (2009). Quorum sensing and quorum quenching: the yin and yang of bacterial communication. *ChemBioChem* 10: 205–216.

158 Dong, Y.H., Wang, L.H., Xu, J.L. et al. (2001). Quenching quorum-sensing-dependent bacterial infection by an *N*-acyl homoserine lactonase. *Nature* 411: 813–817.

159 Varzakas, T.H., Arvanitoyannis, I.S., and Baltas, H. (2007). The politics and science behind GMO acceptance. *Crit. Rev. Food Sci. Nutr.* 47: 335–361.

160 Molina, L., Constantinescu, F., Michel, L. et al. Degradation of pathogen quorum-sensing molecules by soil bacteria: a preventive and curative biological control mechanism. *FEMS Microbiol. Ecol.* 45: 71–81.

161 Diallo, S., Crepin, A., Barbey, C. et al. (2011). Mechanisms and recent advances in biological control mediated through the potato rhizosphere. *FEMS Microbiol. Ecol.* 75: 351–364.

162 Divjot, K., Lata, R.K., Neelam, Y. et al. (2019). Agriculturally and industrially important fungi: current developments and potential biotechnological applications. In: *Recent Advancement in White Biotechnology Through Fungi*, 1–64. Cham: Springer International Publishing.

163 Kumar, M., Kour, D., Yadav, A.N. et al. (2019). Biodiversity of methylotrophic microbial communities and their potential role in mitigation of abiotic stresses in plants. *Biologia* 74: 287–308. https://doi.org/10.2478/s11756-019-00190-6.

164 Nath, Y.A., Divjot, K., Sushma, S. et al. (2019). *Plant Growth Promoting Rhizobacteria for Sustainable Stress Management*. Singapore: Springer Singapore.

165 Jones, K. (1970). Nitrogen fixation in the phyllosphere of the Douglas Fir, Pseudotsuga Douglasii. *Ann. Bot.* 34: 239–244. https://doi.org/10.1093/oxfordjournals.aob.a084358.

166 Furnkranz, M., Wolfgang, W., Andreas, R. et al. (2008). Nitrogen fixation by phyllosphere bacteria associated with higher plants plants and their colonizing epiphytes of a tropical lowland rainforest of Costa Rica. *ISME J.* 2 (5): 561–570. https://doi.org/10.1038/ismej.2008.14.

167 Balint-Kurti, P., Simmons, S.J., Blum, J.E. et al. (2010). Maize leaf epiphytic bacteria diversity patterns are genetically correlated with resistance to fungal pathogen infection. *Mol. Plant-Microbe Interact.* 23: 473–484. https://doi.org/10.1094/MPMI-23-4-0473.

168 Verma, P., Yadav, A.N., Kazy, S.K. et al. (2013). Elucidating the diversity and plant growth promoting attributes of wheat (*Triticum aestivum*) associated acidotolerant bacteria from southern hills zone of India. *Natl. J. Life Sci.* 10: 219–227.

169 Meena, K.K., Kumar, M., Kalyuzhnaya, M.G. et al. (2012). Epiphytic pink-pigmented methylotrophic bacteria enhance germination and seedling growth of wheat (*Triticum aestivum*) by producing phytohormone. *Antonie Van Leeuwenhoek* 101: 777–786. https://doi.org/10.1007/s10482-011-9692-9.

170 Mukhtar, I., Khokhar, I., Sobia, M., and Ali, A. (2010). Diversity of epiphytic and endophytic microorganisms in some dominant weeds. *Pak. J. Weed Sci. Res.* 16: 287–297.

171 Dobrovol'skaya, T., Khusnetdinova, K., Manucharova, N., and Golovchenko, A. (2017). Structure of epiphytic bacterial communities of weeds. *Microbiology* 86: 257–263.

172 Erwin, D.C. and Ribeiro, O.K. (1996). *Phytophthora Diseases Worldwide.* New York: APS Press.

173 Sharma, P.S., D'Souza, F., and Kutner, W. (2012). Molecular imprinting for selective chemical sensing of hazardous compounds and drugs of abuse. *TrAC Trends Anal. Chem.* 34: 59–77. https://doi.org/10.1016/j.trac.2011.11.005.

174 You, C., Zhang, C., Feng, C. et al. (2015). *Myroides odoratimimus*, a biocontrol agent from the rhizosphere of tobacco with potential to control *Alternaria alternata*. *BioControl* 60 (4): 555–564. https://doi.org/10.1007/s10526-015-9654-7.

175 Trabelsi, D. and Mhamdi, R. (2013). Microbial inoculants and their impact on soil microbial communities: a review. *Biomed. Res. Int.* https://doi.org/10.1155/2013/863240.

176 Cha, J.Y., Han, S., Hong, H.J. et al. (2016). Microbial and biochemical basis of a *Fusarium* wilt-suppressive soil. *ISME J.* 10 (1): 119–129. https://doi.org/10.1038/ismej.2015.95.

177 Mikolasch, A., Hammer, E., and Schauer, F. (2003). Synthesis of imidazol-2-yl amino acids by using cells from alkane-oxidizing bacteria. *Appl. Environ. Microbiol.* 69: 1670–1679.

178 Leben, C. and Keitt, G.W. (1954). Antibiotics and plant disease: effects of antibiotics in control of plant diseases. *J. Agric. Food Chem.* 2: 234–239.

179 Nguyen, X.H., Naing, K.W., Lee, Y.S. et al. (2015). Antagonism of antifungal metabolites from *Streptomyces griseus* H7602 against Phytophthora capsici. *J. Basic Microbiol.* 55: 45–53. https://doi.org/10.1002/jobm.201300820.

180 Wang, M. and Ma, Q. (2011). Antagonistic Actinomycete XN-1 from phyllosphere microorganisms of cucumber to control corynespora cassiicola. *Cucurbit Genet. Coop. Rep.* 33–34: 17–21.

181 Sartori, M., Nesci, A., Formento, Á., and Etcheverry, M. (2015). Selection of potential biological control of Exserohilum turcicum with epiphytic microorganisms from maize. *Rev Argent Microbiol* 47: 62–71.

182 Shrestha, B.K., Karki, H.S., Groth, D.E. et al. (2016). Biological control activities of rice-associated *Bacillus* sp. strains against sheath blight and bacterial panicle blight of rice. *PLoS One* 11:e0146764.

183 Michavila, G., Adler, C., De Gregorio, P.R. et al. (2017). *Pseudomonas protegens* CS1 from the lemon phyllosphere as a candidate for citrus canker biocontrol agent. *Plant Biol.* 19: 608–617. https://doi.org/10.1111/plb.12556.

184 Artur, M., Piotr, S., Joanna, P., and Robert, M. (2015). Control of fire blight (Erwinia amylovora) by a novel strain 49M of Pseudomonas graminis from the phyllosphere of apple (Malus spp.). *Eur. J. Plant Pathol.* 145 (2): 265–276. https://doi.org/10.1007/s10658-015-0837-.

185 Scavino, A.F. and Pedraza, R.O. (2013). The role of siderophores in plant growth-promoting bacteria. In: *Bacteria in Agrobiology: Crop Productivity* (ed. D.K. Maheshwari, M. Saraf and A. Aeron), 265–285. Berlin: Springer https://doi.org/10.1007/978-3-642-37241-4-11.

186 Ahmed, E. and Holmström, S.J.M. (2014). Siderophores in environmental research: roles and applications: siderophores in environmental research. *Microb. Biotechnol.* 7: 196–208. https://doi.org/10.1111/1751-7915.12117.

187 Harsonowati, W., Astuti, R.I., and Wahyudi, A.T. (2017). Leaf blast disease reduction by rice-phyllosphere actinomycetes producing bioactive compounds. *J. Gen. Plant Pathol.* 83: 98–108. https://doi.org/10.1007/s10327-017-0700-4.

188 Wensing, A., Braun, S.D., Buttner, P. et al. (2010). Impact of siderophore production by *Pseudomonas syringae* pv. *syringae* 22d/93 on epiphytic fitness and biocontrol activity against *Pseudomonas syringae* pv. *glycinea* 1a/96. *Appl. Environ. Microbiol.* 76: 2704–2711. https://doi.org/10.1128/AEM.02979-09.

189 Walsh, U.F., Morrissey, J.P., and O'Gara, F. (2001). Pseudomonas for biocontrol of phytopathogens: from functional genomics to commercial exploitation. *Curr. Opin. Biotechnol.* 12: 289–295. https://doi.org/10.1016/S0958-1669(00)00212-3.

190 Loper, J.E. and Lindow, S.E. (1987). Lack of evidence for the in situ fluorescent pigment production by *Pseudomonas syringae* pv. *syringae* on bean leaf surfaces. *Phytopathology* 77: 1449–1454. https://doi.org/10.1094/Phyto-77-1449.

191 Wicklow, D.T., Roth, S., Deyrup, S.T., and Gloer, J.B. (2005). A protective endophyte of maize: Acremonium zeae antibiotics inhibitory to Aspergillus flavus and Fusarium verticillioides. *Mycol. Res.* 109: 610–618.

192 Mejía, L.C., Rojas, E.I., Maynard, Z. et al. (2008). Endophytic fungi as biocontrol agents of Theobroma cacao pathogens. *Biol. Control* 46: 4–14.

193 Hanada, R.E., Pomella, A.W.V., Soberanis, W. et al. (2009). Biocontrol potential of Trichoderma martiale against the black-pod disease (Phytophthora palmivora) of cacao. *Biol. Control* 50: 143–149.

194 Núñez-Trujillo, G., Cabrera, R., Burgos-Reyes, R.L. et al. (2012). Endophytic fungi from Vitis vinifera L. isolated in Canary Islands and Azores as potential biocontrol agents of Botrytis cinerea Pers. *Fr. J. Hortic. For. Biotechnol.* 16: 1–6.

195 Videira, S.I.R., Groenewald, J.Z., Verkley, G.J.M. et al. (2015). The rise of Ramularia from the Mycosphaerella labyrinth. *Fungal Biol.* 119: 823–843.

196 Hue, A.G., Voldeng, H.D., Savard, M.E. et al. (2009). Biological control of fusarium head blight of wheat with Clonostachys rosea strain ACM941. *Can. J. Plant Pathol.* 31: 169–179.

197 Rosa, L.H., Tabanca, N., Techen, N. et al. (2012). Antifungal activity of extracts from endophytic fungi associated with Smallanthus maintained in vitro as autotrophic cultures and as pot plants in the greenhouse. *Can. J. Microbiol.* 58 (10): 1202–1221.

198 Zhao, X., Hu, Z., Hou, D. et al. (2020). Biodiversity and antifungal potential of endophytic fungi from the medicinal plant Cornus officinalis. *Symbiosis* 81: 223–233.

199 Cosoveanu, A., Gimenez-Mariño, C., Cabrera, Y. et al. (2014). Endophytic fungi from grapevine cultivars in canary islands and their activity against phytopatogenic fungi. *Int. J. Agric. Crop Sci.* 7: 1497–1503.

200 Brum, M.C.P. (2008). Fungos Endofíticos de Vitis labrusca L. var. Niagara Rosada e o seu Potencial Biotecnológico. Ph.D. thesis. Universidade de Mogi das Cruzes, Mogi das Cruzes, Brazil.

201 Kotasthane, A.S., Agrawal, T., Waris Zaidi, N., and Singh, U.S. (2017). Identification of siderophore producing and cynogenic fluorescent *Pseudomonas* and a simple confrontation assay to identify potential bio-control agent for collar rot of chickpea. *3 Biotech* 7 (2): 137.

202 Dourado, M.N., Neves, A.A.C., Santos, D.S., and Araujo, W.L. (2015). Biotechnological and agronomic potential of endophytic pink-pigmented methylotrophic trophic Methylobacterium spp. *Biomed. Res. Int.* 2015: 909016. https://doi.org/10.1155/2015/909016.

203 Sun, Y., Xiao, J., Jia, X. et al. (2010). The role of wheat Jasmonic acid and ethylene pathways in response to *Fusarium graminearum* infection. *Plant Growth Regul.* 80: 69–77. https://doi.org/10.1007/s10725-016-0147-1.

204 Venkatachalam, S., Ranjan, K., Prasanna, R. et al. (2016). Diversity and functional traits of culturable microbiome members, including cyanobactera in the rice phyllosphere. *Plant Biol.* 18: 627–637. https://doi.org/10.1111/plb.12441.

205 Thapa, S., Ranjan, K., Ramakrishnan, B. et al. (2018). Influence of fertilizers and rice cultivation methods on the abundance and diversity of phyllosphere microbiome. *J. Basic Microbiol.* 58: 172–186. https://doi.org/10.1002/jobm.201700402.

206 Hayat, R., Ali, S., Amara, U. et al. (2010). Soil beneficial bacteria and their role in plant growth promotion: a review. *Ann. Microbiol.* 60: 579–598. https://doi.org/10.1007/s13213-010-0117-1.

207 Nath, Y.A., Divjot, K., Sushma, S. et al. (2019). Psychrotrophic microbes: biodiversity, mechanisms of adaptation, and biotechnological implications in alleviation of cold stress in plants. In: *Plant Growth Promoting Rhizobacteria for Sustainable Stress Management*, 219–253. Singapore: Springer Singapore.

208 Yadav, A.N., Mishra, S., Singh, S., and Gupta, A. (2019). *Recent Advancement in White Biotechnology Through Fungi Volume 1: Diversity and Enzymes Perspectives.* Cham: Springer International Publishing.

209 Lee, M.-K., Singh, C.P., Woo-Jong, Y. et al. (2011). Foliar colonization and growth promotion of red pepper (Capsicum annuum L.) by Methylobacterium oryzae CBMB20. *J. Appl. Biol. Chem.* 54 (2): 120–125. https://doi.org/10.3839/jabc.2011.021.

210 Brandl, M.T., Quinones, B., and Lindow, S.E. (2001). Heterogeneous transcription of an indoleacetic acid biosynthetic gene in Erwinia herbicola on plant surfaces. *Proc. Natl. Acad. Sci.* 98 (6): 3454–3459. https://doi.org/10.1073/pnas.061014498.

211 Singh, R. and Dubey, A.K. (2015). Endophytic actinomycetes as emerging source for therapeutic compounds. *Indo Global J. Pharm. Sci.* 5: 106–116.

212 Parthasarathi, S., Sathya, S., Bupesh, G. et al. (2012). Isolation and characterization of antimicrobial compound from marine *Streptomyces hygroscopicus* BDUS 49. *World J. Fish Mar. Sci.* 4: 268–277.

213 Gouda, S., Das, G., Sen, S.K. et al. (2016). Endophytes: a treasure house of bioactive compounds of medicinal importance. *Front. Microbiol.* 7: 1538. https://doi.org/10.3389/fmicb.2016.01538.

214 Strobel, G.A. and Daisy, B. (2003). Bioprospecting for microbial endophytes and their natural products. *Microbiol. Mol. Biol. Rev.* 67: 491–502. https://doi.org/10.1128/MMBR.67.4.491-502.2003.

215 Joseph, B. and Priya, R.M. (2011). Bioactive compounds from endophytes and their potential in pharmaceutical effect: a review. *Am. J. Biochem. Mol. Biol.* 1: 291–309. https://doi.org/10.3923/ajbmb.2011.291.309.

216 Godstime, O.C., Enwa, F.O., Augustina, J.O., and Christopher, E.O. (2014). Mechanisms of antimicrobial actions of phytochemicals against enteric pathogens – a review. *J. Pharm. Chem. Biol. Sci.* 2: 77–85.

217 Zhang, P., Zhou, P.P., and Yu, L.J. (2009). An endophytic taxol-producing fungus from Taxus media, *Cladosporium cladosporioides* MD2. *Curr. Microbiol.* 59: 227–232. https://doi.org/10.1007/s00284-008-9270-1.

218 Nair Dhanya, N. and Padmavathy, S. (2014). Impact of endophytic microorganisms on plants, environment and humans. *Sci. World J.* 2014: 1–11. https://doi.org/10.1155/2014/250693.

219 Soliman, S.S.M. and Raizada, M.N. (2018). Darkness: a crucial factor in fungal taxol production. *Front. Microbiol.* 9: 353. https://doi.org/10.3389/fmicb.2018.00353.

220 Grobelak, A. and Hiller, J. (2017). Bacterial siderophores promote plant growth: screening of catechol and hydroxamate siderophores. *Int. J. Phytorem.* 19 (9): 825–833. https://doi.org/10.1080/15226514.2017.1290581.

221 Gagné-Bourque, F., Mayer, B.B.F., Charron, J.B.B. et al. (2015). Accelerated growth rate and increased drought stress resilience of the model grass Brachypodium distachyon colonized by Bacillus subtilis B26. *PLoS One* 10: e130456. https://doi.org/10.1371/journal.pone.0130456.

222 Sziderics, A.H., Rasche, F., Trognitz, F. et al. (2007). Bacterial endophytes contribute to abiotic stress adaptation in pepper plants (*Capsicum annuum* L.). *Can. J. Microbiol.* 53: 1195–1202. https://doi.org/10.1139/W07-082.

223 Damodaran, T., Rai, R.B., Jha, S.K. et al. (2014). Rhizosphere and endophytic bacteria for induction of salt tolerance in gladiolus grown in sodic soils. *J. Plant Interact.* 9: 577–584. https://doi.org/10.1080/17429145.2013.873958.

224 Jalgaonwala, R.E., Mohite, B.V., and Mahajan, R.T. (2011). Natural products from plant associated endophytic fungi. *J. Microbiol. Biotechnol. Res.* 1: 21–32.

225 Yadav, A.N., Sachan, S.G., Verma, P. et al. (2015). Culturable diversity and functional annotation of psychrotrophic bacteria from cold desert of Leh Ladakh (India). *World J. Microbiol. Biotechnol.* 31: 95–108. https://doi.org/10.1007/s11274-014-1768-z.

226 Singh, R. and Dubey, A.K. (2018). Diversity and applications of endophytic actinobacteria of plants in special and other ecological niches. *Front. Microbiol.* 9: 1767.

227 Yadav, A.N. (2017). Agriculturally important microbiomes: biodiversity and multifarious PGP attributes for amelioration of diverse abiotic stresses in crops for sustainable agriculture. *Biomed. J. Sci. Tech. Res.* 1: 1–4.

6

Endosphere and Endophyte Communities

CONTENTS

Microbiomics and Sustainable Crop Production, First Edition. Mohammad Yaseen Mir and Saima Hamid.

6.1 Reproduction and Transmission Modes of Microbes

Plants have a throng of microorganisms that dwell on the rhizosphere and phyllosphere, respectively, which are below and above ground plant surfaces. Furthermore, endophytes, which are microorganisms that colonize the inside of plants without causing illness, have received considerable attention in the recent decade [1]. The microbiome within an apparently healthy plant might possibly contain a mix of mutualistic, commensal, and latent pathogenic strains [2], making this a wide and frequently disputed description. A similar role is played by wild plant bacterial endophytes when it comes to stress protection as well as nutrient acquisition [3–7], but in order to fully understand those processes on an individual plant and ecosystem level, we need a better understanding of the endophyte colonization routes and dispersal modes. For example, to understand how much an individual forest tree may profit from endophytic nitrogen fixation [5] or to determine how much nitrogen is transported into a forest ecosystem via this pathway [4], we must understand when and how endophytic communities assemble. We look at the known and prospective pathways of bacterial endophyte transmission and dispersal, as well as the gaps in our knowledge of how bacteria travel between hosts and between the host and the environment.

Host-associated microorganisms can colonize the host in a variety of ways, including horizontally through the environment, vertically from the parent to the offspring, and in a combination of ways [8]. The symbiont in many vertically transmitted symbioses is obligatory, spending its whole life cycle inside the host and unable to live outside [8]. The transmission mode is influenced by ecological and evolutionary connections, and vice versa: Vertical transmission is predicted to emerge when symbiotic partners are mutualistic, as a means of ensuring faithful transfer of the beneficial symbiont from generation to generation [9]. Vertical transmission of bacterial symbionts from parent to offspring is typical in systems where the symbiont plays a critical role, such as the well-studied nutritional symbioses between bacteria and insects [10]. Transmission modality is influenced by ecological and evolutionary connections, and vice versa: When symbiotic partners are mutualistic, vertical transmission is predicted to emerge as a method to assure faithful transfer of the beneficial symbiont from one generation to the next [9]. Vertical transmission of bacterial symbionts from parent to offspring is typical in systems where the symbiont serves an essential role, such as the well-studied bacterial-insect nutritional symbioses [10].

Certain fungal endophytes (e.g. the well-studied Epichlo and Neotyphodium fungal endophytes of grasses) [11] have been shown to transmit vertically through seeds. There are numerous examples of mutualistic horizontally transmitted symbioses [12, 13], and not all mutualists are obligatory. The vast majority of bacterial endophytes are spread horizontally. For starters, the variety of bacteria in seeds and seedlings maintained under sterile conditions is generally smaller than in plants grown in soil [14], suggesting that the bulk of endophytes are obtained from the environment. Subsequent, bacterial endophytes are frequently generalists, as their beneficial characteristics can often be transferred to distantly related plants [15–17]. As a result, bacterial generalists that infect a variety of plant species must travel horizontally among them, and are unlikely to be transmitted exclusively vertically. Beneficial bacteria may be transmitted horizontally, which may benefit the plant. It may be advantageous for plants to be able to recruit symbionts from the environment, as it allows them to adapt to a changing environment [18]. Plants do appear to be capable of harbouring a wide variety of generalist endophytes, whose presence or absence at any one moment appears to be determined by the plant's environment (e.g. soil type) rather than its genotype [19–24]. When comparing more distantly related plant taxa, stronger host impacts have been discovered, although even then, host genotype is less relevant than soil type [25]. Obligate plant–bacteria interactions, on the other hand, appear to be uncommon. Some bacterial endophytes may be transmitted both vertically and horizontally (i.e. mixed-mode transmission), and an endophyte that is helpful to its host in a specific situation (e.g. biotic stress) may be handed down to the offspring through the seed. For bacterial endophytes, we evaluated known and prospective transmission pathways. Rather than plant organ or significance, the review is structured by transmission route (vertical vs. horizontal). We'll start with vertical transmission from the parent plant, such as seed or pollen, and then go on to horizontal transmission from the environment, beginning with the germination environment and finishing with the floral organs. The various transmission mechanisms and routes examined are summarized in Figure 6.1. Bacteria have been found inside the seeds of many different plant species, suggesting that vertical transfer of bacteria via seed and pollen occurs.

Endophytes that are passed down through generations must have a way to get from seeds to reproductive organs, whether by xylem vessels or the shoot apical meristem, which develops into reproductive organs. Soil is thought to be the most common source of bacterial endophytes [26], and soil-to-root transmission is the most well-studied horizontal transmission pathway. Soil bacteria can colonize the core of plants and become endophytes either early, through the spermosphere, or later, through the rhizosphere and into the roots of seedlings and adults. The phyllosphere, or above-ground plant surface, is colonized by a diverse community of microorganisms, providing an alternative, but less-studied, route of entry for bacteria that enter via stomata and are derived from rainwater, bioaerosols from surrounding soil, or dust and other particles in the atmosphere. Stomata are likely to act as a transmission pathway for plants of all life stages, although foliar endophytes of trees may be particularly significant. Moreover, sap-feeders and pollinators, as well as other arthropods, may act as vectors for bacteria that invade the

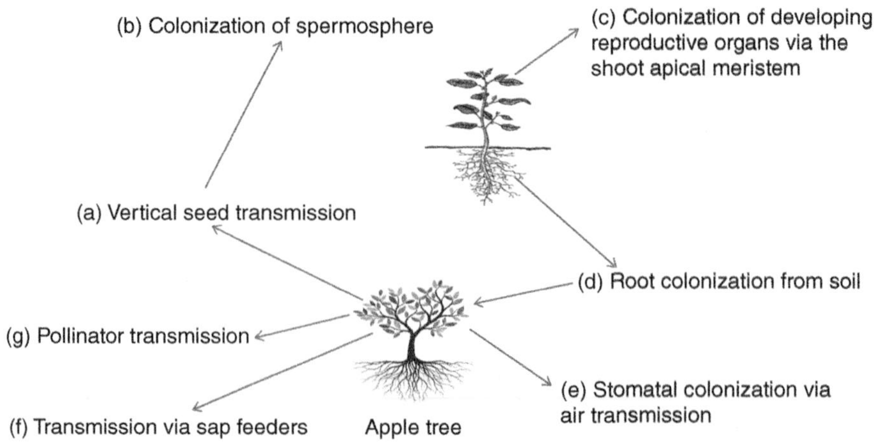

Figure 6.1 Transmission routes for bacterial endophytes across the life cycle of an apple tree. (a) Vertical transmission via seed; (b) Colonization of the spermosphere, depicted as the grey area surrounding the seed; (c) Colonization of developing reproductive organs via the shoot apical meristem as part of vertical transmission; (d) Colonization of root from soil; (e) Colonization of leaves though stomata after transmission via air; (f) Transmission via sap-feeders; (g) Transmission to flowers via pollinators. *Source:* Frank et al. [25]/MDPI/CC BY 4.0.

insides of plants. Bacteria applied to flowers, for example, can be passed through to the following generation [27]. For most plants, however, the relative significance of various transmission pathways is unknown.

To make things even more complicated, numerous endo- and epiphytic bacteria may colonize plants from above and below, inside and out [28, 29], emphasizing the necessity of bacterial mobility within the plant for bacterial transmission between plants. Bacteria have been found to colonize xylem vessels [30], which are considered to be the primary pathway for systemic colonization of inner plant tissues, albeit the process can take several weeks [16].

6.2 Vertical Transmission

6.2.1 Vertical Transfer via Seeds

The seed microbiome is gaining popularity, as seen by numerous recent reviews [31–34]. Bacteria have been found in surface-sterilized seeds of a variety of species, including crop plants such as alfalfa [35], rice [14, 36–40], maize [41, 42], tobacco [43], coffee [44], quinoa [45], common bean [46], grapevine [47], barley [48], and pumpkin [49], as well as wild plants such as the giant cardon cactus (*Pachycereus pringlei*) [50], annual ryegrass (*Lolium rigidum*) [51], various species of eucalyptus (Eucalyptus spp.) [52]. The South American tree *Anadenanthera colubrina* [53] and the Norway spruce (*Picea abies*) [54]. Bacteria have been found in the coat, endosperm, and embryonic tissues of the seed [27, 47, 55]. Truyens et al. [31] examined research on seed endophytes and discovered that bacteria present in seeds are mostly *Bacillus* and *Pseudomonas*, but also *Paenibacillus, Micrococcus, Staphylococcus, Pantoea*, and *Acinetobacter*.

Some seed endophytes have been shown to benefit their hosts. Indigenous endophytes in ryegrass, for example, may help release seed dormancy by producing cytokinins and interacting with bacterial and plant hormones [51]. Other seed endophytes may help plants germinate and develop in difficult environments by providing inorganic nutrients and fixing atmospheric nitrogen [50, 56]. Inoculation with cadmium (Cd) resistant endophytes isolated from seed protects plants from Cd toxicity [43], and seed endophytes can exhibit antifungal activities in vitro [57, 58]. In another research, surface-sterilization and antibiotic treatments to remove rice seed endophytes slowed seedling development compared to control seedlings [40]. Given such advantageous characteristics, it's possible that certain plants established mutualisms with bacteria that are vertically transferred by seed, guaranteeing continuous transmission of beneficial symbionts, similar to the defensive mutualisms between plants and fungal endophytes [59, 60]. Only one vertically transmitted plant–bacterium symbiosis has been described to our knowledge. In the following angiosperm genera *Ardisia, Pavetta, Psychotria*, and *Sericanthe*, leaf-nodulating nitrogen-fixing *Burkholderia* symbionts exist in every vegetative shoot tip and colonize each new leaf [61]. These bacteria are transported into the floral shoot tip, then the developing ovule's embryo sac, and finally the embryo's epicotyl, from whence they are contained in the seedling's shoot tip [61]. Due to mixed-mode transmission involving both vertical inheritance and horizontal transfers from the environment, and frequent host switching, there is no evidence of co-speciation between hosts and leaf-nodulating *Burkholderia* symbionts, despite the fact that they are obligatory and vertically transferred [62]. However, even in this case of verified vertical transmission, it is difficult to identify the symbiont in seeds, due to low levels of bacterial DNA [62]. The presence of bacteria in seed does not imply that they originated in the parent, and not all bacteria seen in seedlings will colonize seedlings. Seed-associated bacterial communities can reveal information about their origins based on their structure. Regardless of environmental factors such as soil type or geographic location, selection by the plant and/or bacterium, as in the case of leaf-nodulating bacteria, should result in long-standing associations and high similarity in seed communities within a plant species and across related plant species, regardless of environmental factors such as soil type or geographic location (i.e. despite the lack of host–microbe co-speciation, the leaf-nodulating symbionts consistently belong to the genus *Burkholderia*). If neutral mechanisms predominate in determining the assembly of a seed-associated community, however, we should expect more diversity among plant individuals, species, and locales. The greatest evidence for vertical endophyte transmission via seed comes from studies that show endophyte taxon overlap between seed and seedling, which is compatible with, but not conclusive, vertical transfer of endophytes via seed [52, 63, 64]. Additional investigations have found that the presence of certain strains in rice and maize persists over generations [58, 65], implying vertical transmission. And, at least in maize, there is evidence of long-term conservation in the seed endophyte community; seeds of genetically related maize hybrids have been found to host similar bacterial taxa [65], and a study using terminal restriction fragment length polymorphism of 16S rDNA revealed the presence of the same genera across several maize genotypes, as well as its ancestor teosinte [42]. A research on the invasive Johnsongrass (*Sorghum halepense*) found that plants grown aseptically from surface-sterilized seed acquired nitrogen from a source other than imported nitrogen, implying that diazotrophs had been transmitted vertically via seeds [56]. Some of the most recent research have investigated how seed microbiome community structure and diversity are affected by numerous variables, including emergence, host genotype, and location. Few endophyte operational taxonomic units (OTUs) were conserved across samples in *Brassica* spp. and common bean (*Phaseolus vulgaris*), and plant genotype did not appear to be an important driver of the bacterial seed endophyte community, implying that neutral processes control seed endophyte assembly in these plant species [66]. Leaf seed communities, on the other hand, appeared to be formed from non-soil environments, such as atmosphere or nutrient solution [67]. Seed endophytes can be transmitted vertically; however, the results of these research show that most endophytes colonize seeds horizontally. A research of *Arabidopsis thaliana* suggests that the plant may choose seed endophytes based on environmental stresses and pass them on to the next generation [68]. However, certain seed endophytes may be transmitted to the following coming generation. The patterns of host and seed endophyte interactions may not leave an evolutionary imprint due to such occasional vertical transmission of endophytes. The existence of identical 16S rRNA sequences between seeds from various genotypes, across seeds of successive generations, or between seed and seedling cannot confirm vertical transmission until strain-level information is known [31]. Furthermore, due to the low quantities of DNA in the seed-transmitted leaf-nodulating symbiosis [62], proving vertical transfer by DNA sequencing may be challenging. Bacteria can colonize seeds horizontally from the outside world via blooms fruit, or after seeds have been disseminated through soil or the germination environment. Microbes can be transported or moved from vegetative portions of the plants to the developing seed via vascular connections to the endosperm, and can also colonize a seed via pollen, according to pathogenic microbe studies [69]. Puente and co-workers found evidence of transfer of cardon cactus seed-borne rock-degrading endophytes [70], indicating that these endophytes play an essential role in their host's establishment on rock surfaces. Endophytic bacteria were found in seedlings sprouted from disinfected seeds' root cortex and vascular system, as well as in surface-disinfected fruit [70]. These bacteria, though, have not been identified and may not be the same as those found in cactus seeds. Similarly, in grapevine, there was overlap in endophyte taxa between seed and fruit, with mostly *Bacillus* spp. found inside berries between pulp cells and xylem, and along cell walls inside seeds [47]. The shoot apical meristem (SAM), which is made up of undifferentiated cells that give birth to all post-embryonic aerial organs [71], is another potential pathway. Tissues derived from the SAM, such as reproductive organs – and hence producing seeds – could pick up bacteria from the meristem. Transfer from the mother plant to the seedling would be ensured via this pathway. Because shoot tip meristems or embryos are frequently utilized as the starting material for tissue culture [72], bacteria from the shoot tip are frequently discovered initially. Poplar trees [73], Norway spruce [74], Scots pine (*Pinus sylvestris*) [75, 76], papaya [77], banana (*Musa* spp.) [78], sour cherry (*Prunus cerasus*) [79], pineapple (*Ananas comosus*), and orchid (*Oncidium flexuosum*) [80] are just a few examples. Many tissue culture or shoot tip endophytes have been identified inside plant cells [81–83], despite the fact that bacterial endophytes are typically found in intercellular gaps. Endophytes were identified inside cells of scale primordia, meristems, and around the resin ducts buds in undamaged Scots pine buds using in situ hybridization [75]. These endophytes' rRNA abundance was measured, and it was discovered to be greatest prior to bud expansion or differentiation [84].

6.2.2 Vertical Transfer via Pollen

As previously stated, one probable route for endophytes to enter seed is through male gametes. Bacteria have been found in and on the surface of pollen from many plant species [49, 85–89]. Pollen grains are exposed to the outside world and can be colonized horizontally from the atmosphere, as well as by pollinators and other animals. Vertical transmission occurs when bacteria in or on pollen come from within the plant and are transferred to seed and seedlings.

The isolation of the endophyte *Enterobacter cloacae* from surface-sterilized pollen of the Mediterranean pines Aleppo pine (*Pinus halepensis*), stone pine (*Pinus pinea*), and Turkish pine (*Pinus brutia*) suggests an origin within the parent plant, and the isolation of the same bacterial species from fertilized *P. brutia* ovules [86] may indicate vertical transfer of *Enterobacter* spp. in pines via pollen. A recent study described the abundance, community structure, diversity, and colonization pattern of bacteria associated with two wind-pollinated and two insect-pollinated species of plant: birch (*Betula pendula*), rye (*Secale cereal*), rape (*Brassica napus*), and autumn crocus (*Colchicum autumnale*) [85]. Bacteria were numerous [90–93] (cultivatable bacteria per gramme of pollen) and appeared as single cells, clusters, or thin biofilms on the outer surface. Because of species-specific differences in pollen shape, nutritional content, or antimicrobial peptides on the pollen coat, the communities vary considerably between plant species [85, 94]. Insect-pollinated species were more in common with each other than wind-pollinated species, suggesting that pollinators may have influenced pollen bacterial community compositions. A plant's microbiome can be horizontally transmitted by pollen in wind-pollinated species.

6.2.3 Horizontal Transmission

6.2.3.1 Colonization of Seed and Root via Soil

Soil has a variety of microorganisms that may live for thousands of years [95]. Endophytes' most significant source of inoculum is soil, which serves as a reservoir for both below- and aboveground plant microbiota [26, 96]. In a recent study of grapevine (*Vitis vinifera*) epi- and endophytic bacteria, the communities associated with leaves, blossoms, and grapes shared a higher proportion of taxa with soil communities than with each other, suggesting that both above- and belowground populations have a soil origin [97]. Victor Gallipe proposed that soil microorganisms penetrate plants more than a century ago [98], and Lorenz Hiltner identified a rich zone of bacteria around plant roots and created the name 'rhizosphere' a few years later [99]. Bacteria from the soil must first travel through the rhizosphere before becoming endophytes via the roots. As a result, some of the same biotic and abiotic variables that impact the organization and assembly of the rhizosphere microbiome are likely to influence the microbe populations transported to the root core.

6.2.4 Endophytic Colonization of the Spermosphere

The spermosphere, a short-lived and microbiologically active zone around a sprouting seed, where bacteria can help with germination [100, 101], is where colonization via the soil begins. Sugars, proteins, and fatty acids are all released by seeds as carbon (reviewed in Nelson [100]). These chemicals have the ability to influence the bacterial makeup of the soil surrounding the seed since they are a likely source of energy for microorganisms [90–92]. These early associations, which the plant may have chosen, might be crucial in establishing positive interactions [101]. The discovery that spermosphere bacteria vary from those found in the rhizosphere suggests that seeds select for microorganisms that may infiltrate the sown seed in a few of hours [100].

6.2.5 Colonization of the Root Endosphere via the Rhizosphere

The rhizosphere, or the millimetres of soil directly around the root, is teeming with bacteria that aid in host growth and development, nutrient acquisition, pathogen protection, systemic resistance, and growth under abiotic stress, such as toxic pollutants or drought [93, 102–105]. Plants cultivated hundreds of kilometres apart from comparable rhizosphere communities [106], demonstrating the plant's ability to control the rhizosphere microbiome's composition, commonly known as the 'rhizosphere effect' [107]. Rhizosphere community composition is influenced by plant genotype [108, 109], although the intensity of the rhizosphere effect varies by plant species [110], and in certain circumstances, soil type can override plant genotype in structuring bacterial communities in the rhizosphere [20, 21]. Plants produce rhizodeposits and exudates that attract soil microorganisms and boost rhizosphere populations around plant roots [107, 110–112], spending a considerable amount of fixed carbon and nitrogen in the process. Plant roots also influence the pH, soil structure, and oxygen availability of the surrounding soil, which influences their microbiota [113, 114]. Plants employ these selection skills at establishment, throughout their lives, and in reaction to other organisms, climate factors, soil microbial composition, and to particularly recruit microorganisms that aid in stress adaption [107, 114]. Plants that are attacked by diseases or insects, for example, can recruit defensive bacteria and boost microbial activity in the rhizosphere to inhibit pathogens [115]. Motility, chemotaxis, and quorum-sensing are all used by bacteria to cluster near plant roots or compete with adjacent bacteria [116–121]. Plant-mediated modulation of the rhizosphere and root endosphere microbiomes is currently being elucidated; it has been demonstrated that defence hormone signalling selects specific bacterial families for colonization from available microbial communities [122] and shapes the root microbiota during plant nutrient stress [123]. According to new research, the rhizosphere has a direct impact on the bacteria that colonize the rhizoplane (root surface) and then the inside as endophytes. Next-generation sequencing of the 16S rRNA gene was utilized in a study of the rice root microbiome to differentiate three separate but overlapping bacterial communities in the rhizosphere, rhizoplane, and endosphere at high resolution and depth [24]. Only a subset of rhizosphere bacteria was detected at the rhizoplane, suggesting that bacterial colonization of rice root surfaces is an active process in which plants select for specific microbial consortia or where some bacteria are stronger at penetrating the root surface, such as by building biofilms [24]. Depletion of additional rhizoplane taxa in the endosphere showed a second selection process for root entry, suggesting that while rhizoplane binding is a necessary requirement for root entry, it is not sufficient. Root exudates and chemical signals from the plant are presumably involved in each of these processes. Microbial penetrance into the endosphere was observed when seedling roots were transplanted from sterile medium to field soil, commencing within a day after transplanting and attaining steady state within two weeks [24]. Previous microscopy investigations back up a two-step hypothesis for endophytic invasion via the rhizoplane. Inoculation of *V. vinifera* with *Burkholderia* spp., for example, has been shown to be effective. Green fluorescent protein (gfp)-tagged strain PsJN showed chronological colonization first on root surfaces, then in root internal tissues, and lastly in xylem vessels of internodes and leaves [16], as well as inside immature berries in a follow-up research [124]. Endoglucanase and endopolygalacturonase, which degrade cell walls, were shown to be secreted by the bacteria, which might explain their penetration into the root endosphere [16]. Bacteria have been found at lateral root emerging locations and root tips in numerous investigations, suggesting that fractures in the root promote entrance and that bacteria enter through root tips [16, 125, 126]. All plant components can be colonized by soil bacteria, according to reports of root inoculants found in the xylem vascular system and aerial plant parts [16, 30, 124, 125, 127]. As revealed by a study of tobacco (*Nicotiana tabacum*) [128], not all root bacteria colonize the rest of the plant; some specialize on roots, or even various parts of roots (fine, secondary, main).

6.2.6 Entry into Aerial Tissues

Endophytes can reach aerial tissues through above-ground surfaces, such as stems, leaves, flowers (anthosphere), and fruits, in addition to entering roots via soil and migrating through xylem vessels (carposphere). The atmosphere, rain, soil, and pollinators or other insects are all potential bacterial source habitats. Bacteria appear to infiltrate leaves and stems through stomata, according to research. Plant diseases have been investigated through the stomatal pathway, while endophytes have received less attention. However, overlap in community composition between the leaf surface and the interior [129] implies that for certain foliar endophytes, the leaf surface constitutes the earliest step of colonization. As a result, we must investigate the transmission and dispersion pathways of plant surface-dwelling bacteria in order to comprehend endophytic colonization of aerial plant parts.

6.2.7 Aerial Dispersal of the Plant Microbiome

The phyllosphere, or aerial surface of plants, is one of the most common microbial homes on the globe [130, 131]. Plant epiphytes can colonize the plant from the inside; for example, seed-borne bacteria have been shown to inoculate plant surfaces via the sprouting seedling [66]. Many phyllosphere bacteria, on the other hand, are believed to be deposited by bioaerosols, which are minute particles such as bacteria, fungus, viruses, or pollen that are discharged into the atmosphere from terrestrial and marine habitats [132]. Bacteria thrive in the atmosphere, with concentrations ranging from 104 to 106 cells/m^3 [133]. Single bacterium cells, bacterial spores, and bacteria aggregated with dust particles, tiny plant debris, or pollen can all be found in bioaerosols [132]. Bioaerosol production is considered a key route of bacterial dispersion and migration on global and continental dimensions [134–136], since many bioaerosol bacteria remain alive and metabolically active in the atmosphere [137–140]. Dust storms may carry tiny particles hundreds of kilometres distant from the source [141–143]. Characterization of microbial communities in airborne dust deposition, for example, shows that dust storms may transport viable bacteria from Saharan soils to high-altitude regions in Europe, without the need for sporulation [144, 145]. Bioaerosols have been investigated for their capacity to disseminate plant- and animal diseases across great distances [146, 147], but their ability to disperse beneficial plant microbiomes has received less attention. Particles in the atmosphere are thought to be mostly produced by plant canopies, which cover the majority of Earth's terrestrial surface. Plant canopies were discovered to have greater quantities of viable bacteria than soil in early attempts to measure airborne bacteria [148], suggesting that plants canopies are a key source of airborne bacteria. Some of the bacteria in the near-surface atmosphere have been traced back to soil and plant surfaces in agricultural fields, suburban areas, and woods in more recent studies [149]. Similar to tropical soils and leaf surfaces, fungal communities in the Amazon rainforest's atmosphere were discovered to be quite similar [150]. 'A single drop of water splashed on the ground may aerosolize millions of germs', a recent research found [151]. A pattern of strong bacterial upward flow from dry leaves on sunny days and a high downward flux and lateral movement of bacteria on wet days has been observed in aerial dispersal of epiphytes from bean plants [152]. Rain and sprinkler watering have been shown to aid in the dispersion of plant diseases [153, 154]. Bacteria can disperse laterally to the surfaces of different leaves, individuals, or species of plants [155, 156], or upwards by air currents, where they can be transported by wind much faster and farther than other mechanisms, such as diffusion through soil, before being deposited again by precipitation or dry deposition [135]. Soil bacteria, on the other hand, have a limited dispersion that allows for regional endemism to develop [157, 158]. It's unclear how much of the endophytic microbiome gets distributed via the atmosphere, if any at all, but overlap between endophytic and airborne populations implies it's a possibility worth considering. The *Alphaproteobacterial* order *Rhodospirillales*, for example, dominated the viable airborne bacterial population above an Oregon mountainside [140], as well as the community deposited on California subalpine snow [159]. In California and Colorado, the *Rhodospirillales* have been regularly identified in the foliar endophytic community of various species of subalpine conifers [4, 160, 161], perhaps indicating interactions between the air and conifer canopy microbiome in subalpine areas.

6.2.8 Endophytic Leaf Colonization via Stomata

Bacterial plant pathogens can enter through all these openings, but the stomata dominate in number and are considered the main route of entry to the interior of plant aboveground parts [162–164]. The stomata are formed by a pair of guard cells that control the stomatal pore in response to light intensity, carbon dioxide concentration, and relative humidity. In addition, the stomata represent a first line of defence against bacterial pathogens, as the guard cells can sense microbe-associated molecular patterns (MAMPs) and close the stomatal pore in response, in a process mediated by phytohormones [165]. Bacteria colonizing the surface of leaves, unlike epiphytic fungi, are not known to penetrate the leaf cuticle [162, 166, 167], but studies on bacterial pathogens (and, to a lesser extent, endophytes) suggest that they may use openings in the plant epidermis, such as the stomata (openings in the aerial part of plants that allow and control gas exchange and water transpiration between the plant interior and the environment) [162, 166, 167]. In exchange, certain pathogens have developed virulence factors that inhibit MAMP-triggered stomatal closure [168–171] and effectors produced by type III secretion systems that either prevent or promote stomatal pore opening [172, 173]. Stomatal resistance has also been demonstrated against the human pathogen *Escherichia coli* O157:H7, implying that plants actively police not just plant pathogens but also human pathogens from entering the stomata. *Salmonella enterica serovar Typhimurium* SL1344S, another human pathogen, employs chemotaxis to move towards stomata, where it induces a brief stomatal closure but colonizes the apoplast, implying that it can escape or subvert plant immunity [174, 175]. These findings imply that stomatal response to phyllosphere bacteria varies based on bacterial MAMPs' natural variance [176]. If phyllosphere commensal or helpful bacteria are permitted to infiltrate the stomata and colonize the above-ground portions as endophytes, and if bacteria evolve methods to resist such responses, the issue arises as to whether they are recruited to do so. Bacteria can penetrate stomata and cause harm to the plant, according to some research (i.e. as endophytes). *Herbaspirillum seropedicae*, a nitrogen-fixing endophyte that promotes pineapple (*A. comosus*) development, has been shown to penetrate the stomata [177]. To begin with, large numbers of bacteria were detected on trichomes, epidermal cell wall junctions, and in the vicinity of stomata; this was followed by stomatal chamber penetration and colonization of leaf mesophyll intercellular spaces. While imaging natural bacteria on maize leaves, a similar finding of bacterium clusters around stomata was made [178]. On the basis of indirect evidence, bacteria were found on the top surface of leaves, indicating colonization of the host following stomatal penetration. This *Bacillus amyloliquefaciens* endophyte was discovered on the vanilla orchid (*Vanilla phaeantha*) shoot meristems, stomatal regions, and guard cells of *Amaranthus caudatus* [179].

A study of the diazotroph *Azospirillum brasiliense* provides more evidence for stomatal colonization by beneficial bacteria. When sprayed to maize and wheat by leaf spray inoculation (with controls for entrance via soil), the bacteria did not survive in the phyllosphere, but colonized the interior of leaves and stems, most likely via the stomata [180].

6.2.9 Floral Transmission of Bacterial Endophytes

Bacterial populations are found on the surfaces of various floral organs [2]. Bacteria have been found in nectar [181–184], petals [49, 184], pistils [49, 185], pollen [86, 87], and fruit [55, 124, 186] in both culture-dependent and independent investigations. As with leaf surfaces, bacteria that colonize floral surfaces may have the same origins (i.e. air, dust, wind, rain splash, and surrounding plants and soil, or from pollinators and other insect visiting flowers [187]). According to a study of apple blossom microbial populations, wind is likely a dispersion mechanism to and from the blooms [188]. It is unknown if environmental bacteria may colonize flowers on the endophytic level, but we do know from investigations of the apple and pear fire blight agent *Erwinia amylovora*, which infects flowers predominantly [189–191] that they are able to enter floral tissue. *Epiamomum epiphyticum* is a plant that grows on other plants. Amylovora infections appear on the stigma or hypanthium (where nectar is produced), and the bacteria enter the plant via nectarthodes [190, 192, 193]. Petals also have stomata, via which epiphytes may theoretically enter, although this has not been documented to our knowledge. It is known, however, that bacteria acquired horizontally via flowers may penetrate developing seeds and populate the progeny. Mitter and colleagues [27] sprayed a bacterial inoculant directly on the mother plant's blossom to transfer endophytes into maize, pepper, and soybean seeds. They were then able to identify the bacteria in the embryo/cotyledons, seedling's demonstrating that the bacteria may multiply and colonize the root and stem of the progeny. Flowers may provide two predictable transmission channels for the aerial tissue microbiome to plants: insect vectors and pollen in the case of wind-pollinated species. Flowers are visited by predators searching for prey [194], herbivores that feed on floral tissue [195], and insects looking for a partner [196], in addition to pollinators. Flowers, in fact, are invertebrate biodiversity hotspots, with concentrations ten to ten thousand times higher than adjacent foliage [197]. Pollinator visitation may impact floral microbiomes, according to surveys of flower and pollinator microbiomes [181, 198, 199]. Flowers may also function as hubs for pollinator bacterium transfer. Pollen from both insect-pollinated and wind-pollinated plants is home to a varied bacterial population, as previously mentioned. Pollen grains may be an efficient vector not only for vertical transmission of endophytes, but also for canopy-to-canopy horizontal transmission of plant microbiomes, because they are released in large quantities into the environment – Molina et al. [200] reported the release of up to 500 000 million grains for a single tree. Pollen-mediated microbe dispersion does not require viable pollen or fertilization, and it can occur over long distances. Tree pollen has been shown to travel 600 km for viable pollen and 3000 km for possibly viable pollen [201]. Pollen transmission may be used by fungal plant diseases, according to some evidence [202].

6.2.10 Endophyte Transmission by Plant-Feeding Insects

Sap-feeding insects of the insect order *Hemiptera*, such as leafhoppers, planthoppers, and psyllids, can spread plant disease, including viruses and bacteria like *Phytoplasma* (an obligatory plant pathogen) and *Xylella* [203–205]. These insects have piercing–sucking mouthparts that allow them to penetrate phloem or xylem cells and suck out the contents, which can occasionally spread plant diseases. Interestingly, *Cardinium*, an intracellular symbiont of many sap-feeding insects, can be horizontally transmitted between different phloem sap-feeding insect species through plants [170], and *Wolbachia* associated with the phloem-feeding whitefly (*Bemisia tabaci*), can be horizontally transmitted via multiple species of plants, where it has been visualized in situ in phloem vessels and 'reservoir' spherules along the phloem [206]. These results suggest that *Cardinium* and *Wolbachia* can reside in the plant as endophytes, at least temporarily, and that non-pathogens can be transmitted between plants via sap-feeders. The possible transfer of whole population endophytes was examined in a recent research of the American grapevine leafhopper *Scaphoideus titanus*, a phloem-feeder [207]. In the study, insects were given the option of feeding on source plants grown in normal circumstances and containing typical microbial communities, or axenically micropropagated grapevine plantlets. The experiment was repeated four times, demonstrating that sink plants were colonized by a microbiome remarkably similar to that of the source plant, implying that most, if not all, endophytes were transferred this way. The endophytic population was discovered throughout the sink plant, including the roots [207], not simply in the stems where the insects feed. As a result, sap-feeding insects can spread the beneficial or commensal plant microbiome within and between plant species. Beneficial phloem endophytes, such as the *Bacillus pumilus* strain isolated from lodgepole pine (*Pinus ponderosa*) and discovered to be antagonistic towards a fungal symbiont of the mountain pine beetle (*Dendroctonus ponderosae*) [6], might be transferred in this way. However, many sap-feeders appear to have very sparse bacterial populations [205]. Sap-feeders are important vectors of plant microbiomes, although additional study is needed to confirm this.

6.3 Endophyte Genomes and Metagenomes

6.3.1 Genome Analysis

Endophyte–host interactions are poorly understood on a physiological level. To effectively manipulate their mutualistic relationship, it is crucial to identify, isolate, and characterize the genes involved in such beneficial interactions. Genome study of the endophyte has offered a new method for examining the endophytism in detail and identifying the characteristics required to harbour plants as a home. These genes include those involved in nitrogen fixation, phytohormone synthesis (IAA, GA, etc.), mineral acquisition (Fe, P, etc.), stress tolerance and adhesion, among others [208–210]. As a result of these characteristics, microorganisms play an important role in nutrient cycling and can invade the plant endosphere. The genetic characteristics that directly or indirectly impact different bioactivities as well as colonization preferences have been discovered through whole genome research of endophytic bacteria. It helps identify specific genes involved in antibiotic resistance mechanisms, antibiotic synthesis, plant growth promotion, endophytic secretory system, surface attachment and insertion elements, transport system, and other metabolic systems. Endophyte ecology and evolution have been better understood thanks to such research. The presence of genes producing *N*-acyl homoserine lactone synthases and hydrolases, as well as hyperadherence factors and fusaric acid resistance proteins, highlights *Pantoea ananatis'* biotechnological potential [211]. Endophytic members of the fungal order Sebacinales have piqued interest due to their ability to promote plant development and stress tolerance [212]. *Pirifomospora indica* (order sebacinales) is a model organism for studying symbiotic relationships. *P. indica's* genome sequence study has indicated its potential as a plant probiotic agent [213]. Many bacterial and fungal endophytes have had their genomes sequenced (Table 6.1), and the list is growing. Plant–microbe and microbe–microbe interactions are studied using endophyte genomes as model systems. Individual genome sequences can aid data processing in plant-associated microorganism metaomics (metagenomics, transcriptomics, and proteomics) investigations. Endophytic colonization affects the evolution of the host plant genome [232]; however, Zgadzaj et al. [233] found that host genetic variables influence the establishment of both the endophyte and the symbiont within root-nodules, thus host genome studies are also necessary for a fuller picture.

6.3.2 Multigenome Analysis

Comparative multigenome analysis is very useful in determining the genetic and metabolic diversity of similar or related microbes that interact with plants and animals in various ways. The colonization of a host plant by a microbe as a

Table 6.1 Endophytes whose genomes have been sequenced.

Endophyte	Reference
Microdochium bolleyi J235TASD1	[214]
Rhodotorula graminis WP1	[209]
Penicillium aurantiogriseum NRRL 62431	[215]
Xylona heveae	[216]
Shiraia sp. slf14	[217]
Enterobacter cloacae subsp. Cloacae strainENHKU01	[65]
Stenotrophomonas maltophilia RR-10	[218]
Bacillus pumilus	[219]
Rhizobium sp. strain IRBG74	[220]
Variovorax paradoxus	[221]
Enterobacter sp. strain638	[222]
Pantoea agglomerans	[223]
Bacillus subtilis	[224]
Pseudomonas fluorescens PICF7	[210]
Raoultella terrigena R1Gly	[225]
Staphylococcus epidermidis strains SE2.9; 4.6; 4.7 and 4.8	[226]
Paenibacillus dauci	[227]
Pantoea ananatis AMG521	[211]
Bacillus thuringiensis KB1	[228]
Stenotrophomonas Strain 169	[229]
Burkholderia sp. Strain MS389	[230]
Bacillus cereus T4S	[231]

symptomless endophyte or as a pathogen appears to be separated by a fine line. The genome of *Piriformospora indica* has been compared to the genomes of fungi from various classes, revealing the presence of genes related to both saprotropism and biotropism. Surprisingly, the bacterium has all of the necessities for both life modes, therefore there must be some external forces influencing the microbe's decision. The genetic factors responsible for this differentiation can be discovered by comparing the genomes of closely similar animals living in various microbiomes. Differences in metabolic, secretory, transport, and surface attachment proteins are primarily responsible for habitat selection, according to comparative genomics research. According to Monteiro et al. [234], lipopolysaccharide and adhesins are probable molecular components responsible for the phenotypic differences between closely related species when colonizing a host plant as a symbiotic endophyte or as a phytopathogen. The distribution of key genes linked to signalling, surface attachment, secretion, and transport differed between the two *Klebsiella pneumoniae* strains, revealing differences in the preferred lifestyles of Kp342 as a plant endophyte and MGH78578 as a human pathogen [208]. These findings imply that Kp342 is well-adapted to evade plant defensive responses and establishes itself successfully inside a plant. Using suppression subtractive hybridization, Monteiro et al. [234] discovered the genes responsible for variations in colonization patterns between two closely related species, endophytic *H. seropedicae* Smr1 and phytopathogenic *Herbaspirillum rubrisubalbicans* M1 (SSH). SSH is a useful method for determining the genetic diversity of microorganisms [235, 236]. SSH libraries are used to find DNA segments that are present in one species but not in another. Genome comparisons of endophytic and non-endophytic isolates show characteristics that are believed to be required for establishing and sustaining plant microbial relationships. Furthermore, the research aids in understanding the non-culturable characteristic of many endophytes [237]. Dinsdale et al. [237] used a metagenomics method to uncover the potential of uncultured microbial communities [endophytes], providing information beyond the genetic information of particular species. This method involves extracting DNA from the entire population and analysing it for gene content. Metagenome analysis to deduce the potential functional features of rice root endophytes. They discovered various metabolic adaptations of endophytes to their microhabitat, suggesting that the endophyte community has a lot of promise in terms of plant growth promotion, stress tolerance, disease management, and bioremediation. Microbial communities living in many settings have preserved functional diversity. The functional diversity of endophytes from the same or distinct host plants may be studied using a comparative metagenomics method. To explain the differences in functional potential of nine distinct microbiomes, Dinsdale et al. [237] employed a comparative metagenomics method. Next-generation sequencing (NGS), a type of high-throughput sequencing, has made metagenomic research a lot easier and accelerated the characterization of microbiomes to new levels [238]. It has given researchers a versatile tool for studying DNA sequences from environmental samples in a fast and cost-effective manner [239]. The use of 454 sequencing has made it easier to characterize fungal populations. Toju et al. [240] studied the diversity of root-associated fungus in a Japanese temperate forest. Using 454 pyrosequencing methods, they revealed the coexistence of mycorrhizal fungi and endophytic fungi in the roots of several plant species. Coexistence would very certainly include intricate interactions between the two ecotypes, which might be investigated further utilizing metaproteomics, metatranscriptomics, or metaproteogenomics. Before adopting NGS technology for experimental investigations, one should be aware of its limitations [241, 242]. One of the primary constraints of metagenomic investigations is the high ratio of sequences with no homolog in public databases. To some degree, genome sequencing investigations of strains obtained from the same habitat might circumvent this restriction.

6.3.3 Metagenomics

Metagenomics involves analysis of sequence information from microbial members of various ecological communities. It evades the need for isolation and cultivation of individual species. In order to understand and manipulate the contribution of endophytes to the host plant, it is important to uncover their metabolic potential and beneficial characteristics. However, determination of endophytic microbial functions is impeded by the non-culturable feature of many endophytes [237]. Coexistence would very certainly include intricate interactions between the two ecotypes, which might be investigated further utilizing metaproteomics, metatranscriptomics, or metaproteogenomics. Before adopting NGS technology for experimental investigations, one should be aware of its limitations [241, 242]. One of the primary constraints of metagenomic investigations is the high ratio of sequences with no homolog in public databases. To some degree, genome sequencing investigations of strains obtained from the same habitat might circumvent this restriction. Dinsdale et al. [237] found that a metagenomics method is very useful in unlocking the potential of uncultured microbial communities (endophytes), providing information beyond the genetic information of particular species. This method involves extracting DNA from the entire population and analysing it for gene content. Based on metagenome research, Sessitsch et al. [243] deduced the probable functional features of rice root endophytes. Many metabolic adaptations of endophytes were documented to their microhabitat, indicating a great potential for plant growth promotion, stress resistance enhancement, pathogen management, and bioremediation by the endophyte community. The functional diversity of microorganisms living in diverse settings has been maintained. It is possible to examine functional diversity among endophytes of the same or different host plants using a comparative metagenomics method. To characterize the functional potential of nine distinct microbiomes, Dinsdale et al. [237] employed a comparative metagenomic method. Microbiome characterization investigations have been accelerated by the use of next-generation sequencing (NGS), a high-throughput sequencing technique [238]. DNA sequences from environmental samples may now be studied quickly and affordably using this technology. One of the most convenient ways to characterize fungal communities is through the use of 454 sequencing. In a temperate forest in Japan, Toju et al. [240] studied the community composition of root-associated fungus. Using 454 pyrosequencing methods, they revealed the coexistence of mycorrhizal and endophytic fungi in the roots of several plant species. Coexistence would very certainly include intricate interactions between the two ecotypes, which might be investigated further utilizing metaproteomics, metatranscriptomics, or metaproteogenomics. Before adopting NGS technology for experimental investigations, one should be aware of its limitations [241, 242]. One of the primary constraints of metagenomic investigations is the high ratio of sequences with no homolog in public databases. To some degree, genome sequencing investigations of strains obtained from the same habitat might circumvent this restriction.

6.3.4 Advanced Fundamental Research on Microbe Interactions in the Endosphere

After infection with *Fusarium oxysporum* f.sp.*pisi*, endophytic *Bacillus pumilus* strain SE34 caused ultrastructural alterations in pea (*Pisum sativum*) plants [244]. The deposition of callose and the influx of phenolic chemicals strengthened the cortical and epidermal cell walls. Only when the pathogen infected pre-bacterized plants did cell wall changes appear. The bacterial endophyte's so-called 'priming action' was required to prevent the infection from reaching the roots' core stele. The Gram-negative bacteria endophyte *Serratia plymuthica* R1GC4 elicited ultrastructural changes that protected against invasion by the oomycete pathogen *Pythium ultimum* [245]. Callose deposition in cell walls in response to pathogen penetration and the buildup of osmiophilic material in infected areas provided protection. The endophytic bacterial strain *P. fluorescens* PICF7 colonized olive (*Olea europaea* L.) roots, causing systemic resistance to *Verticillium dahliae* Kleb. pathogens not only in the root system, but also in the aerial portions of the host plant [246]. Transcriptional activation of genes involved in phytohormone production, phenylpropanoids biosynthesis, and the generation of oxidases and reductases were among the plant defensive responses. Both bacterial (*Pseudomonas syringae*) and fungal (*F. oxysporum*) pathogens influence the host plant's protein production, according to proteomics research. Protein fingerprints for both pathogens vary considerably between virulent and avirulent strains. In all experiments, virulent strains caused greater superoxide dismutase production in plants than attenuated ones. Plants produce super oxide dismutase, which inhibits the oxidation of cellular components. Plant defence has been shown to be stimulated by effector molecules generated by bacterial and fungal pathogens. Resistance to a variety of bacterial, fungal, oomycetal, and nematode diseases was induced by the colonization of plant roots by arbuscular mycorrhizal (AM) fungi [247]. In this regard, it's noteworthy that AM fungi suppress salicylic acid-dependent responses in plants during initial colonization to prevent plant defence responses from preventing further colonization of the hyphae, and then stimulate jasmonate-dependent responses later in infection to increase resistance to necrotrophic pathogens and herbivores [247]. The SYM route is used by bacteria (nitrogen-fixing *Rhizobium* species) and AM fungus in higher plants. AM fungus (Myc factors) and *Rhizobium* species (Nod factors) produce lipochitooligosaccharides, which are detected by plants through LysM receptor kinases. Mycorrhiza fungus symbiotic interactions cause changes in bacterial and fungal groups in the mycorrhizosphere [248]. AM fungi interact with several types of Proteobacteria, Firmicutes, Actinobacteria, and fungi [249], resulting in synergism in plant growth stimulation [250]. The development of biofilms on fungal hyphae by these so-called 'mycorrhiza helper bacteria' was thought to be stimulated by fungal synthesis of particular compounds such as trehalose [251]. It has previously been established that bacterial impacts on AM fungal colonization and plant interactions are particular rather than universal [252]. The effects of *P. fluorescens* SBW25 and *Paenibacillus brasiliensis* PB177 on *Glomus mosseae* and *Glomus intraradices* colonization of wheat (*Triticum aestivum*) roots were distinct. Only *P. brasiliensis* PB177 promoted colonization of wheat roots by *G. intraradices*, but both strains had favourable impacts on *G. mossae* colonization. *G. intraradices* in combination with *P. fluorescens* PB177, on the other hand, had a synergistic effect on *Microdochium nivale*, a fungal pathogen for wheat, resulting in higher shoot dry weight in infected plants treated with *G. intraradices* and *P. fluorescens* PB177 compared to control plants treated with *M. nivale* alone [253]. The colonization of the symbiotic fungus *Laccaria bicolor* was improved by 19 *Pseudomonas* strains isolated from the *Populus deltoides* microbiome [254]. When fungi are present as endophytes in aboveground portions of plants, selectivity for bacterial populations arises. *Epichlo coenophiala*, which was found in tall fescue (*Schedonorus arundinaceus*) plants, promoted the growth of *Burkholderia ambifaria*, epiphytic bacteria that were able to use the fungus's loline alkaloid as a food source on infected leaves [255]. This bacterium, however, was not found in *E. coenopliala*-free plants growing near infected fescue plants (*Festuca arundinacea*). Because of the presence of the loline alkaloid, the metacommunity of plants growing near infected fescue plants experienced so-called 'niche partitioning'. Endophytic fungus, according to recent studies, can alter nutrition, water, and sugar levels, hormone signalling, metabolic pathways, and oxidative balances in plants, therefore affecting bacteria at various sites inside or around plants [256–258].

6.3.5 Fungal Hyphae as Vehicles for Bacterial Colonization of the Endosphere

Bacteria that live in close proximity to fungus benefit not only from the nutrients supplied by fungi, but also from the physical protection and transportation provided by fungal hyphae to locations that are inaccessible to bacterial cells alone. Leveau and Preston [259] proposed three types of bacterial–fungal interactions: (i) necrotrophic interactions; (ii) extracellular biotrophic interactions; and

(iii) endocellular biotrophic interactions. Because the first contact causes fungal cell death, the latter two are the more relevant for bacterial penetration of plants. Planktonic, mixed-biofilm, and intrahyphal colonization are all examples of physical interactions between bacteria and fungus [259]. Different types of connection, including as competition, adhesion, signalling, secretion, attraction, metabolic conversion, and even genetic exchanges, occur in these relationships [259]. Only intrahyphal colonization by bacteria and mixed biofilm development are explored further since physical interaction between bacterial and fungal cells is most significant in the context of this article.

6.3.6 Bacterial Intrahyphal Colonization

The cytoplasm of spores and mycelium of the AM fungus *Gigaspora margarita* included a bacterium-like creature [260]. This bacterium shares a close relationship with the taxa *Burkholderia* and *Ralstonia*, according to taxonomic analysis [261]. The symbiont, known as *Candidatus Glomeribacter gigasporarum*, has a tiny genome and is metabolically dependent on the host environment, as well as an obligatory endosymbiotic lifestyle, according to genomic research. This organism's ability to generate vitamin B12, which has been identified in the metagenomes of mixed-algae biofilms and lung lichen [259–262], is thought to be an important nutrition for the fungal host. Due to its intracytoplasmic lifestyle, *Candidatus G. gigasporarum* is limited to its fungal host, and it is unlikely that this bacterium would spread to other endosphere locations following the colonization of the plant by AM fungi. The ability of culturable bacteria from the Burkholderia clade (*Burkholderia cepacia, Burkholderia pseudomallei*, and *Bacillus vietnamensis*) to penetrate the spores of the AM fungus *Gigaspora decipiens* has been proven [263–265]. Proteobacteria (alpha, beta, and gamma) and Firmicutes were found in the hyphae of fungal endophytes [266]. The endobacterial strain *Agrobacterium tumefaciens* was discovered in *P. indica* (order Sebacinales) [267]. Endobacteria may colonize plants by employing their fungal host as a vector for transfer to the plants. Given that some of these endobacteria are opportunistic pathogens to plants or people [266, 267], this route of bacterial transfer to the endosphere must have implications for plant and food production.

6.4 Bacteria and Fungi in Mixed Biofilms in Plants

Wilting is caused by the formation of bacterial extracellular polymeric substances (EPS) inside plants, which is caused by vascular pathogens that impede the movement of water and nutrients to higher regions of the plant [268, 269]. Proteobacteria (*A. tumefaciens, E. amylovora, Pantoea stewartii, Ralstonia solanacearum,*

P. syringae, and *Xanthomonas campestris*) and Actinobacteria (*Clavibacter michiganensis*) are among the bacterial phytopathogens that produce EPS. The easI and esaR quorum-sensing systems were found to be important for colonization of secondary cell wall thickening in the protoxylem of sweet corn (*Zea mays*) [270], and precise regulation of EPS production in *Pantoea stewartii* (subspecies stewartii) was found to be important for colonization of secondary cell wall thickening in the protoxylem of sweet corn (*Z. mays*). In *Xylella fastidiosa*, disruption of genes involved in EPS synthesis (homologous to gumD and gumH in *X. campestris*) led in avirulent mutants unable to move inside plants and whose transmission to new plants through insect vectors was hampered [270]. Biofilms require bacterial EPS to function properly. Bacteria and fungus have been discovered in the same biofilm in a variety of settings, including food, medical equipment, and soil [271]. Cell clustering on the surface of hyphae of the AM fungus *G. intraradices* was caused by the production of EPS by a mutant derivative of *P. fluorescens* strain CHA0 [272]. Biofilm development on fungal hyphae is necessary for bacterial translocation to plant interior tissues, but biofilm creation by endophytes is also important for effective apoplast colonization. In a maize model, *S. enterica* cells were shown to form biofilms on *Aspergillus niger* hyphae [273]. Plants were co-inoculated with both species, which resulted in a greater reduction in plant growth than either species alone. When flag leaves were co-injected with the fungus, no *S. enterica* colony-forming units were identified, and *S. enterica* colonies were only found when the bacterium was inoculated alone. Both partners are required for a considerable decrease in plant development; however, the fungus prevented the bacteria from colonizing the plant [274]. As a result, this example demonstrates that the paradigm of fungal endophytes functioning as vehicles for non-endophyte bacteria colonization is not universal for all bacteria–fungi pairings.

Plant-associated bacteria, including endophytes, from various taxonomic clades have been shown to colonize plant surfaces and occasionally form biofilms, including *A. brasilense* [275], *Methylobacterium mesophilicum* [276], *Rhizobium leguminosarum* [277], *Gluconacetobacter diazotrophicus* [278], and *H. seropedicae* [279]. In all cases, EPS generation was shown to be significant in the creation of biofilms, albeit in the case of *H. seropedicae*, it was only important for the formation of a biofilm on an inert (glass) surface and not on maize (*Z. mays*) roots [243]. Balsanell et al. [279] presented a scenario in which bacterial cells adhere to the maize root surface only weakly and do not form a biofilm. Individual cells would be free to migrate to find optimal penetration sites on root surfaces if they were loosely attached to the root surface. This would have ecological consequences for endophytes. The relatively high number of hits with genes encoding flagella monomers and phosphodiesterase A (responsible for the degradation of cyclic-di-GMP and, thus, leading to transitions of bacteria from biofilm to planktonic forms) [280] in the rice metagenome [279] may explain why biofilm formation may not be an advantageous property for endophytes in all circumstances. Although biofilm development may be an effective strategy for surviving the frequently harsh conditions found within the plant's internal environment, transitioning to more mobile planktonic forms may be required for the occupancy of new areas, such as vertical transmission through seeds [281]. Bacteria embedded in an EPS matrix are more drought resistant and have a longer lifespan in soils [282]. Antimicrobial drugs such sodium dodecyl sulphate and tobramycin were more resistant to bacteria in a mixed-species biofilm including *Pseudomonas aeruginosa, Pseudomonas protegens,* and *K. pneumoniae* than bacteria in single-species biofilms [283]. *Microbacterium oxydans, Paenibacillus amylolyticus, Stenotrophomonas rhizophila*, and *Xanthomonas retroflexus*, all of which were isolated from the same soil, coexisted in a mixed biofilm [284]. When grown in monoculture, these species produced greater biofilm mass, but one species, *X. retroflexus*, dominated in cell number over time. Biofilms containing several species are more prevalent in nature than biofilms containing only one species. Cooperation, competition, communication, and reciprocal interactions at various gene expression and physiological levels are all suspected to occur in mixed biofilms [285]. Different species can be geographically dispersed over the biofilm matrix, but when working together, they are more resistant to host defensive systems than when working alone. Antibiosis, which causes nonlethal cell damage, has been shown to be a significant trigger in biofilm development, as evidenced by the treatment of *P. aeruginosa* cells with narrow- and broad-spectrum antibiotics [286]. When bacteria compete with other microbial species, biofilms are formed. They collaborate with other microbial consortia members to defend themselves from external influences such host defensive responses, and they metabolically complement each other [287]. Endophyte survival is aided by biofilms, which are microenvironments in plants.

The biofilm matrix may be thought of as a molecular filter that keeps macromolecules such extracellular enzymes, other proteins, and DNA together [288]. It's unclear where the DNA in biofilms comes from, whether it's created from scratch or released from lysed cells. It's worth noting, though, that phages have been discovered in biofilms [222, 289], suggesting that biofilms might be a hotspot for gene transfer. The presence of a plasmid with aberrant codon usage in plasmid backbone genes in the endophyte *Enterobacter* sp. 638 [290], mobile genetic elements in the *P. ananatis* accessory genome [289] and the bacterial origin of PKS genes in the genome of the fungal endophyte *Pestalotiopsis fici* [290] all indicate that horizontal gene transfer events in endophytes have occurred in the past. We can't imagine the huge potential of biofilms in genetic recombination in microorganisms in diverse natural settings, given their cellular compactness and the newly discovered unanticipated wide transfer of plasmids to soil community members [288].

6.5 Conclusion and Future Perspectives

Before mass manufacturing can begin, detailed study on a specific population of endophytes functioning in a host is necessary, which frequently necessitates the use of modern technologies. Plant-specific endophyte inoculum dosages are also needed, which assist will reduce bulk production while simultaneously increasing productivity by lowering our reliance on synthetic fungicides, insecticides, and fertilizers. Acceptance of endophytes in integrated pest control will be aided by the development of endophytes that can be sprayed like most chemical pesticides. For more research on how endophytes colonize plants and plant–microbe interactions, future studies will need to take into consideration the development of genomic and metabolomics techniques. The chemicals generated by endophytes, as well as their role in disease reduction, need to be studied further. This will aid in the development of effective markers for a variety of essential and unique biocontrol agents, as well as the evaluation of the impacts of plant genotypes, innate microbial communities, and, most crucially, the environment. This method will also aid in the discovery of novel endophytes with essential characteristics. It is critical to do molecular research on these endophytes in order to improve medication development. Furthermore, a metagenomics research will be critical in demonstrating the variety of endophytes and the activities they may perform by a comprehensive examination of their genes. Molecular biology techniques may be used to isolate and identify the many types of genes found in biosynthetic pathways, which will lead to the discovery of new bioactive chemicals both commercially and in the laboratory. Future research should concentrate on the biochemical pathways that may be responsible for endophytes secreting a variety of essential bioactive chemicals. Potential research might also look at the production of endophytic nanoparticles, which could aid plant growth. The use of gene transfer can also be used to detect more efficient species. Manipulation of genes can aid host plants in generating new characteristics such as phytoremediation and herbicide resistance, among others, that can better control metabolism. There is no such thing as a successful microbial technology unless it has been marketed. One of the constraints in large-scale production is the specialization of endophytes inside a plant.

References

1 Hallman, J., Quadt-Hallmann, A., Mahaffee, W.F., and Kloepper, J.W. (1997). Bacterial endophytes in agricultural crops. *Can. J. Microbiol.* 43: 895–914. https://doi.org/10.1139/m97-131.

2 Hardoim, P.R., van Overbeek, L.S., Berg, G. et al. (2015). The hidden world within plants: ecological and evolutionary considerations for defining functioning of microbial endophytes. *Microbiol. Mol. Biol. Rev.* 79: 293–320. https://doi.org/10.1128/MMBR.00050-14.

3 Naylor, D., DeGraaf, S., Purdom, E., and Coleman-Derr, D. (2017). Drought and host selection influence bacterial community dynamics in the grass root microbiome. *ISME J.* https://doi.org/10.1038/ismej.2017.118.

4 Moyes, A.B., Kueppers, L.M., Pett-Ridge, J. et al. (2016). Evidence for foliar endophytic nitrogen fixation in a widely distributed subalpine conifer. *New Phytol.* 210: 657–668. https://doi.org/10.1111/nph.13850.

5 Doty, S.L., Sher, A.W., Fleck, N.D. et al. (2016). Variable nitrogen fixation in wild populus. *PLoS One* 11: e0155979. https://doi.org/10.1371/journal.pone.0155979.

6 Adams, A.S., Six, D.L., Adams, S.M., and Holben, W.E. (2008). In vitro interactions between yeasts and bacteria and the fungal symbionts of the mountain pine beetle (Dendroctonus ponderosae). *Microb. Ecol.* 56: 460–466. https://doi.org/10.1007/s00248-008-9364-0.

7 Elbeltagy, A., Nishioka, K., Sato, T. et al. (2001). Endophytic colonization and in planta nitrogen fixation by a Herbaspirillum sp. isolated from wild rice species. *Appl. Environ. Microbiol.* 67: 5285–5293. https://doi.org/10.1128/AEM.67.11.5285-5293.2001.

8 Bright, M. and Bulgheresi, S. (2010). A complex journey: transmission of microbial symbionts. *Nat. Rev. Microbiol.* 8: 218–230. https://doi.org/10.1038/nrmicro2262.

9 Herre, E.A., Knowlton, N., Mueller, U.G., and Rehner, S.A. (1999). The evolution of mutualisms: exploring the paths between conflict and cooperation. *Trends Ecol. Evol.* 14: 49–53. https://doi.org/10.1016/S0169-5347(98)01529-8.

10 Moran, N.A. (2006). Symbiosis. *Curr. Biol.* 16: R866–R871. https://doi.org/10.1016/j.cub.2006.09.019.

11 Schardi, C.L. (2001). Epichloë festucae and related mutualistic symbionts of grasses. *Fungal Genet. Biol.* 33: 69–82. https://doi.org/10.1006/fgbi.2001.1275.

12 Wilkinson, D.M. and Sherratt, T.N. (2001). Horizontally acquired mutualisms, an unsolved problem in ecology? *Oikos* 92: 377–384. https://doi.org/10.1034/j.1600-0706.2001.920222.x.

13 Foster, K.R. and Wenseleers, T. (2006). A general model for the evolution of mutualisms. *J. Evol. Biol.* 19: 1283–1293. https://doi.org/10.1111/j.1420-9101.2005.01073.x.

14 Hardoim, P.R., Hardoim, C.C., Van Overbeek, L.S., and Van Elsas, J.D. (2012). Dynamics of seed-borne rice endophytes on early plant growth stages. *PLoS One* 7: e30438. https://doi.org/10.1371/journal.pone.0030438.

15 Ma, Y., Rajkumar, M., Luo, Y., and Freitas, H. (2011). Inoculation of endophytic bacteria on host and non-host plants – effects on plant growth and Ni uptake. *J. Hazard. Mater.* 195: 230–237. https://doi.org/10.1016/j.jhazmat.2011.08.034.

16 Compant, S., Reiter, B., Sessitsch, A. et al. (2005). Endophytic colonization of Vitis vinifera L. by plant growth-promoting bacterium Burkholderia sp. strain PsJN. *Appl. Environ. Microbiol.* 71: 1685–1693. https://doi.org/10.1128/AEM.71.4.1685-1693.2005.

17 Khan, Z., Guelich, G., Phan, H. et al. (2012). Bacterial and yeast endophytes from poplar and willow promote growth in crop plants and grasses. *ISRN Agron.* 2012: 890280. https://doi.org/10.5402/2012/890280.

18 Carroll, G. (1988). Fungal endophytes in stems and leaves: from latent pathogen to mutualistic symbiont. *Ecology* 69: 2–9. https://doi.org/10.2307/1943154.
19 Schlaeppi, K. and Bulgarelli, D. (2015). The plant microbiome at work. *Mol. Plant-Microbe Interact.* 212: 212–217. https://doi.org/10.1094/MPMI-10-14-0334-FI.
20 Bulgarelli, D., Rott, M., Schlaeppi, K. et al. (2012). Revealing structure and assembly cues for Arabidopsis root-inhabiting bacterial microbiota. *Nature* 488: 91–95. https://doi.org/10.1038/nature11336.
21 Lundberg, D.S., Lebeis, S.L., Paredes, S.H. et al. (2012). Defining the core Arabidopsis thaliana root microbiome. *Nature* 488: 86–90. https://doi.org/10.1038/nature11237.
22 Peiffer, J.A., Spor, A., Koren, O. et al. (2013). Diversity and heritability of the maize rhizosphere microbiome under field conditions. *Proc. Natl. Acad. Sci. U. S. A.* 110: 6548–6553. https://doi.org/10.1073/pnas.1302837110.
23 Schlaeppi, K., Dombrowski, N., Oter, R.G. et al. (2014). Quantitative divergence of the bacterial root microbiota in Arabidopsis thaliana relatives. *Proc. Natl. Acad. Sci. U. S. A.* 111: 585–592. https://doi.org/10.1073/pnas.1321597111.
24 Edwards, J., Johnson, C., Santos-Medellín, C. et al. (2015). Structure, variation, and assembly of the root-associated microbiomes of rice. *Proc. Natl. Acad. Sci. U. S. A.* 112: E911–E920. https://doi.org/10.1073/pnas.1414592112.
25 Frank, A., Guzmán, J., and Shay, J. (2017). Transmission of bacterial endophytes. *Microorganisms* 5: 70. https://doi.org/10.3390/microorganisms5040070.
26 Hardoim, P.R., van Overbeek, L.S., and Elsas, J.D. (2008). Properties of bacterial endophytes and their proposed role in plant growth. *Trends Microbiol.* 16: 463–471. https://doi.org/10.1016/j.tim.2008.07.008.
27 Mitter, B., Pfaffenbichler, N., Flavell, R. et al. (2017). A new approach to modify plant microbiomes and traits by introducing beneficial bacteria at flowering into progeny seeds. *Front. Microbiol.* 8. https://doi.org/10.3389/fmicb.2017.00011.
28 Bai, Y., Müller, D.B., Srinivas, G. et al. (2015). Functional overlap of the Arabidopsis leaf and root microbiota. *Nature* 528: 364–369. https://doi.org/10.1038/nature16192.
29 Ji, K.-X., Chi, F., Yang, M.-F. et al. (2010). Movement of rhizobia inside tobacco and lifestyle alternation from endophytes to free-living rhizobia on leaves. *J. Microbiol. Biotechnol.* 20: 238–244.
30 James, E.K., Gyaneshwar, P., Mathan, N. et al. (2002). Infection and colonization of rice seedlings by the plant growth-promoting bacterium Herbaspirillum seropedicae Z67. *Mol. Plant-Microbe Interact.* 15: 894–906. https://doi.org/10.1094/MPMI.2002.15.9.894.
31 Truyens, S., Weyens, N., Cuypers, A., and Vangronsveld, J. (2015). Bacterial seed endophytes: genera, vertical transmission and interaction with plants: bacterial seed endophytes. *Environ. Microbiol. Rep.* 7: 40–50. https://doi.org/10.1111/1758-2229.12181.

32 Shade, A., Jacques, M.-A., and Barret, M. (2017). Ecological patterns of seed microbiome diversity, transmission, and assembly. *Curr. Opin. Microbiol.* 37: 15–22. https://doi.org/10.1016/j.mib.2017.03.010.

33 Nelson, E.B. (2017). The seed microbiome: origins, interactions, and impacts. *Plant Soil.* https://doi.org/10.1007/s11104-017-3289-7.

34 Rodríguez, C.E., Mitter, B., Barret, M. et al. (2017). Commentary: seed bacterial inhabitants and their routes of colonization. *Plant Soil.* https://doi.org/10.1007/s11104-017-3368-9.

35 Charkowski, A.O., Sarreal, C.Z., and Mandrell, R.E. (2001). Wrinkled alfalfa seeds harbor more aerobic bacteria and are more difficult to sanitize than smooth seeds. *J. Food Prot.* 64: 1292–1298. https://doi.org/10.4315/0362-028X-64.9.1292.

36 Cottyn, B., Regalado, E., Lanoot, B. et al. (2001). Bacterial populations associated with rice seed in the tropical environment. *Phytopathology* 91: 282–292. https://doi.org/10.1094/PHYTO.2001.91.3.282.

37 Bacilio-Jiménez, M., Aguilar-Flores, S., del Valle, M.V. et al. (2001). Endophytic bacteria in rice seeds inhibit early colonization of roots by Azospirillum brasilense. *Soil Biol. Biochem.* 33: 167–172. https://doi.org/10.1016/S0038-0717(00)00126-7.

38 Kaga, H., Mano, H., Tanaka, F. et al. (2009). Rice seeds as sources of endophytic bacteria. *Microbes Environ.* 24: 154–162. https://doi.org/10.1264/jsme2.ME09113.

39 Okunishi, S., Sako, K., Mano, H. et al. (2005). Bacterial flora of endophytes in the maturing seed of cultivated rice (Oryza sativa). *Microbes Environ.* 20: 168–177. https://doi.org/10.1264/jsme2.20.168.

40 Verma, S.K., Kingsley, K., Irizarry, I. et al. (2017). Seed vectored endophytic bacteria modulate development of rice seedlings. *J. Appl. Microbiol.* 122: 1680–1691. https://doi.org/10.1111/jam.13463.

41 Liu, Y., Zuo, S., Zou, Y. et al. (2013). Investigation on diversity and population succession dynamics of endophytic bacteria from seeds of maize (Zea mays L., Nongda108) at different growth stages. *Ann. Microbiol.* 63: 71–79. https://doi.org/10.1007/s13213-012-0446-3.

42 Johnston-Monje, D. and Raizada, M.N. (2011). Conservation and diversity of seed associated endophytes in Zea across boundaries of evolution, ethnography and ecology. *PLoS One* 6: e20396. https://doi.org/10.1371/journal.pone.0020396.

43 Mastretta, C., Taghavi, S., van Der Lelie, D. et al. (2009). Endophytic bacteria from seeds of Nicotiana tabacum can reduce cadmium phytotoxicity. *Int. J. Phytoremediation* 11: 251–267. https://doi.org/10.1080/15226510802432678.

44 Vega, F.E., Pava-Ripoll, M., Posada, F., and Buyer, J.S. (2005). Endophytic bacteria in Coffea arabica L. *J. Basic Microbiol.* 45: 371–380. https://doi.org/10.1002/jobm.200410551.

45 Pitzschke, A. (2016). Developmental peculiarities and seed-borne endophytes in quinoa: omnipresent, robust bacilli contribute to plant fitness. *Front. Microbiol.* 7. https://doi.org/10.3389/fmicb.2016.00002.

46 López-López, A., Rogel, M.A., Ormeno-Orrillo, E. et al. (2010). Phaseolus vulgaris seed-borne endophytic community with novel bacterial species such as Rhizobium endophyticum sp. nov. *Syst. Appl. Microbiol.* 33: 322–327. https://doi.org/10.1016/j.syapm.2010.07.005.

47 Compant, S., Mitter, B., Colli-Mull, J.G. et al. (2011). Endophytes of grapevine flowers, berries, and seeds: identification of cultivable bacteria, comparison with other plant parts, and visualization of niches of colonization. *Microb. Ecol.* 62: 188–197. https://doi.org/10.1007/s00248-011-9883-y.

48 Zawoznik, M.S., Vázquez, S.C., Díaz Herrera, S.M., and Groppa, M.D. (2014). Search for endophytic diazotrophs in barley seeds. *Braz. J. Microbiol.* 45: 621–625. https://doi.org/10.1590/S1517-83822014000200033.

49 Fürnkranz, M., Lukesch, B., Müller, H. et al. (2012). Microbial diversity inside pumpkins: microhabitat-specific communities display a high antagonistic potential against phytopathogens. *Microb. Ecol.* 63: 418–428. https://doi.org/10.1007/s00248-011-9942-4.

50 Puente, M.E., Li, C.Y., and Bashan, Y. (2009). Endophytic bacteria in cacti seeds can improve the development of cactus seedlings. *Environ. Exp. Bot.* 66: 402–408. https://doi.org/10.1016/j.envexpbot.2009.04.007.

51 Goggin, D.E., Emery, R.J.N., Kurepin, L.V., and Powles, S.B. (2015). A potential role for endogenous microflora in dormancy release, cytokinin metabolism and the response to fluridone in Lolium rigidum seeds. *Ann. Bot.* 115: 293–301. https://doi.org/10.1093/aob/mcu231.

52 Ferreira, A., Quecine, M.C., Lacava, P.T. et al. (2008). Diversity of endophytic bacteria from Eucalyptus species seeds and colonization of seedlings by Pantoea agglomerans. *FEMS Microbiol. Lett.* 287: 8–14. https://doi.org/10.1111/j.1574-6968.2008.01258.x.

53 Alibrandi, P., Cardinale, M., Rahman, M.M. et al. (2017). The seed endosphere of Anadenanthera colubrina is inhabited by a complex microbiota, including Methylobacterium spp. and Staphylococcus spp. with potential plant-growth promoting activities. *Plant Soil* 1–19. https://doi.org/10.1007/s11104-017-3182-4.

54 Cankar, K., Kraigher, H., Ravnikar, M., and Rupnik, M. (2005). Bacterial endophytes from seeds of Norway spruce (Picea abies L. Karst). *FEMS Microbiol. Lett.* 244: 341–345. https://doi.org/10.1016/j.femsle.2005.02.008.

55 Glassner, H., Zchori-Fein, E., Yaron, S. et al. (2017). Bacterial niches inside seeds of Cucumis melo L. *Plant Soil* 1–13. https://doi.org/10.1007/s11104-017-3175-3.

56 Rout, M.E., Chrzanowski, T.H., Westlie, T.K. et al. (2013). Bacterial endophytes enhance competition by invasive plants. *Am. J. Bot.* 100: 1726–1737. https://doi.org/10.3732/ajb.1200577.

57 Herrera, S.D., Grossi, C., Zawoznik, M., and Groppa, M.D. (2016). Wheat seeds harbour bacterial endophytes with potential as plant growth promoters and biocontrol agents of Fusarium graminearum. *Microbiol. Res.* 186: 37–43. https://doi.org/10.1016/j.micres.2016.03.002.

58 Mukhopadhyay, K., Garrison, N.K., Hinton, D.M. et al. (1996). Identification and characterization of bacterial endophytes of rice. *Mycopathologia* 134: 151–159. https://doi.org/10.1007/BF00436723.

59 Saikkonen, K., Saari, S., and Helander, M. (2010). Defensive mutualism between plants and endophytic fungi? *Fungal Divers.* 41: 101–113. https://doi.org/10.1007/s13225-010-0023-7.

60 Hodgson, S., de Cates, C., Hodgson, J. et al. (2014). Vertical transmission of fungal endophytes is widespread in forbs. *Ecol. Evol.* 4: 1199–1208. https://doi.org/10.1002/ece3.953.

61 Miller, I.M. (1990). Advances in botanical research. In: *Bacterial Leaf Nodule Symbiosis*, vol. 17 (ed. J.-C. Kader and M. Delseny), 163–234. Amsterdam, The Netherlands: Elsevier.

62 Lemaire, B., Janssens, S., Smets, E., and Dessein, S. (2012). Endosymbiont transmission mode in bacterial leaf nodulation as revealed by a population genetic study of Psychotria leptophylla. *Appl. Environ. Microbiol.* 78: 284–287. https://doi.org/10.1128/AEM.06130-11.

63 Gagne-Bourgue, F., Aliferis, K.A., Seguin, P. et al. (2013). Isolation and characterization of indigenous endophytic bacteria associated with leaves of switchgrass (Panicum virgatum L.) cultivars. *J. Appl. Microbiol.* 114: 836–853. https://doi.org/10.1111/jam.12088.

64 Ringelberg, D., Foley, K., and Reynolds, C.M. (2012). Bacterial endophyte communities of two wheatgrass varieties following propagation in different growing media. *Can. J. Microbiol.* 58: 67–80. https://doi.org/10.1139/w11-122.

65 Liu, Y., Zuo, S., Xu, L. et al. (2012). Study on diversity of endophytic bacterial communities in seeds of hybrid maize and their parental lines. *Arch. Microbiol.* 194: 1001–1012. https://doi.org/10.1007/s00203-012-0836-8.

66 Barret, M., Briand, M., Bonneau, S. et al. (2015). Emergence shapes the structure of the seed microbiota. *Appl. Environ. Microbiol.* 81: 1257–1266. https://doi.org/10.1128/AEM.03722-14.

67 Truyens, S., Beckers, B., Thijs, S. et al. (2016). The effects of the growth substrate on cultivable and total endophytic assemblages of Arabidopsis thaliana. *Plant Soil* 405: 325–336. https://doi.org/10.1007/s11104-015-2761-5.

68 Truyens, S., Weyens, N., Cuypers, A., and Vangronsveld, J. (2013). Changes in the population of seed bacteria of transgenerationally Cd-exposed Arabidopsis thaliana. *Plant Biol.* 15: 971–981. https://doi.org/10.1111/j.1438-8677.2012.00711.x.

69 Agarwal, V.K. and Sinclair, J.B. (1997). *Principles of Seed Pathology*, 2e. Boca Raton, FL: Lewis Publishers.

70 Puente, M.E., Li, C.Y., and Bashan, Y. (2009). Rock-degrading endophytic bacteria in cacti. *Environ. Exp. Bot.* 66: 389–401. https://doi.org/10.1016/j.envexpbot.2009.04.010.

71 Clark, S.E. (1997). Organ formation at the vegetative shoot meristem. *Plant Cell* 9: 1067–1076. https://doi.org/10.1105/tpc.9.7.1067.

72 Pirttilä, A.M. (2011). Endophytic bacteria in tree shoot tissues and their effects on host. In: *Endophytes of Forest Trees: Biology and Applications*, vol. 80 (ed. A.M. Pirttila and A.C. Frank), 139–149. New York: Springer.

73 Van Aken, B., Peres, C.M., Doty, S.L. et al. (2004). Methylobacterium populi sp. nov., a novel aerobic, pink-pigmented, facultatively methylotrophic, methane-utilizing bacterium isolated from poplar trees (Populus deltoids × nigra DN34). *Int. J. Syst. Evol. Microbiol.* 54: 1191–1196. https://doi.org/10.1099/ijs.0.02796-0.

74 Ulrich, K., Ulrich, A., and Ewald, D. (2008). Paenibacillus – a predominant endophytic bacterium colonizing tissue cultures of woody plants. *Plant Cell Tissue Organ Cult.* 93: 347–351. https://doi.org/10.1007/s11240-008-9367-z.

75 Pirttilä, A.M., Laukkanen, H., Pospiech, H. et al. (2000). Detection of intracellular bacteria in the buds of Scotch pine (Pinus sylvestris L.) by in situ hybridization. *Appl. Environ. Microbiol.* 66: 3073–3077. https://doi.org/10.1128/AEM.66.7.3073-3077.2000.

76 Laukkanen, H., Soini, H., Kontunen-Soppela, S. et al. (2000). A mycobacterium isolated from tissue cultures of mature Pinus sylvestris interferes with growth of Scots pine seedlings. *Tree Physiol.* 20: 915–920. https://doi.org/10.1093/treephys/20.13.915.

77 Thomas, P., Kumari, S., Swarna, G.K. et al. (2007). Ubiquitous presence of fastidious endophytic bacteria in field shoots and index-negative apparently clean shoot-tip cultures of papaya. *Plant Cell Rep.* 26: 1491–1499. https://doi.org/10.1007/s00299-007-0363-2.

78 Thomas, P., Swarna, G.K., Roy, P.K., and Patil, P. (2008). Identification of culturable and originally non-culturable endophytic bacteria isolated from shoot tip cultures of banana cv. Grand Naine. *Plant Cell Tissue Organ Cult.* 93: 55. https://doi.org/10.1007/s11240-008-9341-9.

79 Kamoun, R., Lepoivre, P., and Boxus, P. (1997). Evidence for the occurrence of endophytic prokaryotic contaminants in micropropagated plantlets of Prunus cerasus cv. 'Montmorency'. In: *Pathogen and Microbial Contamination Management in Micropropagation* (ed. A.C. Cassells), 145–148. New York: Springer.

80 Esposito-Polesi, N.P., de Abreu-Tarazi, M.F., de Almeida, C.V. et al. (2017). Investigation of endophytic bacterial community in supposedly axenic cultures of pineapple and orchids with evidence on abundant intracellular bacteria. *Curr. Microbiol.* 74: 103–113. https://doi.org/10.1007/s00284-016-1163-0.

81 Thomas, P. and Sekhar, A.C. (2014). Live cell imaging reveals extensive intracellular cytoplasmic colonization of banana by normally non-cultivable endophytic bacteria. *AoB Plants* 6. https://doi.org/10.1093/aobpla/plu002.

82 Thomas, P., Kumari, S., Swarna, G.K., and Gowda, T.K.S. (2007). Papaya shoot tip associated endophytic bacteria isolated from in vitro cultures and host-endophyte interaction in vitro and in vivo. *Can. J. Microbiol.* 53: 380–390. https://doi.org/10.1139/W06-141.

83 Koskimäki, J.J., Pirttilä, A.M., Ihantola, E.-L. et al. (2015). The intracellular scots pine shoot symbiont methylobacterium extorquens DSM13060 aggregates around the host nucleus and encodes eukaryote-like proteins. *MBio* 6: e00039-15. https://doi.org/10.1128/mBio.00039-15.

84 Pirttilä, A.M., Pospiech, H., Laukkanen, H. et al. (2005). Seasonal variations in location and population structure of endophytes in buds of Scots pine. *Tree Physiol.* 25: 289–297. https://doi.org/10.1093/treephys/25.3.289.

85 Ambika, M.B., Ratering, S., Rusch, V. et al. (2016). Bacterial microbiota associated with flower pollen is influenced by pollination type, and shows a high degree of diversity and species-specificity. *Environ. Microbiol.* 18: 5161–5174. https://doi.org/10.1111/1462-2920.13524.

86 Madmony, A., Chernin, L., Pleban, S. et al. (2005). Enterobacter cloacae, an obligatory endophyte of pollen grains of Mediterranean pines. *Folia Microbiol. (Praha)* 50: 209–216. https://doi.org/10.1007/BF02931568.

87 Heydenreich, B., Bellinghausen, I., König, B. et al. (2012). Gram-positive bacteria on grass pollen exhibit adjuvant activity inducing inflammatory T cell responses. *Clin. Exp. Allergy* 42: 76–84. https://doi.org/10.1111/j.1365-2222.2011.03888.x.

88 Obersteiner, A., Gilles, S., Frank, U. et al. (2016). Pollen-associated microbiome correlates with pollution parameters and the allergenicity of pollen. *PLoS One* 11: e0149545. https://doi.org/10.1371/journal.pone.0149545.

89 Jojima, Y., Mihara, Y., Suzuki, S. et al. (2004). Saccharibacter floricola gen. nov., sp. nov., a novel osmophilic acetic acid bacterium isolated from pollen. *Int. J. Syst. Evol. Microbiol.* 54: 2263–2267. https://doi.org/10.1099/ijs.0.02911-0.

90 Roberts, D.P., Baker, C.J., McKenna, L. et al. (2009). Influence of host seed on metabolic activity of Enterobacter cloacae in the spermosphere. *Soil Biol. Biochem.* 41: 754–761. https://doi.org/10.1016/j.soilbio.2009.01.010.

91 Kageyama, K. and Nelson, E.B. (2003). Differential inactivation of seed exudate stimulation of *Pythium ultimum* sporangium germination by Enterobacter cloacae influences biological control efficacy on different plant species. *Appl. Environ. Microbiol.* 69: 1114–1120. https://doi.org/10.1128/AEM.69.2.1114-1120.2003.

92 Simon, H.M., Smith, K.P., Dodsworth, J.A. et al. (2001). Influence of tomato genotype on growth of inoculated and indigenous bacteria in the spermosphere. *Appl. Environ. Microbiol.* 67: 514–520. https://doi.org/10.1128/AEM.67.2.514-520.2001.

93 Mendes, R., Kruijt, M., de Bruijn, I. et al. (2011). Deciphering the rhizosphere microbiome for disease-suppressive bacteria. *Science* 332: 1097–1100. https://doi.org/10.1126/science.1203980.

94 Zasloff, M. (2017). Pollen has a microbiome: implications for plant reproduction, insect pollination and human allergies: pollen has a microbiome. *Environ. Microbiol.* 19: 1–2. https://doi.org/10.1111/1462-2920.13661.

95 Lennon, J.T. and Jones, S.E. (2011). Microbial seed banks: the ecological and evolutionary implications of dormancy. *Nat. Rev. Microbiol.* 9: 119–130. https://doi.org/10.1038/nrmicro2504.

96 Turner, T.R., James, E.K., and Poole, P.S. (2013). The plant microbiome. *Genome Biol.* 14: 209. https://doi.org/10.1186/gb-2013-14-6-209.

97 Zarraonaindia, I., Owens, S.M., Weisenhorn, P. et al. (2015). The soil microbiome influences grapevine-associated microbiota. *MBio* 6: e02527-14. https://doi.org/10.1128/mBio.02527-14.

98 Galippe, V. (1887). Note sur la présence de micro-organismes dans les tissus végétaux. *C. R. Seances Soc. Biol. Fil.* 39: 410–416.

99 Hiltner, L. (1904). Über neue Erfahrungen und Probleme auf dem Gebiet der Bodenbakteriologie und unter besonderer Ber- ücksichtigung der Gründüngung und Brache. *Arb. Deust. Landw. Ges.* 98: 59–78.

100 Nelson, E.B. (2004). Microbial dynamics and interactions in the spermosphere. *Annu. Rev. Phytopathol.* 42: 271–309. https://doi.org/10.1146/annurev.phyto.42.121603.131041.

101 Schiltz, S., Gaillard, I., Pawlicki-Jullian, N. et al. (2015). A review: what is the spermosphere and how can it be studied? *J. Appl. Microbiol.* 119: 1467–1481. https://doi.org/10.1111/jam.12946.

102 Rolli, E., Marasco, R., Vigani, G. et al. (2015). Improved plant resistance to drought is promoted by the root-associated microbiome as a water stress-dependent trait. *Environ. Microbiol.* 17: 316–331. https://doi.org/10.1111/1462-2920.12439.

103 Vergani, L., Mapelli, F., Zanardini, E. et al. (2017). Phyto-rhizoremediation of polychlorinated biphenyl contaminated soils: an outlook on plant-microbe beneficial interactions. *Sci. Total Environ.* 575: 1395–1406. https://doi.org/10.1016/j.scitotenv.2016.09.218.

104 Mendes, R., Garbeva, P., and Raaijmakers, J.M. (2013). The rhizosphere microbiome: significance of plant beneficial, plant pathogenic, and human pathogenic microorganisms. *FEMS Microbiol. Rev.* 37: 634–663. https://doi.org/10.1111/1574-6976.12028.

105 Schuhegger, R., Ihring, A., Gantner, S. et al. (2006). Induction of systemic resistance in tomato by N-acyl-L-homoserine lactone-producing rhizosphere bacteria. *Plant Cell Environ.* 29: 909–918. https://doi.org/10.1111/j.1365-3040.2005.01471.x.

106 Nuccio, E.E., Anderson-Furgeson, J., Estera, K.Y. et al. (2016). Climate and edaphic controllers influence rhizosphere community assembly for a wild annual grass. *Ecology* 97: 1307–1318. https://doi.org/10.1890/15-0882.1.

107 Philippot, L., Raaijmakers, J.M., Lemanceau, P., and van der Putten, W.H. (2013). Going back to the roots: the microbial ecology of the rhizosphere. *Nat. Rev. Microbiol.* 11: 789–799. https://doi.org/10.1038/nrmicro3109.

108 Haichar, F.Z., Marol, C., Berge, O. et al. (2008). Plant host habitat and root exudates shape soil bacterial community structure. *ISME J.* 2: 1221–1230. https://doi.org/10.1038/ismej.2008.80.

109 Berg, G. and Smalla, K. (2009). Plant species and soil type cooperatively shape the structure and function of microbial communities in the rhizosphere. *FEMS Microbiol. Ecol.* 68: 1–13. https://doi.org/10.1111/j.1574-6941.2009.00654.x.

110 Bulgarelli, D., Schlaeppi, K., Spaepen, S. et al. (2013). Structure and functions of the bacterial microbiota of plants. *Annu. Rev. Plant Biol.* 64: 807–838. https://doi.org/10.1146/annurev-arplant-050312-120106.

111 Compant, S., Clement, C., and Sessitsch, A. (2010). Plant growth-promoting bacteria in the rhizo- and endosphere of plants: their role, colonization, mechanisms involved and prospects for utilization. *Soil Biol. Biochem.* 42: 669–678. https://doi.org/10.1016/j.soilbio.2009.11.024.

112 Bais, H.P., Weir, T.L., Perry, L.G. et al. (2006). The role of root exudates in rhizosphere interactions with plants and other organisms. *Annu. Rev. Plant Biol.* 57: 233–266. https://doi.org/10.1146/annurev.arplant.57.032905.105159.

113 Marschner, H., Römheld, V., Horst, W.J., and Martin, P. (1986). Root-induced changes in the rhizosphere: importance for the mineral nutrition of plants. *J. Plant Nutr. Soil Sci.* 149: 441–456. https://doi.org/10.1002/jpln.19861490408.

114 Dennis, P.G., Miller, A.J., and Hirsch, P.R. (2010). Are root exudates more important than other sources of rhizodeposits in structuring rhizosphere bacterial communities? *FEMS Microbiol. Ecol.* 72: 313–327. https://doi.org/10.1111/j.1574-6941.2010.00860.

115 Berendsen, R.L., Pieterse, C.M.J., and Bakker, P.A.H.M. (2012). The rhizosphere microbiome and plant health. *Trends Plant Sci.* 17: 478–486. https://doi.org/10.1016/j.tplants.2012.04.001.

116 Persello-Cartieaux, F., Nussaume, L., and Robaglia, C. (2003). Tales from the underground: molecular. *Plant Cell Environ.* 26: 189–199. https://doi.org/10.1046/j.1365-3040.2003.00956.x.

117 Müller, H., Westendorf, C., Leitner, E. et al. (2009). Quorum-sensing effects in the antagonistic rhizosphere bacterium *Serratia plymuthica* HRO-C48. *FEMS Microbiol. Ecol.* 67: 468–478. https://doi.org/10.1111/j.1574-6941.2008.00635.x.

118 Podile, A.R., Vukanti, R., Sravani, A. et al. (2013). Root colonization and quorum sensing are the driving forces of plant growth promoting rhizobacteria (PGPR) for growth promotion. *Proc. Natl. Acad. Sci. India Sect. B Biol. Sci.* 80: 407–413. https://doi.org/10.16943/ptinsa/2014/v80i2/55117.

119 Wei, H.-L. and Zhang, L.-Q. (2006). Quorum-sensing system influences root colonization and biological control ability in Pseudomonas fluorescens

2P24. *Antonie Van Leeuwenhoek* 89: 267–280. https://doi.org/10.1007/s10482-005-9028-8.

120 De Weger, L.A., van der Vlugt, C.I., Wijfjes, A.H. et al. (1987). Flagella of a plant-growth-stimulating Pseudomonas fluorescens strain are required for colonization of potato roots. *J. Bacteriol.* 169: 2769–2773. https://doi.org/10.1128/jb.169.6.2769-2773.1987.

121 Scharf, B.E., Hynes, M.F., and Alexandre, G.M. (2016). Chemotaxis signaling systems in model beneficial plant-bacteria associations. *Plant Mol. Biol.* 90: 549–559. https://doi.org/10.1007/s11103-016-0432-4.

122 Lebeis, S.L., Paredes, S.H., Lundberg, D.S. et al. (2015). PLANT MICROBIOME. Salicylic acid modulates colonization of the root microbiome by specific bacterial taxa. *Science* 349: 860–864. https://doi.org/10.1126/science.aaa8764.

123 Castrillo, G., Teixeira, P.J.P.L., Paredes, S.H. et al. (2017). Root microbiota drive direct integration of phosphate stress and immunity. *Nature* 543: 513–518. https://doi.org/10.1038/nature21417.

124 Compant, S., Kaplan, H., Sessitsch, A. et al. (2008). Endophytic colonization of Vitis vinifera L. by Burkholderia phytofirmans strain PsJN: from the rhizosphere to inflorescence tissues. *FEMS Microbiol. Ecol.* 63: 84–93. https://doi.org/10.1111/j.1574-6941.2007.00410.x.

125 Hurek, T., Reinhold-Hurek, B., van Montagu, M., and Kellenberger, E. (1994). Root colonization and systemic spreading of Azoarcus sp. strain BH72 in grasses. *J. Bacteriol.* 176: 1913–1923. https://doi.org/10.1128/jb.176.7.1913-1923.1994.

126 Reinhold-Hurek, B. and Hurek, T. (1998). Life in grasses: diazotrophic endophytes. *Trends Microbiol.* 6: 139–144. https://doi.org/10.1016/S0966-842X(98)01229-3.

127 Chi, F., Shen, S.-H., Cheng, H.-P. et al. (2005). Ascending migration of endophytic rhizobia, from roots to leaves, inside rice plants and assessment of benefits to rice growth physiology. *Appl. Environ. Microbiol.* 71: 7271–7278. https://doi.org/10.1128/AEM.71.11.7271-7278.2005.

128 Saleem, M., Law, A.D., and Moe, L.A. (2016). Nicotiana roots recruit rare rhizosphere taxa as major root-inhabiting microbes. *Microb. Ecol.* 71: 469–472. https://doi.org/10.1007/s00248-015-0672-x.

129 Bodenhausen, N., Horton, M.W., and Bergelson, J. (2013). Bacterial communities associated with the leaves and the roots of Arabidopsis thaliana. *PLoS One* 8: e56329. https://doi.org/10.1371/journal.pone.0056329.

130 Vorholt, J.A. (2012). Microbial life in the phyllosphere. *Nat. Rev. Microbiol.* 10: 828–840. https://doi.org/10.1038/nrmicro2910.

131 Lindow, S.E., Hecht-Poinar, E.I., and Elliott, V.J. (ed.) (2002). *Phyllosphere Microbiology*. St. Paul, MN: APS Press/American Phytopathological Society.

132 Fröhlich-Nowoisky, J., Kampf, C.J., Weber, B. et al. (2016). Bioaerosols in the Earth system: climate, health, and ecosystem interactions. *Atmos. Res.* 182: 346–376. https://doi.org/10.1016/j.atmosres.2016.07.018.

133 Lighthart, B. (2000). Mini-review of the concentration variations found in the alfresco atmospheric bacterial populations. *Aerobiologia* 16: 7–16. https://doi.org/10.1023/A:1007694618888.

134 Burrows, S.M., Butler, T., Jöckel, P. et al. (2009). Bacteria in the global atmosphere – part 2: modeling of emissions and transport between different ecosystems. *Atmos. Chem. Phys.* 9: 9281–9297. https://doi.org/10.5194/acp-9-9281-2009.

135 Burrows, S.M., Elbert, W., Lawrence, M.G., and Pöschl, U. (2009). Bacteria in the global atmosphere – part 1: review and synthesis of literature data for different ecosystems. *Atmos. Chem. Phys.* 9: 9263–9280. https://doi.org/10.5194/acp-9-9263-2009.

136 Womack, A.M., Bohannan, B.J.M., and Green, J.L. (2010). Biodiversity and biogeography of the atmosphere. *Philos. Trans. R. Soc. B Biol. Sci.* 365: 3645–3653. https://doi.org/10.1098/rstb.2010.0283.

137 Hill, K.A., Shepson, P.B., Galbavy, E.S. et al. (2007). Processing of atmospheric nitrogen by clouds above a forest environment. *J. Geophys. Res.* 112. https://doi.org/10.1029/2006JD008002.

138 Sattler, B., Puxbaum, H., and Psenner, R. (2001). Bacterial growth in supercooled cloud droplets. *Geophys. Res. Lett.* 28: 239–242. https://doi.org/10.1029/2000GL011684.

139 DeLeon-Rodriguez, N., Lathem, T.L., Rodriguez-R, L.M. et al. (2013). Microbiome of the upper troposphere: species composition and prevalence, effects of tropical storms, and atmospheric implications. *Proc. Natl. Acad. Sci. U. S. A.* 110: 2575–2580. https://doi.org/10.1073/pnas.1212089110.

140 Klein, A.M., Bohannan, B.J.M., Jaffe, D.A. et al. (2016). Molecular evidence for metabolically active bacteria in the atmosphere. *Front. Microbiol.* 7: 772. https://doi.org/10.3389/fmicb.2016.00772.

141 Bovallius, A., Bucht, B., Roffey, R., and Anäs, P. (1978). Long-range air transmission of bacteria. *Appl. Environ. Microbiol.* 6: 1231–1232.

142 Yamaguchi, N., Ichijo, T., Sakotani, A. et al. (2012). Global dispersion of bacterial cells on Asian dust. *Sci. Rep.* 2. https://doi.org/10.1038/srep00525.

143 Prospero, J.M., Blades, E., Mathison, G., and Naidu, R. (2005). Interhemispheric transport of viable fungi and bacteria from Africa to the Caribbean with soil dust. *Aerobiologia* 21: 1–19. https://doi.org/10.1007/s10453-004-5872-7.

144 Barberán, A., Henley, J., Fierer, N., and Casamayor, E.O. (2014). Structure, inter-annual recurrence, and global-scale connectivity of airborne microbial communities. *Sci. Total Environ.* 487: 187–195. https://doi.org/10.1016/j.scitotenv.2014.04.030.

145 Meola, M., Lazzaro, A., and Zeyer, J. (2015). Bacterial composition and survival on Sahara dust particles transported to the European Alps. *Front. Microbiol.* 6. https://doi.org/10.3389/fmicb.2015.01454.
146 McCartney, H.A., Fitt, B.D.L., and Schmechel, D. (1997). Sampling bioaerosols in plant pathology. *J. Aerosol Sci.* 28: 349–364. https://doi.org/10.1016/S0021-8502(96)00438-7.
147 Brown, J.K.M. (2002). Aerial dispersal of pathogens on the global and continental scales and its impact on plant disease. *Science* 297: 537–541. https://doi.org/10.1126/science.1072678.
148 Lindemann, J., Constantinidou, H.A., Barchet, W.R., and Upper, C.D. (1982). Plants as sources of airborne bacteria, including ice nucleation-active bacteria. *Appl. Environ. Microbiol.* 44: 1059–1063.
149 Bowers, R.M., McLetchie, S., Knight, R., and Fierer, N. (2011). Spatial variability in airborne bacterial communities across land-use types and their relationship to the bacterial communities of potential source environments. *ISME J.* 5: 601–612. https://doi.org/10.1038/ismej.2010.167.
150 Womack, A.M., Artaxo, P.E., Ishida, F.Y. et al. (2015). Characterization of active and total fungal communities in the atmosphere over the Amazon rainforest. *Biogeosciences* 12: 6337–6349. https://doi.org/10.5194/bg-12-6337-2015.
151 Joung, Y.S., Ge, Z., and Buie, C.R. (2017). Bioaerosol generation by raindrops on soil. *Nat. Commun.* 8: 14668. https://doi.org/10.1038/ncomms14668.
152 Lindemann, J. and Upper, C.D. (1985). Aerial dispersal of epiphytic bacteria over bean plants. *Appl. Environ. Microbiol.* 50: 1229–1232.
153 Quinn, C.E., Sells, I.A., and Graham, D.C. (1980). Soft rot Erwinia bacteria in the atmospheric bacterial aerosol. *J. Appl. Bacteriol.* 49: 175–181. https://doi.org/10.1111/j.1365-2672.1980.tb01055.x.
154 Graham, D.C., Quinn, C.E., and Bradley, L.F. (1977). Quantitative Studies on the generation of aerosols of Erwinia carotovora var. atroseptica by simulated raindrop impaction on blackleg-infected potato stems. *J. Appl. Bacteriol.* 43: 413–424. https://doi.org/10.1111/j.1365-2672.1977.tb00768.x.
155 Ercolani, G.L. (1974). Epiphytic survival of Pseudomonas syringae on hairy vetch in relation to epidemiology of bacterial brown spot of bean in Wisconsin. *Phytopathology* 64: 1330. https://doi.org/10.1094/Phyto-64-1330.
156 Cafati, C.R. (1980). Role of Nonhost species as alternate inoculum sources of Xanthomonas phaseoli. *Plant Dis.* 64: 194. https://doi.org/10.1094/PD-64-194.
157 Martiny, J.B.H. (2016). History leaves its mark on soil bacterial diversity. *MBio* 7: e00784-16. https://doi.org/10.1128/mBio.00784-16.
158 Andam, C.P., Doroghazi, J.R., Campbell, A.N. et al. (2016). A latitudinal diversity gradient in terrestrial bacteria of the genus streptomyces. *MBio* 7: e02200-15. https://doi.org/10.1128/mBio.02200-15.

159 Carey, C.J., Hart, S.C., Aciego, S.M. et al. (2016). Microbial community structure of subalpine snow in the Sierra Nevada, California. *Arct. Antarct. Alp. Res.* 48: 685–701. https://doi.org/10.1657/AAAR0015-062.

160 Carrell, A.A. and Frank, A.C. (2014). Pinus flexilis and Piceae engelmannii share a simple and consistent needle endophyte microbiota with a potential role in nitrogen fixation. *Front. Microbiol.* 5. https://doi.org/10.3389/fmicb.2014.00333.

161 Carrell, A.A., Carper, D.L., and Frank, A.C. (2016). Subalpine conifers in different geographical locations host highly similar foliar bacterial endophyte communities. *FEMS Microbiol. Ecol.* 92. https://doi.org/10.1093/femsec/fiw124.

162 Melotto, M., Underwood, W., and He, S.Y. (2008). Role of stomata in plant innate immunity and foliar bacterial diseases. *Annu. Rev. Phytopathol.* 46: 101–122. https://doi.org/10.1146/annurev.phyto.121107.104959.

163 Scott, R., Chard, J., Hocart, M. et al. (1996). Penetration of potato tuber lenticels by bacteria in relation to biological control of black leg disease. *Potato Res.* 39: 333–344. https://doi.org/10.1007/BF02357937.

164 Hugouvieux, V., Barber, C.E., and Daniels, M.J. (1998). Entry of Xanthomonas campestris pv. campestris into hydathodes of Arabidopsis thaliana leaves: a system for studying early infection events in bacterial pathogenesis. *Mol. Plant-Microbe Interact.* 11: 537–543. https://doi.org/10.1094/MPMI.1998.11.6.537.

165 Arnaud, D. and Hwang, I. (2015). A sophisticated network of signaling pathways regulates stomatal defenses to bacterial pathogens. *Mol. Plant* 8: 566–581. https://doi.org/10.1016/j.molp.2014.10.012.

166 Arnold, E.A. (2008). Endophytic fungi: hidden components of tropical community ecology. In: *Tropical Forest Community Ecology* (ed. W. Carson and S. Schnitzer), 254–271. Oxford: Blackwell Scientific, Inc.

167 Mendgen, K., Hahn, M., and Deising, H. (1996). Morphogenesis and mechanisms of penetration by plant pathogenic fungi. *Annu. Rev. Phytopathol.* 34: 367–386. https://doi.org/10.1146/annurev.phyto.34.1.367.

168 Melotto, M., Underwood, W., Koczan, J. et al. (2006). Plant stomata function in innate immunity against bacterial invasion. *Cell* 126: 969–980. https://doi.org/10.1016/j.cell.2006.06.054.

169 Zhang, W., He, S.Y., and Assmann, S.M. (2008). The plant innate immunity response in stomatal guard cells invokes G-protein-dependent ion channel regulation. *Plant J. Cell Mol. Biol.* 56: 984–996. https://doi.org/10.1111/j.1365-313X.2008.03657.x.

170 Montillet, J.-L., Leonhardt, N., Mondy, S. et al. (2013). An abscisic acid-independent oxylipin pathway controls stomatal closure and immune defense in Arabidopsis. *PLoS Biol.* 11: e1001513. https://doi.org/10.1371/journal.pbio.1001513.

171 Panchal, S., Roy, D., Chitrakar, R. et al. (2016). Coronatine facilitates Pseudomonas syringae infection of Arabidopsis leaves at night. *Front. Plant Sci.* 7: 880. https://doi.org/10.3389/fpls.2016.00880.

172 Zhou, Z., Wu, Y., Yang, Y. et al. (2015). An Arabidopsis plasma membrane proton ATPase modulates JA signaling and is exploited by the Pseudomonas syringae effector protein AvrB for stomatal invasion. *Plant Cell* 27: 2032–2041. https://doi.org/10.1105/tpc.15.00466.

173 Hurley, B., Lee, D., Mott, A. et al. (2014). The Pseudomonas syringae type III effector HopF2 suppresses Arabidopsis stomatal immunity. *PLoS One* 9: e114921. https://doi.org/10.1371/journal.pone.0114921.

174 Roy, D., Panchal, S., Rosa, B.A., and Melotto, M. (2013). Escherichia coli O157:H7 induces stronger plant immunity than Salmonella enterica Typhimurium SL1344. *Phytopathology* 103: 326–332. https://doi.org/10.1094/PHYTO-09-12-0230-FI.

175 Kroupitski, Y., Golberg, D., Belausov, E. et al. (2009). Internalization of Salmonella enterica in leaves is induced by light and involves chemotaxis and penetration through open stomata. *Appl. Environ. Microbiol.* 75: 6076–6086. https://doi.org/10.1128/AEM.01084-09.

176 Melotto, M., Zhang, L., Oblessuc, P.R., and He, S.Y. (2017). Stomatal defense a decade later. *Plant Physiol.* 174: 561–571. https://doi.org/10.1104/pp.16.01853.

177 Baldotto, L.E.B., Olivares, F.L., and Bressan-Smith, R. (2011). Structural interaction between GFP-labeled diazotrophic endophytic bacterium Herbaspirillum seropedicae RAM10 and pineapple plantlets "Vitória". *Braz. J. Microbiol.* 42: 114–125. https://doi.org/10.1590/S1517-83822011000100015.

178 De Souza, R.O., Pamphile, J.A., Lúcia, C. et al. (2004). Plant-microbe interactions between maize (Zea mays L.) and endophytic microrganisms observed by Scanning Electron Microscopy. *Acta Sci. Biol. Sci.* 26: 357–359.

179 White, J.F., Torres, M.S., Sullivan, R.F. et al. (2014). Occurrence of Bacillus amyloliquefaciens as a systemic endophyte of vanilla orchids: B. amyloliquefaciens as a systemic endophyte of vanilla orchids. *Microsc. Res. Tech.* 77: 874–885. https://doi.org/10.1002/jemt.22410.

180 Fukami, J., Nogueira, M.A., Araujo, R.S., and Hungria, M. (2016). Accessing inoculation methods of maize and wheat with Azospirillum brasilense. *AMB Express* 6: 3. https://doi.org/10.1186/s13568-015-0171-y.

181 Aizenberg-Gershtein, Y., Izhaki, I., and Halpern, M. (2013). Do honeybees shape the bacterial community composition in floral nectar? *PLoS One* 8: e67556. https://doi.org/10.1371/journal.pone.0067556.

182 Álvarez-Pérez, S., Herrera, C.M., and de Vega, C. (2012). Zooming-in on floral nectar: a first exploration of nectar-associated bacteria in wild plant communities. *FEMS Microbiol. Ecol.* 80: 591–602. https://doi.org/10.1111/j.1574-6941.2012.01329.x.

183 Fridman, S., Izhaki, I., Gerchman, Y., and Halpern, M. (2012). Bacterial communities in floral nectar. *Environ. Microbiol. Rep.* 4: 97–104. https://doi.org/10.1111/j.1758-2229.2011.00309.x.

184 Jacquemyn, H., Lenaerts, M., Brys, R. et al. (2013). Among-population variation in microbial community structure in the floral nectar of the bee-pollinated forest herb Pulmonaria officinalis L. *PLoS One* 8: e56917. https://doi.org/10.1371/journal.pone.0056917.

185 Baruzzi, F., Cefola, M., Carito, A. et al. (2012). Changes in bacterial composition of zucchini flowers exposed to refrigeration temperatures. *Sci. World J.* 2012: 127805. https://doi.org/10.1100/2012/127805.

186 Glassner, H., Zchori-Fein, E., Compant, S. et al. (2015). Characterization of endophytic bacteria from cucurbit fruits with potential benefits to agriculture in melons (Cucumis melo L.). *FEMS Microbiol. Ecol.* 91. https://doi.org/10.1093/femsec/fiv074.

187 Aleklett, K., Hart, M., and Shade, A. (2014). The microbial ecology of flowers: an emerging frontier in phyllosphere research. *Botany* 92: 253–266. https://doi.org/10.1139/cjb-2013-0166.

188 Shade, A., McManus, P.S., and Handelsman, J. (2013). Unexpected diversity during community succession in the apple flower microbiome. *MBio* 4: e00602-12. https://doi.org/10.1128/mBio.00602-12.

189 Vanneste, J. (2000). *Fire Blight: The Disease and Its Causative Agent* (ed. E. Amylovora and J.L. Vanneste). Wallingford/New York: CABI Pub.

190 Spinelli, F., Ciampolini, F., Cresti, M. et al. (2005). Influence of stigmatic morphology on flower colonization by Erwinia amylovora and Pantoea agglomerans. *Eur. J. Plant Pathol.* 113: 395–405. https://doi.org/10.1007/s10658-005-4511-7.

191 Wilson, M. (1993). Interactions between the biological control agent Pseudomonas fluorescens A506 and Erwinia amylovora in pear blossoms. *Phytopathology* 83: 117. https://doi.org/10.1094/Phyto-83-117.

192 Buban, T., Orosz-Kovacs, Z., and Farkas, A. (2003). The nectary as the primary site of infection by Erwinia amylovora (Burr.) Winslow et al.: a mini review. *Plant Syst. Evol.* 238: 183–194. https://doi.org/10.1007/s00606-002-0266-1.

193 Farkas, Á., Mihalik, E., Dorgai, L., and Bubán, T. (2012). Floral traits affecting fire blight infection and management. *Trees* 26: 47–66. https://doi.org/10.1007/s00468-011-0627-x.

194 Louda, S.M. (1982). Inflorescence spiders: a cost/benefit analysis for the host plant, Haplopappus venetus Blake (Asteraceae). *Oecologia* 55: 185–191. https://doi.org/10.1007/BF00384486.

195 McCall, A.C. and Irwin, R.E. (2006). Florivory: the intersection of pollination and herbivory. *Ecol. Lett.* 9: 1351–1365. https://doi.org/10.1111/j.1461-0248.2006.00975.x.

196 Pellmyr, O. and Thien, L.B. (1986). Insect reproduction and floral fragrances: keys to the evolution of the angiosperms? *Taxon* 35: 76. https://doi.org/10.2307/1221036.

197 Wardhaugh, C.W., Stork, N.E., Edwards, W., and Grimbacher, P.S. (2012). The overlooked biodiversity of flower-visiting invertebrates. *PLoS One* 7: e45796. https://doi.org/10.1371/journal.pone.0045796.

198 Ushio, M., Yamasaki, E., Takasu, H. et al. (2015). Microbial communities on flower surfaces act as signatures of pollinator visitation. *Sci. Rep.* 5. https://doi.org/10.1038/srep08695.

199 McFrederick, Q.S., Thomas, J.M., Neff, J.L. et al. (2017). Flowers and wild megachilid bees share microbes. *Microb. Ecol.* 73: 188–200. https://doi.org/10.1007/s00248-016-0838-1.

200 Molina, R.T., Rodríguez, A.M., Palaciso, I.S., and López, F.G. (1996). Pollen production in anemophilous trees. *Grana* 35: 38–46. https://doi.org/10.1080/00173139609430499.

201 Campbell, I.D., McDonald, K., Flannigan, M.D., and Kringayark, J. (1999). Long-distance transport of pollen into the Arctic. *Nature* 399: 29–30. https://doi.org/10.1038/19891.

202 Marques, J.P.R., Amorim, L., Spósito, M.B. et al. (2013). Infection of citrus pollen grains by Colletotrichum acutatum. *Eur. J. Plant Pathol.* 136: 35–40. https://doi.org/10.1007/s10658-012-0078-2.

203 Nault, L.R. (1997). Arthropod transmission of plant viruses: a new synthesis. *Ann. Entomol. Soc. Am.* 90: 521–541. https://doi.org/10.1093/aesa/90.5.521.

204 Nault, L.R. and Ammar, E.D. (1989). Leafhopper and planthopper transmission of plant viruses. *Annu. Rev. Entomol.* 34: 503–529. https://doi.org/10.1146/annurev.en.34.010189.002443.

205 Weintraub, P.G. and Beanland, L. (2006). Insect vectors of phytoplasmas. *Annu. Rev. Entomol.* 51: 91–111. https://doi.org/10.1146/annurev.ento.51.110104.151039.

206 Li, S.-J., Ahmed, M.Z., Lv, N. et al. (2017). Plantmediated horizontal transmission of Wolbachia between whiteflies. *ISME J.* 11: 1019–1028. https://doi.org/10.1038/ismej.2016.164.

207 Lòpez-Fernàndez, S., Mazzoni, V., Pedrazzoli, F. et al. (2017). A phloem-feeding insect transfers bacterial endophytic communities between grapevine plants. *Front. Microbiol.* 8. https://doi.org/10.3389/fmicb.2017.00834.

208 Fouts, D.E., Tyler, H.L., DeBoy, R.T. et al. (2008). Complete genome sequence of the N2- fixing broad host range endophyte Klebsiella pneumoniae 342 and virulence predictions verified in mice. *PLoS Genet.* 4: e1000141. https://doi.org/10.1371/journal.pgen.1000141.

209 Firrincieli, A., Otillar, R., Salamov, A. et al. (2015). Genome sequence of the plant growth promoting endophytic yeast Rhodotorula graminis WP1. *Front. Microbiol.* 6: 978. https://doi.org/10.3389/fmicb.2015.00978.

210 Martinez-Garcia, P.M., Ruano-Rosa, D., Schiliro, E. et al. (2015). Complete genome sequence of Pseudomonas fluorescens strain PICF7, an indigenous root endophyte from olive (Olea europaea L.) and effective biocontrol agent against Verticillium dahlia. *Stand. Genomic Sci.* 10: 10. https://doi.org/10.1186/1944-3277-10-10.

211 Megias, E., Megias, M., Ollero, F.J., and Hungria, M. (2016). Draft genome sequence of Pantoea ananatis strain AMG521, a rice plant growthpromoting bacterial endophyte isolated from the Guadalquivir marshes in southern Spain. *Genome Announc.* 4: e01681-15. https://doi.org/10.1128/genomeA.01681-15.

212 Weiss, M., Sýkorová, Z., Garnica, S. et al. (2011). Sebacinales everywhere: previously overlooked ubiquitous fungal endophytes. *PLoS One* 6: e16793. https://doi.org/10.1371/journal.pone.0016793.

213 Qiang, X., Weiss, M., Kogel, K.H., and Schafer, S. (2012). Piriformospora indicaa mutualistic basidiomycete with an exceptionally large plant host range. *Mol. Plant Pathol.* 13: 508–518. https://doi.org/10.1111/j.1364-3703.2011.00764.x.

214 David, A.S., Haridas, S., LaButti, K. et al. (2016). Draft genome sequence of microdochium bolleyi, a dark septate fungal endophyte of beach grass. *Genome Announc.* 4: e00270-16. https://doi.org/10.1128/genomeA.00270-16.

215 Yang, Y., Zhao, H., Barrero, R.A. et al. (2014). Genome sequencing and analysis of the paclitaxel-producing endophytic fungus Penicillium aurantiogriseum NRRL 62431. *BMC Genomics* 15: 69. https://doi.org/10.1186/1471-2164-15-69.

216 Gazis, R., Kuo, A., Riley, R. et al. (2016). The genome of Xylona heveae provides a window into fungal endophytism. *Fungal Biol.* 120: 26–42. https://doi.org/10.1016/j.funbio.2015.10.002.

217 Yang, H., Wang, Y., Zhang, Z. et al. (2014). Whole-genome shotgun assembly and analysis of the genome of Shiraia sp. strain Slf14, a novel endophytic fungus producing huperzine A and hypocrellin A. *Genome Announc.* 2: e00011-14. https://doi.org/10.1128/genomeA.00011-14.

218 Zhu, B., Liu, H., Tian, W.X. et al. (2012). Genome sequence of Stenotrophomonas maltophilia RR-10, isolated as an endophyte from rice root. *J. Bacteriol.* 194: 1280–1281. https://doi.org/10.1128/JB.06702-11.

219 Jeong, H., Choi, S.K., Kloepper, J.W., and Ryu, C.M. (2014). Genome sequence of the plant endophyte Bacillus pumilus INR7, triggering induced systemic resistance in field crops. *Genome Announc.* 2: e01093-14. https://doi.org/10.1128/genomeA.01093-14.

220 Crook, M.B., Mitra, S., Ane, J.M. et al. (2013). Complete genome sequence of the Sesbania symbiont and rice growth-promoting endophyte Rhizobium sp. strain IRBG74. *Genome Announc.* 1: e00934-13. https://doi.org/10.1128/genomeA.00934-13.

221 Han, J., Choi, H.K., Lee, S.W. et al. (2011). Complete genome sequence of the metabolically versatile plant growthpromoting endophyte Variovorax paradoxus S110. *J. Bacteriol.* 193: 1183–1190. https://doi.org/10.1128/JB.00925-10.

222 Taghavi, S., van der Lelie, D., Hoffman, A. et al. (2010). Genome sequence of the plant growth promoting endophytic bacterium Enterobacter sp. 638. *PLoS Genet.* 6: e1000943. https://doi.org/10.1371/journal.pgen.1000943.

223 Gan, H.Y., Gan, H.M., Savka, M.A. et al. (2014). Whole-genome sequences of 13 endophytic bacteria isolated from shrub willow (Salix) grown in Geneva, New York. *Genome Announc.* 2: e00288-14. https://doi.org/10.1128/genomeA.00288-14.

224 Deng, Y., Zhu, Y., Wang, P. et al. (2011). Complete genome sequence of Bacillus subtilis BSn5, an endophytic bacterium of Amorphophallus konjac with antimicrobial activity for the plant pathogen Erwinia carotovora subsps. carotovora. *J. Bacteriol.* 193: 2070–2071. https://doi.org/10.1128/JB.00129-11.

225 Schicklberger, M., Shapiro, N., Loqué, D. et al. (2015). Draft genome sequence of Raoultella terrigena R1Gly, a diazotrophic endophyte. *Genome Announc.* 3: e607–e615. https://doi.org/10.1128/genomeA.00607-15.

226 Chaudhry, V. and Prabhu, B.P. (2016). Genomic investigation reveals evolution and lifestyle adaptation of endophytic Staphylococcus epidermidis. *Sci. Rep.* 6: 19263. https://doi.org/10.1038/srep19263.

227 Wu, Q., Zhu, L., Jiang, L. et al. (2015). Draft genome sequence of Paenibacillus dauci sp. nov., a carrot-associated endophytic actinobacteria. *Genom. Data* 5: 241–253. https://doi.org/10.1016/j.gdata.2015.06.010.

228 Jeong, H., Jo, S.H., Hong, C.E., and Park, J.M. (2016). Genome sequence of the endophytic bacterium Bacillus thuringiensis strain KB1, a potential biocontrol agent against phytopathogens. *Genome Announc.* 4: e00279-16. https://doi.org/10.1128/genomeA.00279-16.

229 Ulrich, K., Kube, M., Becker, R. et al. (2021). Genomic analysis of the endophytic *Stenotrophomonas* strain 169 reveals features related to plant-growth promotion and stress tolerance. *Front. Microbiol.* 12: 687463. https://doi.org/10.3389/fmicb.2021.687463.

230 Jia, J., Ford, E., Baird, S., and Lu, S. (2021). Complete genome sequence resource for the endophytic Burkholderia sp. Strain MS389 isolated from a healthy soybean growing adjacent to charcoal rot disease patch. *Plant Dis.* 105: 2704–2707. https://doi.org/10.1094/PDIS-02-21-0377-A.

231 Adeleke, B., Ayangbenro, A., and Babalola, O. (2021). Genomic analysis of endophytic Bacillus cereus T4S and its plant growth-promoting traits. *Plants* 2021 (10): 1776. https://doi.org/10.3390/plants10091776.

232 Guo, L., Qiu, J., Han, Z. et al. (2015). A host plant genome (Zizania latifolia) after a century long endophyte infection. *Plant J.* 83: 600–609. https://doi.org/10.1111/tpj.12912.

233 Zgadzaj, R., James, E.K., Kelly, S. et al. (2015). A legume genetic framework controls infection of nodules by symbiotic and endophytic bacteria. *PLoS Genet.* 11: e1005280. https://doi.org/10.1371/journal.pgen.1005280.

234 Monteiro, R.A., Balsanelli, E., Tuleski, T. et al. (2012). Genomic comparison of the endophyte Herbaspirillum seropedicae SmR1 and the phytopathogen Herbaspirillum rubrisubalbicans M1 by suppressive subtractive hybridization and partial genome sequencing. *FEMS Microbiol. Ecol.* 80: 441–451. https://doi.org/10.1111/j.1574-6941.2012.01309.

235 Winstanley, C. (2002). Spot the difference: applications of subtractive hybridisation to the study of bacterial pathogens. *J. Med. Microbiol.* 51: 459–467. https://doi.org/10.1099/0022-1317-51-6-459.

236 Galbraith, E.A., Antonopoulos, D.A., and White, B.A. (2004). Suppressive subtractive hybridization as a tool for identifying genetic diversity in an environmental metagenome: the rumen as a model. *Environ. Microbiol.* 6: 928–937. https://doi.org/10.1111/j.1462-2920.2004.00575.x.

237 Dinsdale, E.A., Edwards, R.A., Hall, D. et al. (2008). Functional metagenomic profiling of nine biomes. *Nature* 452: 629–632. https://doi.org/10.1038/nature06810.

238 Akinsanya, M.A., Goh, J.K., Lim, S.P., and Ting, A.Y. (2015). Metagenomics study of endophytic bacteria in Aloe vera using next-generation technology. *Genom. Data* 6: 159–163. https://doi.org/10.1016/j.gdata.2015.09.004.

239 Jones, W.J. (2010). High throughput sequencing and metagenomics. *Estuaries Coast* 33: 944–952. https://doi.org/10.1007/s12237-009-9182-8.

240 Toju, H., Yamamoto, S., Sato, H. et al. (2013). Community composition of root-associated fungi in a Quercusdominated temperate forest: "codominance" of mycorrhizal and rootendophytic fungi. *Ecol. Evol.* 3: 1281–1293. https://doi.org/10.1002/ece3.546.

241 Jumpponen, A., Jones, K.L., Mattox, J.D., and Yaege, C. (2010). Massively parallel 454-sequencing of fungal communities in Quercus spp. ectomycorrhizas indicates seasonal dynamics in urban and rural sites. *Mol. Ecol.* 19: 41–53. https://doi.org/10.1111/j.1365-294X.2009.04483.x.

242 Daniel, G.H., Cristopher, P.F., and Annelise, E.B. (2008). Advantages and limitations of next-generation sequencing technologies: a comparison of electrophoresis and non-electrophoresis method. *Electrophoresis* 29: 4618–4626. https://doi.org/10.1002/elps.200800456.

243 Sessitsch, A., Hardoim, P., Döring, J. et al. (2012). Functional characteristics of an endophyte community colonizing rice roots asrevealed by metagenomic analysis. *Mol. Plant Microbe Interact.* 25 (1): 28–36. https://doi.org/10.1094/MPMI-08-11-0204.

244 Benhamou, N., Kloepper, J.W., Quadt-Hallman, A., and Tuzun, S. (1996). Induction of defense-related ultrastructural modifications in pea roottissues inoculated with endophytic bacteria. *Plant Physiol.* 112 (3): 919–929. https://doi.org/10.1104/pp.112.3.919.

245 Benhamou, N., Gagné, S., Le Quéré, D. et al. (2000). Bacterial-mediated induced resistance in cucumber: beneficial effect of the endophytic bacterium *Serratia plymuthica* on the protection against infection by *Pythium ultimum*. *Phytopathology* 90 (1): 45–56. https://doi.org/10.1094/PHYTO.2000.90.1.45.

246 Gómez-Lama Cabanás, C., Schilirò, E., Valverde-Corredor, A., and Mercado-Blanco, J. (2014). The biocontrol endophytic bacterium Pseudomonasfluorescens PICF7 induces systemic defense responses in aerial tissuesupon colonization of olive roots. *Front. Microbiol.* 5: 427. https://doi.org/10.3389/fmicb.2014.00427.

247 Pozo, M.J. and Azcón-Aguilar, C. (2007). Unraveling mycorrhizainduced resistance. *Curr. Opin. Plant Microbiol.* 10: 393–398.

248 Venkateswaran, M., Volkening, J.D., Sussman, M.R., and Ane, J.-M. (2013). Symbiosis and the social network of higher plants. *Curr. Opin. Plant Biol.* 16: 1–10. https://doi.org/10.1016/j.pbi.2012.11.007.

249 Johansson, J.F., Paul, L.R., and Finlay, R.D. (2004). Microbialinteractions in the mycorrhizosphere and their significance for sustainableagriculture. *FEMS Microbiol. Ecol.* 48 (1): 1–13. https://doi.org/10.1016/j.femsec.2003.11.012.

250 Miransari, M. (2011). Interactions between arbuscular mycorrhizal fungi and soil bacteria. *Appl. Environ. Microbiol.* 89: 917–930.

251 Artursson, V., Finlay, R.D., and Jansson, R.D. (2006). Interactions between arbuscular mycorrhizal fungi and bacteria and their potential for stimulating plant growth. *Environ. Microbiol.* 8 (1): 1–10. https://doi.org/10.1111/j.1462-2920.2005.00942.x.

252 Frey-Klett, P., Garbaye, J., and Tarkka, M. (2007). The mycorrhiza helperbacteria revisited. *New Phytol.* 176: 22–36. https://doi.org/10.1111/j.1469-8137.2007.02191.x.

253 Jaderlund, L., Arthurson, V., Granhall, U., and Jansson, J.K. (2008). Specific interactions between arbuscular mycorrhizal fungiand plant growth-promoting bacteria: as revealed by different combinations. *FEMS Microbiol. Lett.* 287 (2): 174–180. https://doi.org/10.1111/j.1574-6968.2008.01318.x.

254 Labbe, J.L., Weston, D.J., Dunkirk, N. et al. (2014). Newly identified helper bacteria stimulate ectomycorrhizal formation in Populus. *Front. Plant Sci.* 5: 579. https://doi.org/10.3389/fpls.2014.00579.

255 Roberts, E. and Lindow, S. (2013). Loline alkaloid production by fungal endophytes of fescue species select for particular epiphytic bacterial microflora. *ISME J.* 8: 359–368.

256 Saikkonen, K., Mikola, J., and Helander, M. (2015). Endophytic phyllospere fungi and nutrientcycling in terrestrial ecosystems. *Current Science* 109: 121–126.

257 Hamilton, C.E., Pedro, G., Helander, M., and Saikkonen, K. (2012). Endophytic mediation of reactive oxygen species and antioxidantactivity in plants: a review. *Fungal Divers.* 54. https://doi.org/10.1007/s13225-012-0158-9.

258 Saikkonen, K., Gundel, P.E., and Helander, M. (2013). Chemical ecology mediated by fungal endophytes in grasses. *J. Chem. Ecol.* 39: 962–968. https://doi.org/10.1007/s10886-013-0310-3.

259 Krohn-Molt, I., Wemheuer, B., Alawi, M. et al. (2013). Metagenome survey of a multispecies and alga-associated biofilm revealed keyelements of bacterial-algal interactions in photobioreactors. *Appl. Environ. Microbiol.* 79 (20): 6196–6206. https://doi.org/10.1128/AEM.01641-13.

260 Leveau, J. and Preston, G.M. (2007). Bacterial mycophagy: definition and diagnosis of a unique bacterial-fungal interaction. *New Phytopathol.* 177: 859–876.

261 Frey-Klett, P., Burlinson, P., Deveau, A. et al. (2011). Bacterial-fungal interactions: hyphens between agricultural, clinical, environmental and food microbiologists. *Microbiol. Mol. Biol. Rev.* 75: 583–609.

262 Bianciotto, V., Bandi, C., Minerdi, D. et al. (1996). An obligately endosymbiotic mycorrhizal fungus itself harbors obligately intracellular bacteria. *Appl. Environ. Microbiol.* 62: 3005–3010. https://doi.org/10.1128/aem.62.8.3005-3010.1996.

263 Bonfante, P. (2003). Plants, mycorrhizal fungi and endobacteria: a dialog among cells and genomes. *Biol. Bull.* 204: 215–220. https://doi.org/10.2307/1543562.

264 Ghignone, S., Salvioli, A., Anca, I. et al. (2012). The genome of the obligate endobacterium of an AM fungus reveals an interphylum network of nutritional interactions. *ISME J.* 6: 136–145. https://doi.org/10.1038/ismej.2011.110.

265 Levy, A., Chang, B.J., Abbott, L.K. et al. (2003). Invasion of spores of the arbuscular mycorrhizal fungus Gigaspora decipiens by Burkholderia spp. *Appl. Environ. Microbiol.* 69: 6250–6256. https://doi.org/10.1128/AEM.69.10.6250-6256.2003.

266 Hoffman, M.T. and Arnold, A.E. (2010). Diverse bacteria inhabit living hyphae of phylogenetically diverse fungal endophytes. *Appl. Environ. Microbiol.* 76: 4063–4074.

267 Bonfante, P. and Anca, I. (2009). Plants, mycorrhizal fungi, and bacteria: a network of interactions. *Annu. Rev. Microbiol.* 63: 363–383.

268 Denny, T.P. (1995). Involvement of bacterial polysaccharides in plant pathogenesis. *Annu. Rev. Phytopathol.* 33: 173–197.

269 Danhorn, T. and Fuqua, C. (2007). Biofilm formation by plantassociated bacteria. *Annu. Rev. Microbiol.* 61: 401–422.

270 Koutsoudis, M.D., Tsaltas, D., Minogue, T.D., and von Bodman, S.B. (2006). Quorum-sensing regulation governs bacterial adhesion, biofilm development, and host colonization in Pantoea stewartii subspecies stewartii. *Proc. Natl. Acad. Sci. U. S. A.* 103: 5983–5988.

271 Killiny, N., Martinez, R.H., Dumenyo, C.K. et al. (2013). The exopolysaccharide of Xylella fastidiosa is essential for biofilm formation, plant virulence, and vector transmission. *Mol. Plant Microbe Interact.* 26: 1044–1053. https://doi.org/10.1094/MPMI-09-12-0211-R.

272 Wargo, M.J. and Hogan, D.A. (2006). Fungal-bacterial interactions: a mixed bag of mingling microbes. *Curr. Opin. Microbiol.* 9: 359–364.

273 Bianciotto, V., Andreotti, S., Balestrini, R. et al. (2001). Mucoid mutants of the biocontrol strain Pseudomonas fluorescens CHA0 show increased ability in biofilm formation on mycorrhizal and nonmycorrhizal carrot roots. *Mol. Plant Microbe Interact.* 14: 255–260. https://doi.org/10.1094/MPMI.2001.14.2.255.

274 Balbontin, R., Vlamakis, H., and Kolter, R. (2014). Mutualistic interaction between Salmonella enterica and Aspergillus niger and its effects on Zea mays colonization. *Microb. Biotechnol.* 7: 589–600. https://doi.org/10.1111/1751-7915.12182.

275 Jofré, E., Lagares, A., and Mori, G. (2004). Disruption of dTDP-rhamnose biosynthesis modifies lipopolysaccharide core, exopolysaccharide production, and root colonization in Azospirillum brasilense. *FEMS Microbiol. Lett.* 231: 267–275. https://doi.org/10.1016/S0378-1097(04)00003-5.

276 Andreote, F.D., Lacava, P.T., Gai, C.S. et al. (2006). Model plants for studying the interaction between Methylobacterium mesophilicum and Xylella fastidiosa. *Can. J. Microbiol.* 52: 419–426. https://doi.org/10.1139/w05-142.

277 Janczarek, M., Kutkowska, J., Piersiak, T., and Skorupska, A. (2010). Rhizobium leguminosarum bv. trifolii rosR is required for interaction with clover, biofilm formation and adaptation to the environment. *BMC Microbiol.* 10: 284. https://doi.org/10.1186/1471-2180-10-284.

278 Meneses, C.H.S.G., Rouws, L.F., Simoes-Araujo, J.L. et al. (2011). Exopolysaccharide production is required for biofilm formation and plant colonization by the nitrogen-fixing endophyte Gluconacetobacter diazotrophicus. *Mol. Plant Microbe Interact.* 24: 1448–1458. https://doi.org/10.1094/MPMI-05-11-0127.

279 Balsanelli, E., de Baura, V.A., Pedrosa Fde, O. et al. (2014). Exopolysaccharide biosynthesis enables mature biofilm formation on abiotic surface by Herbaspirillum seropedicae. *PLoS One* 9: e110392. https://doi.org/10.1371/journal.pone.0110392.

280 Cotter, P.A. and Stibitz, S. (2007). c-di-GMP-mediated regulation of virulence and biofilm formation. *Curr. Opin. Microbiol.* 10: 17–23.

281 Truyens, S., Cuypers, N.W.A., and Vangronsveld, J. (2015). Bacterial seed endophytes: genera, vertical transmission and interaction with plants. *Environ. Microbiol. Rep.* 7: 40–50.

282 Roberson, E.B. and Firestone, M.K. (1992). Relationship between desiccation and exopolysaccharide production in a soil Pseudomonas sp. *Appl. Environ. Microbiol.* 58: 1284–1291.

283 Lee, K.W.K., Periasamy, S., Mukherjee, M. et al. (2014). Biofilm development and enhanced stress resistance of a model, mixed-species community biofilm. *ISME J.* 8: 894–907. https://doi.org/10.1038/ismej.2013.194.

284 Ren, D., Madsen, J.S., Sørensen, S.J., and Burmølle, M. (2015). High prevalence of biofilm synergy among bacterial soil isolates in cocultures indicates bacterial interspecific cooperation. *ISME J.* 9: 81–89. https://doi.org/10.1038/ismej.2014.96.

285 Elias, S. and Banin, E. (2012). Multi-species biofilms: living with friendly neighbors. *FEMS Microbiol. Rev.* 36: 990–1004.

286 Oliveira, N.M., Martinez-Garcia, E., Xavier, J. et al. (2015). Biofilm formation as a response to ecological competition. *PLoS Biol.* 13: e1002191.

287 Flemming, H. and Wingender, J. (2010). The biofilm matrix. *Nat. Rev. Microbiol.* 8: 623–633.

288 Klümper, U., Riber, L., Dechesne, A. et al. (2015). Broad host range plasmids can invade an unexpectedly diverse fraction of a soil bacterial community. *ISME J.* 9: 934–945. https://doi.org/10.1038/ismej.2014.191.

289 Sheibani-Tezerji, R., Naveed, M., Jehl, M.A. et al. (2015). The genomes of closely related Pantoea ananatis maize seed endophytes having different effects on the host plant in secretion system genes and mobile genetic elements. *Front. Microbiol.* 6: 440. https://doi.org/10.3389/fmicb.2015.00440.

290 Wang, X., Zhang, X., Liu, L. et al. (2015). Genomic and transcriptomic analysis of the endophytic fungus Pestalotiopsis fici reveals its lifestyle and high potential for synthesis of natural products. *BMC Genomics* 16: 28. https://doi.org/10.1186/s12864-014-1190-9.

291 Martin Grube, Tomislav Cernava, Jung Soh. et al. (2015). Exploring functional contexts of symbiotic sustain within lichen-associated bacteria by comparative omics. *ISME J.* 9: 412–424.

7

Core Microbiomes

For Sustainable Agroecosystems

CONTENTS

Microbiomics and Sustainable Crop Production, First Edition. Mohammad Yaseen Mir and Saima Hamid.

7.1 Core Microbiome for Agriculture: A Taxonomic and Functional Aspect

The identification of the components that drive the dynamics of ecosystems is an important problem. It is much more difficult to manipulate them in order to create practical advantages. Agroecosystems are among the most complicated objects of human manipulation as they are the interweaving of macroorganism (such as plants and insects) and innumerable microorganisms (e.g. Bacteria, archaea, protists, and fungus) interaction ties [1–3]. As novel pest/pathogen strains, climate change, and limited availability (and environmental costs) of chemical fertilizers threaten steady agricultural output, scientists are now faced with the unprecedented job of optimizing the functions of these microbial communities (microbiomes) [4–8]. With global crop demand expected to double by 2050, achieving this objective would necessitate quick and urgent solutions. Using microorganisms to improve plant nutrient uptake [9–11] and tolerance to biotic and abiotic stresses [12–16] is one of the few unexplored reservoirs of possibilities for addressing agricultural sustainability challenges [1–3]. The intricacy of plant–microorganism and microorganism–microorganism interactions, as well as their sensitivity to environmental variables, makes optimizing plant–microbial relationships a difficult undertaking. One of the master regulators of the plant phosphate starvation response is also involved in several plant defence responses, indicating that changes in plant-associated microbiome structure can be induced [11, 17]. Plant signalling mechanisms have developed to interact with mycorrhizal fungi and nitrogen-fixing bacteria, as well as to fight off a variety of diseases, and in certain circumstances, microbes are directly recruited or repelled by them [1, 18, 19]. As a result, a greater knowledge of the intricate relationships between plant genotypes, environmental circumstances, and microbiome structure is critical information for breeding efforts. Novel sensing technologies, in addition to traditional breeding methods, provide real-time diagnostics and predictions of plant physiological states based on climatic and soil physiochemical data [20, 21]. Despite the enormous potential, we are just starting to investigate how microbiome data might be incorporated into 'smart farming' practices [20, 21]. The goal of this study was to suggest multidisciplinary research techniques for improving microbiome functioning in agroecosystems. Despite the discovery of a variety of plant-growth-promoting microbes, microbial populations introduced to farmlands frequently decline [22, 23]. Furthermore, they may occasionally outcompete native (resident) microbiomes, which might have unanticipated negative consequences for ecosystem functions [24, 25]. Beyond conventional research that has concentrated on direct, paired plant–microorganism interactions in controlled experimental circumstances, we offer an interdisciplinary method for improving plant–microbiome interactions at the agroecosystem level. We first discuss the importance of manipulating microbiome assembly during the early stages of plant development, and then introduce the concept of 'core microbiomes', which are defined as collections of microorganisms that form cores of interactions and can be used to optimize microbial functions at the individual plant and ecosystem levels. Rather than encouraging host plant development, these organisms are crucial in coordinating the building of plant-associated microbiomes. Next, we demonstrate how informatics pipelines can be used to identify core microbiomes that, if properly deployed, will mediate and organize plant–microorganism and microorganism–microorganism interactions by recruiting indigenous microorganisms with diverse functions and suppressing high pathogen loads in the field. Finally, we consider how recent developments in medical science, microfluidics, robotics, and computer science may be used to disseminate and control plant-associated core microbiomes in agroecosystems. Core microbiomes are a collection of microorganisms that form interaction hubs and may be utilized to enhance microbial activities at the individual plant and ecosystem level. Various functional indigenous microorganisms may be recruited, and high pathogenic stressors in the field may be regulated utilizing a variety of informatics methods, in order to better coordinate plant–microbial and microbe–microbe interactions. The major focus is the distribution and control of plant-related microbiomes in agroecosystems [26, 27], taking into account major advancements in medical, microfluidic, robotics, and informatics. Plant-associated microbiomes are frequently categorized in several categories based on crop types [6, 28, 29] and farming management [29–31]. In summary, these researches have suggested that microbiomes constitute local stability equilibriums, alternative community or ecosystem dynamics, and alternative stable states in ecology. Unlike common perception, moving a biological population to alternate solutions necessitates some environmental damage [32–34]. As a result, a disease-related microbiome can become a beneficial microbiome. Almost all microbiomes are strong and resilient, and they revert to a similar group structure following internal disruption or new species [22, 34, 35]. Throughout the microbiome, the core microbiome comprises species that are widely considered to serve a vital part in the ecosystem's functioning. With modest modifications in ribosomal genes, the core microbiome of the human gut [36] and plants [37] was identified at the operational taxonomic unit (OTU) level. Both human and plant illnesses are weakened and destructive as a result of changes in the core microbiome. Farmers are widely known for transferring dirt from one field to another in order to utilize disease to suppress microbial populations in the soil [38]. Microbial soil research has mostly relied on cultivation-dependent approaches since the identification of numerous bacteria. Plant growth-promoting rhizobacteria (PGPR), a microbiome treatment based on the *Pseudomonas* spp. species, has the ability to eradicate this illness in such soils [26, 39, 40]. Studies now indicate that a microbiome, rather than a single or a few microorganisms, determines the complete microbial community assembly and activities in a rhizosphere [41]. Advanced technologies were used in the research [42, 43].

7.1.1 Core Microbiome Identification

Microbial products help in the identification of plant-related microorganisms in most plant specimens, allowing them to be prioritized for future investigation, including collecting and research testing. Although the microbiota associated with plants is varied, not all of them play significant roles in host biology. The primary microbiome concept allows researchers to filter away transient populations and concentrate on those that are more likely to impact host phenotypic stability. Culturally independent trials, such as internal transcribed spacer (ITS) sequencing and 16S rRNA amplicons of huge volumes of similar microbiome species, can help us get closer to our aim [3, 44–52]. Instead, sequencing multiple plant microbiomes would help with cultivation. Microbiota interact with and colonies taxonomic microbial plants in a broad range of ways. They have an influence on the parent plant's and the environment's nature and development. Furthermore, holobionts and their associated microbiome have an impact on the system's overall strength and fitness [4, 5]. Recently, a core array of microbiota species linked with a certain species or plant variety was discovered. Biogeographic studies, on the other hand, show that soil-dependent microbial reservoirs vary, casting doubt on the universal distribution of microbios in taxonomic core circumstances. In various settings, the microbiota attracted by a genotype appear to be functionally more similar than taxonomic similarities [53–55]. The ability of community members to metabolize root exudates is used to recruit the 16S rRNA gene deep-sequencing in *Arabidopsis thaliana* roots in a typical core microbiome [55, 56]. Methodological advances in defining microbial variants fuelled this focus, which was aided by large-scale polymorphism analyses in target taxonomic sequences. Proteobacteria, actinobacteria, firmicutes, and *Bacteroides* were found in order to discover groups typically linked with plants. Though this was a major breakthrough for research on microbial ecology, it did not seek to integrate the evolution, interaction, metabolism, and the fitness of working issues [57–59]. This was a major stride ahead. Therefore, in the microbial reservoir where plants recruit their mates the soil-based changes in their taxonomic centre might occur [60–62].

7.1.2 Functional Core Microbiome

Due to phylogenetic variations, it may be adaptive and necessary to define a functional core in a taxonomic significant microbiome. This functional core may be characterized using metagenomic and metatranscriptomic methods, which are likely to be important for the specific crop population under consideration [63]. Genes using phylogenetic markers such as extrapolation by ITS provide limited details. Functional DNA metagenomics and shotgun metagenomics, on the other hand, allow us to better understand community functional capability [64], and their phenotype is recognized by metatranscriptomics, metaproteomics, and metabolomics. The major community functions can be discovered by integrating multi-omic methods between large plant samples and the breadth of taxonomic functional redundancy. Comparing core or metagenomic microbiomes from important plants and genotypes of the same plant species may reveal variations in host-driven microbiomas. Although seed-borne vertical transmission [65] has happened and the atmosphere also contributes to the plant microbiota overground, soil is the primary microbial source for the plant microbiome [4, 5, 11, 20, 48, 49, 66–68]. Furthermore, cultivars that reliably build diverse microbiomes, or cultivars that reliably assemble multiple microbiomes, may rely on microbiome to satisfy diverse needs, especially in difficult settings [69]. Finally, crop productivity is influenced by plant microbiomes, phenotype, and climate interaction. Drought, heat, salt, heavy metals, sickness, and herbivores, among other stressors, can be prevented by common bacteria, particularly in difficult environs. For example, core meta-transcriptomes in dry soil crown plants may differ from those in well-watered soil crown plants, providing insight into the possible contribution of microbial genes to drought resistance [70]. Microbiome subsets that play a key role in host functioning in agriculture systems might be identified using metagenomic wide combination studies [71]. Roots influence the rhizosphere through altering pH, rhizobiome structure, and the availability of other chemicals than quorum sensor imitations, as well as by supplying dead roots and high-carbon exudates as a source of energy [6, 7]. It might be helpful to the microbiota that live in this niche, but it might be detrimental to the health of the plants. Mutual microorganisms can provide the plant with physiologically accessible phytohormones and other bioactive compounds, improve crop growth, eliminate crop pathogens, and help crops survive heat, humidity, and drought [8, 72]. Rhizosphere societies are a subset of soil microorganisms that enter the endophytic compartment through niche utilization and host interactions [9, 10]. Microbial community structure varies with plant species [11, 73], as do microbial patterns related to host genotype [12, 13]. High-performance sequencing has been used to sample the root microbiome many times utilizing reliable experimental methods. Because of the controlled experimental design and capability of deep sequencing, the results of many of the diverse research offer a crucial step in identifying the functional capability of this microbiome and the host genes, which may contribute to the connection of microbial phenotypes. The use of 454 inbred pyrosequencing in DNA sequence 16S rRNA gene amplicons of *A. thaliana's* accessions would be one of the key agronomic aims for such plant genes. Exploration of additional pairings of core microbiomes that might interact in a complementary or enabling way is an important component in the creation of core microbiomes [74]. In plant research, the creation of ideal pairing or sets of microbes greatly reduces the unpredictable nature of plant microorganisms [20, 21], their persistence, and the possible harmful consequences of microbial inoculation on ecosystems [22, 23]. Collecting and categorizing microbial activities is certainly a daunting task, and it is the most difficult aspect of developing informatics pipelines to construct core microbiomes. Many identified PGPs have been classed as sensitive to bias in intentional annotations [69], and some species in the databases have not yet been classed, including numerous endophytic organisms and fungi. More experiments are being conducted to assess genome-based microbial activity [75] as well as experimental microbe inoculation on plant crops [13, 73, 76]. New computer pipelines, however, have been developed to optimize synergistic genome-specific microbial interactions. Bacteria are thought to be generated by diverse taxonomic communities of crop and soil-related organisms that are known to be free of competition from endospheres, phyllospheres, and rhizospheres. Similarly, by using jasmonic acid (JA) or salicylic acid (SA), early occupants might alter the levels of host plants, which may alter their immune responses to late invaders [48, 49, 77]. The fundamental goal of plant microbiome management isn't always to transform preexisting structures into alternative microbiomes. Because the majority of agricultural plant species are annual and easily accessible in their early stages of growth, more attention may be focused on the formation and maintenance of a beneficial microbiome structure [44].

7.1.3 Conservative Approaches to Core Plant Microbiomes

By combining fundamental microbiome and abundance ubiquity test ideas, we may identify persistent, possibly symbiotic relationships without relying on sequencing or sample preference. Bacterial community patterns fluctuate more as a result of events than as a result of abundance. Through metagenomics and Illumina sequencing, a more comprehensive examination of bacteria associated to corals, especially the uncommon ones, has become feasible. Bacteria and their nutritional environment are well characterized in terms of their identification, location, and possible role. By combining fundamental microbiome and abundance ubiquity test ideas, we may identify persistent, possibly symbiotic relationships without relying on sequencing or sample preference. Bacterial community patterns fluctuate more as a result of events than as a result of abundance. Through metagenomics and Illumina sequencing, a more comprehensive examination of bacteria associated to corals, especially the uncommon ones, has become feasible. Bacteria and their nutritional environment are well characterized in terms of their identification, location, and possible role. Several biogeochemical processes are governed by microorganisms. Many studies have found that the microbial community composition varies geographically and with environmental changes [6, 78]. Despite this, the vast array of microbial diversity and possible functional redundancy provides little information on microbial structure at the microbial system level. In plant biology, predicting changes in the microbial composition of the environment and ecosystem processes based on functional features has received a lot of interest. According to studies, a crop community's phylogenetic range is a significant determinant of crop performance and that a specific plant trait is communicated. In support of this theory, a microcosmic research found that closer Protistans share more characteristics and that the association has influenced competition amongst organisms. Due of substantial heterogeneity in functional capacity among allied lines, phylogenetic organism relatedness is required instead of species wealth or variety to define community composition. Both community assembly and functioning may be influenced by phylogenetic distribution and functional clustering. Salt tolerance and metabolic activity at different temperatures are continuous qualities of microbes, but the metabolic ability to fix C or N is discrete. However, when the traits are well maintained, they are shared throughout deep microorganism groups and found in tiny phylogenetic classes. Multiple phylogenetic groupings spreading various traits emerge through gene loss, convergent evolution, and gene transfer, resulting in random correlations between phylogenetic and functional connections [74, 79]. This might be related to the delayed growth and lower likelihood of central gene systems [6, 19–25, 28, 35, 79, 80], which are composed of numerous proteins that interact amongst species. Functional features based on the organism's integrated complex genetic systems have also been shown to evolve slowly and to be retained more phylogenetically, resulting in deep-seated clüsters' common features. Several techniques for estimating the evolutionary rate and phylogenetic dispersion of characteristics have been proposed, including character evolution evaluation using either the highest probability or Bayesian approaches. For the phylogenetic assignment of a characteristic, several techniques are compared to random or Brownian motion evolution models. These methods fail to detect the phylogenetic combination of feature-like microbe groups because they measure the rate of development or the significance of phylogenetic clusters. ConsenTRAIT, a novel phylogenetic technique that determines the sequence similarity of organisms collecting linear features, was developed to detect the morphological sustainability of certain features in microorganisms. This study was successful in determining the distribution and degree of phylogenetic clustering of distinct microbial characteristics. The presence of genomic subsystems underlying numerous ecologically relevant processes, including as photosynthesis, CO_2 fixation, nitrogen-cycling, and the utilization of C sources for this test, gives a wide overview of the metabolic capabilities of different variety in various phylogenetic groupings. Many features, such as regulatory differences or related gene alterations, appear to be linked to genetic changes beyond the frequency of a route at about the same period. The random importance of a biotic or abiotic factor's activity in distributing and operating microbial differences is specified in general. By connecting an OTU with techniques like ARISA, T-RFLP, or a phospholipid fatty acid profile, or by combining lines with 97% sequential sequences, an operational taxonomical unit (OTU) can be defined with cluster groups or by connecting an OTU with techniques like ARISA, T-RFLP, or a phospholipid fatty acid profile.

When a microbial taxon's sequences are compared to the phylogenies of the associated characteristic, ConsenTRAIT can determine if the similarity is significant. The phylogenetic level can't always be used as a conventional definition of taxa. With contrast, in a consenTRAIT metric, it is critical to detect the evolutionary distance between the tips and the last common ancestors, whereas grouped methods like Mothr are interlinear based. Diversity may be important for community functioning and diverse roles linked with distinct phylogenetic cluster dimensions, according to Martiny et al. As a result, large changes in the Community's taxonomy may have an impact on the functioning of deep-cluster characteristics, but not on low-lying microdiversity clusters, and vice versa. Although numerous other factors, in addition to complication and likely complexity, may influence the phylogenetic distribution of the features, the dispersion and depth supported by the genomic data set analysed can also have an impact on the feature's genetic complexity. Moreover, since of its capacity to develop in both low and high light, the Prochlorococcus marine cyanobacterium may be divided into two categories. Other abiotic variables, such as light adjustment, are also likely to be complicated due to the presence of numerous interacting proteins. In high/low heat acclimatizing cells, it is also possible to separate the highly adaptable cover into a low and high iron group. The metabolic genes for NO_3 and PO_4 are also linked to clades. As a few genes are included in their clusters, it may be simpler to absorb nutrients. As a result, particular Prochlorococcus traits are clearly phylogenetically distinct, and this variance may be attributed to variations in feature complexity rather than bacteria. Extreme soil humidity tolerance in plant temperature or light was related with higher taxonomic levels in Europe. The relationship between features and complexity is a paradigm for analysing the evolutionary distribution of specific traits as well as the stability and robustness of a micro-mediated ecosystem. However, as there are millions of microbial taxa and several critical traits, it is unlikely that all relevant characteristics are split into all phylogenetic lines of microorganisms. Furthermore, given the critical role of microorganisms in many ecosystem processes, more research can help anticipate how changes in microbial biodiversity might influence a community's well-being, as well as help anticipate future changes in global biogeochemical cycles. COREMIC is now able to identify particular samples of essential habitat-specific microorganisms. Switch grass-associated bacteria are closely linked to the *Mesorhizobium* sp., *Chitinophagaceae,* and *Lysobacter* spp., among others. Antibiotics, on the other hand, help plants grow and flourish. It's unclear how effective the bacteria are at switching grass. As a result, the analysis of a large range of data yields information that considerably narrows the range of possibilities, setting the scene for controlled experiments to determine if the crucial microbiota supports or contradicts its function. The new toolkit is simple to use and supports a large number of researchers, and it is particularly significant in the field of bioinformatics, despite its limited background. This technique is used to identify microbiomic biota in general. COREMIC conducts a new study of current techniques utilizing frequency data, which is viewed as a possible benefit because it is uncertain whether a large amount of sequence information reveals true comparative differences between populations. Sample plots were put in various settings in the current investigation, which helped to generate the heterogeneity background. Despite the fact that the possibility for plant–microbial interaction is taken into consideration in this diversity of habitat circumstances, this diversity uses a conservative method to identify a very accurate and true core microbiome, regardless of heterogeneity. Researchers can investigate the impacts of habitat characteristics while constructing and evaluating COREMIC. The general conclusions that various grass species have diverse microbial communities were also implied in plant taxonomy. This method helped to support it. Plant taxonomy is also critical for detecting microbial populations in the root zone. While it is necessary to carefully evaluate experimental issues of interest while using COREMIC, it is a frequent, if not omnipresent, basis for all tests and investigations. ACC deaminase activities and IAA generation capability are also found in core discovered microbiome species from the *Chitinophagaceae* family of soil dwellers [26]. The current method should be compared to scientific studies conducted prior to the advent of metagenomics. Prior to it, finding and identifying antagonistic plants that encourage microorganisms was time-consuming and labour intensive. Hundreds of studies have been conducted to produce and identify potential microorganisms that can grow on crops. Even when acknowledged under ambient circumstances, the positive impacts of excellent microorganisms have seldom become plant-friendly under circumstances of plant development. With new data showing the importance of plant–microbial relationships in plant setting and root crop interactions for the creation of growth-supporting plant–microbial relationships, the approach used in that study shifts its focus to the search for microbes that appear to have been fully tailored for root crops. This method streamlines the search for appropriate microorganisms in a daunting pool of thousands of bacterial species. The COREMIC programme helps researchers to create working hypotheses for finding environmentally friendly microorganisms or hosts. It may also be used to double-check laboratory experiments with known target microorganisms that might be essential symbionts or are connected to a certain environment. The discovery of diazo-trophic bacteria that aid in the growth and design of bioenergy crops is an example of this. These low-nutrient waters, as well as other environmental gradients, photosynthesize endosymbiotic dinoflagellates. Furthermore, current estimations suggest that thousands of distinct phylotypes of coral have been identified as a significant component in the acclimatization, adjustment, and adaptation of eukaryotes to environmental changes in a metaorganism that has not been investigated previously. Although these interactions are meant to have an ecological effect on corals, it is difficult to discern between special and compulsory coral-bacterial symbiosis in this complex system. To assess the role and hosting contribution of bacterial communities, it is necessary to pinpoint exactly where prokaryotic interactions occur with their activities. Differentiation of the core microbioids from a host microhabitat or ecological niche is essential to create stable and consistent community connection. This technique separates microbial gut–host mammalian interactions with the environment from interactions between plant roots and comparable mammalian niche habitats. As a consequence, distinguishing between the fundamental pathways and metabolic aspects of the interaction between host and microbe will allow researchers to identify the essential components of a metaorganism. Photosynthesis and methanogenesis, for example, have been found to have complicated features encoded by many deep clusters. In the current context, the complexity of phylogenetic functional characteristics affects their retention. This approach allows for the prediction of microbial composition changes and the use of phylogenetic patterns and biogeographic models to impact microbial driven ecosystem process.

7.2 Core Microorganisms and Priority Effects in Initial Assembly

In medical research, significant progress has been achieved in connecting microbiome structure to host health, with growing awareness of common characteristics in microbial dynamics [35, 80] leading to successful manipulation of gut microbial communities [79] (e.g. faecal transplants) between human individuals [79]. However, in agriculture, we are just beginning to develop a broad framework for regulating the complexity of plant-associated microbiomes [1, 16, 28]. The ultimate objective is to find effective ways for transforming microbiomes from uncontrolled black box components to rich sources of functions in agroecosystems.

7.2.1 Microbiome Types

If microorganisms in the environment colonized plants at random, the possible combinations of coexisting species on plant shoots and roots would be enormous. Microbiomes, on the other hand, are constructed in such a manner that within a local population, individuals of a plant species may be categorized into a small number of types (hereinafter, 'microbiome types'). Plant-associated microbiomes are typically divided into a few categories based on agricultural plant varieties [30–32] and even cropping practices [32, 33]. In general, perturbing a biological community's equilibrium to create alternatives [78] necessitates particular environmental disturbance. This is why changing a disease-related microbiome type to a healthy microbiome can be difficult: microbiomes are resistant and robust, reverting to similar community structure following external disruption or the introduction of new species [24, 78, 79]. Bifidobacterium or Lactobacillus, for example, are frequently replaced by other microorganisms in the human stomach, thus their 'probiotic' benefits are short-lived unless the host-beneficial bacteria are continuously supplied by food or medicine [44]. Even if changes from unfavourable (host-damaging) microbiome types occur as a result of significant perturbations (such as antibiotic or fungicide treatment), such shifts frequently result in unanticipated microbiome structure fluctuations [45, 46]. As a result, changing established microbiome types remains a significant issue in human microbiome management [79]. Providentially, converting existing microbiome structures into new ones isn't often the primary objective of plant microbiome management. Because most agricultural plant species are annuals with easily accessible early developmental phases, we may focus our efforts on the construction and maintenance of benign microbiome structures rather than the treatment of sick host people as is required in medical science. Furthermore, whereas human intestines are fluid habitats with regular disturbances (inflow of food and accompanying microbial invaders), plant tissue environments, particularly those within root endospheres, offer some insulation from physical perturbations. Early microbial colonists are likely to survive on host plant individuals in such conditions. Microorganisms began colonizing the seedling root endospheres within 24 hours of first soil contact, according to an experimental research on rice hosts [32], and achieved a stable microbiome condition within two weeks [32].

7.2.2 Priority Effects in Initial Assembly

Small changes in early colonization processes, such as the sequence of species introduction, can result in significant variances in community structure in plant microbiomes [48, 49] (i.e. priority effects) [50]. *Pantoea agglomerans* inhibits *Pseudomonas syringae* (which causes basal kernel blight of barley) only when it is introduced to hosts before the pathogen [49]. Early colonizers have an advantage in species assembly because they can utilize space and resources sooner than other microbes, and/or they can generate physical barriers and/or antibiotics that inhibit future bacteria' colonization of the plants [63] (a process known as niche pre-emption and modification) [50]. Antibiotics are known to be produced by several taxonomic groupings of plant-associated and soil microorganisms [64, 65], which are used to keep rivals out of endospheres, phyllospheres, and rhizospheres. Seeds and seedlings with established resident communities in endospheres and rhizospheres are predicted to establish core microbial species more reliably than mature plants with established resident communities in endospheres and rhizospheres [47]. Controlling early microbiome assembly might have a substantial influence on agricultural production [64], because pathogen infection during early growth stages is one of the most damaging causes lowering yields of key crop plants (e.g. soybean) [67]. How may priority effects be used to regulate otherwise stochastic processes in early microbiome successions.

7.2.3 Deploying Core Microorganisms

Three mechanisms may be used to regulate the plant microbiome based on priority effects: (i) functional species recruitment, (ii) pathogen/pest blocking, and (iii) core reinforcement. To begin, beneficial species must be recruited first and foremost. This is aided by microbial species that may attract a large number of additional microbial species with beneficial physiological functions [1, 73]. These facilitator microorganisms would be the first to be injected as pioneer symbionts in seeds or seedlings, therefore prioritizing the recruitment of communities from natural pools of prospective microbial symbionts. *Bacillus amyloliquefaciens* has been shown to promote soybean nodulation by *Bradyrhizobium japonicum* [68], a nitrogen-fixing bacterium, making it a potential option for use as an early colonizer. Second, preferred microbiome formation offers a foundation for avoiding pathogen and pest infestation. Pathogen/pest blocking is made easier by priority effects, which allow resident (early colonizing) microorganisms to slow or prevent hostile late colonists from entering [15, 63]. While screening for plant-growth-promoting bacteria [81] will remain a popular research topic, we suggest a new approach to microbial diversity research that focuses on the functions of core microorganisms in orchestrating community-scale events. Core microorganisms (and core microbiomes) may not always support plant development directly, according to our concept (see ref. [77] for a more basic explanation). Rather, they are predicted to play an important role in the organization of plant-associated microbiomes within and surrounding host plants. When a core microbe is the only inoculum in a plant performance experiment, it might be a mutualist, commensalist, or even antagonist. We focus on microbiome functioning because microbes without direct advantages to hosts can play critical roles in particular community situations.

7.2.4 Prioritizing a Core Microbiome over Space

In macroecology, abundance-occupancy distributions are often used to think about changes in diversity over time [14–16], but they are less frequently used to think about 'core' taxa, a notion that is more actively used and researched in microbiome ecology [5, 76]. There are many research types in microbiome ecology that have a clear geographical component and allow direct application of the abundance-occupancy approach to the data set. Because of the significant environmental variability that is often imposed by the host, spatial components are inherent in plant–microbiome investigations in particular (e.g. soils chemistry, root zones) [17, 72]. Efforts are made in certain plant–microbiome experimental designs to reduce pot-to-pot, row-to-row, or field-to-field variability so that interesting characteristics to the factor of interest (e.g. plant genotype, management) may be separated [18]. Characterizing plant–microbiome alterations in response to spatial environmental gradients is a biogeographic topic in which space is the element of interest in previous investigations [19, 20]. Using abundance-occupancy distributions, core microbiomes may be inferred from any of these geographical research designs. However, for the computation of taxon occupancy, the geographical research design is critical. The most straightforward method for calculating occupancy is to have all discrete samples contribute equally to the total, represented as a proportion of one (or a percentage out of 100%). This method is the most conservative, as it limits the core to just those taxa that are found in every sample, and so favours more numerous taxa (see neutral model discussion, ahead). When it comes to replication over time, it's sometimes more ecological to look at the replicates as a whole and exclude taxa that are incompatible. Alternatively, occupancy might be thought of as detection inside a location/treatment, with the taxon being considered as occurring there as long as it was represented in each location (albeit not necessarily in all replicates within that location). Because this method is less cautious, there may be false positive taxa that are later removed from the core set after functional testing. Though, depending on the community structure, this method is more likely to include species with a medium to low abundance. To establish the ecologically supported core membership and threshold, a thorough investigation of the core is recommended [21], but it should be supplemented with ecological information gained through the experimental design and research question addressed. To assess the core membership's explanatory value for beta diversity, we suggest an iterative exploration. After ranking taxa by occupancy, the aggregate beta diversity contribution of each ranked taxon is determined, and finally, a threshold is applied that implies declining benefits when adding the next ranked-taxon another geographical concern is the range of the biogeography area included in the study, due to the considerable environmental variation among soils. With occupancy of 1, a greenhouse research using a normal potting soil mix should expect higher variety and fewer taxa than a greenhouse study using a standard potting soil mix. The adoption of a core set based on abundance-occupancy distributions can provide a consistent strategy for selecting biologically relevant species throughout place and time, even when there are no one-size-fits-all inclusion criteria.

7.2.5 Prioritizing a Core Microbiome over Time

Longitudinal studies in microbiome ecology are becoming more common, and are of particular interest for plants-microbiome because of known changes in phytohormone and root exudate production over plant development [22–24]. Another interesting temporal component is how plants with perennial or biannual life histories may differently recruit and maintain core microbiome members from plants with annual strategies [25]. In addition, some plant compartments, like epiphytic phyllosphere surfaces, may be expected to experience greater environmental fluctuations over time (seasons) than other compartments, like root endophytes [19].

However, there is much work to be done to understand how the microbiota that live in these compartments change temporally, and, because destructive sampling is typically required, there can be logistical challenges in executing temporal replication within a study's design. Therefore, quantitative temporal studies of core plant–microbiota are at their onset and the field is relatively open. Abundance-occupancy distributions have been applied in other systems to identify core microbiota over time. 'Persistent' and 'transient' were used to categorize microbiome members by occupancy bins over a daily time series in the human gut [35].

A neutral model was applied to abundance-occupancy distributions that were calculated from data collected over discrete time points in zebrafish development [80]. Given the need to understand microbiome changes over plant development or with seasonal changes, these approaches are conceptually aligned with goals in plant–microbiome research.

As for calculating occupancy over space, there are considerations for collapsing the occupancy axis to include only the most ecologically relevant series for the question rather than the entire sample set. This is, arguably, more important for defining core microbiome over time because occupancy is indicative of persistence: which taxa are most consistently detected in association with the plant, even as the environment changes over time. Persistence is a key ecological consideration for the maintenance of system stability, which is often a translational outcome desired in understanding core microbiota. In addition, the temporal dynamics of core plant microbiota offer important baseline series against which to compare disturbances (e.g. for crops, growing seasons with flood, drought, or other management changes).

We have recently used abundance-occupancy distributions to prioritize a core in the perennial switchgrass phyllosphere (epiphytes), grown in the field for up to two seasons in a replicated block design [18]. We discuss this work in detail to show how the occupancy calculation can be ecologically informed. Because we were interested in the seasonal trends in microbiome structure, replicated plots were crucial for identifying consistent core members, and occupancy was conservatively defined to include taxa detected in all replicate fields at a given time point. Defining the core set as having an occupancy of 1 at any time point accounted for the seasonal aspect of the data set, and then we were able to observe which of those members were also persistent over the growing season and across two years. We found that much of the core membership had high mean abundance and also were persistent seasonally, and that it was the collective dynamics of these core members that defined the temporal patterns in betadiversity for the study. In summary, setting a core microbiome occupancy threshold of 1 for a longitudinal study is likely overly conservative when environmental changes occur, but a replicated time series can offer support for prioritizing core taxa that are key at particular times and under particular seasonal conditions.

7.2.6 Neutral Model to Inform Core Taxa That Are Deterministically Assembled

The general shape of the abundance-occupancy distribution is expected to be an 'S', with the most abundant taxa exhibiting highest occupancies and the rare taxa exhibiting lowest occupancies. This is the 'neutral' expectation of the data that assumes unlimited dispersal of microbial cells and no fitness differences among taxa. This neutral model was discussed in detail and brought to the attention of many microbial ecologists by Sloan and Curtis in their theoretical explorations and applications of abundance-occupancy distributions from wastewater treatment microbial communities [26, 79]. The neutral model is used as a null hypothesis of community assembly, and is often tested in parallel with deterministic models to ask which model best describes the overall community assembly. However, the neutral model can offer an interesting extension to prioritizing core microbiota because it can discriminate taxa that are expected to be deterministically selected from taxa that are expected to be neutral (e.g. occupancy as expected, given mean abundance). Taxa that fall outside of the confidence intervals meet a deterministic expectation. In particular, taxa that fall above the neutral model prediction are in higher occupancy than expected by their mean abundance in the data set, and taxa that fall below the prediction are in higher abundance than expected by their occupancy. For plant microbiomes, we may interpret this to form a hypothesis that, in the former case, the plant environment selects for these taxa, and, in the latter case, that these taxa are dispersal limited. Parallel interpretation was previously applied in the developing zebra gut, human lungs and skin, as well as from a wider range of hosts and environments [28–34]. Among the core set of taxa selected by abundance and occupancy, those that are deterministically selected above the model fit (selected by the plant environment) can be given heighten priority for robust colonization efficiency and/or the potential to have interactions with the plant. Therefore, we recommend exploring the assembly mechanisms of core taxa to glean deeper ecological insights into their distributions and to guide functional follow-up studies.

7.3 Informatics of Microbial Networks

While a number of studies have demonstrated priority effects in plant microbiome assembly [48, 50, 63] finding potential core microorganisms with preferable priority effects is a challenge. To explore candidate core microorganisms, it is important to understand how microbial species in endospheres and rhizospheres interact with each other [74]. Network theory provides a 'bird's-eye' view of potential microorganism–microorganism interactions [51, 63, 82] by allowing us to understand which species are most likely to mediate interactions among many others within microbiomes [16, 69, 70, 80, 83]. Combined with 'bottom-up' approaches for evaluating the persistence and ecological effects of synthetic communities [17, 28, 71, 81] studies targeting in natural dynamics of microbial interactions will help us optimize relationships between introduced and resident microbiomes in agroecosystems.

7.3.1 Microbial Networks

Diverse functional groups of microorganisms interact with each other within and around plants, forming complex webs of interactions. Recent studies have demonstrated that the exact physiological effects of microorganisms on host plants will vary considerably depending on the community structure of plant microbiomes [84]. Such background networks of microorganisms can be complex, not just in diversity but also in structure. For example, plant–microorganism interactions can involve symbionts that are physically nested within another symbiont. These partners can play critical roles: bacteria found inside hyphae of mycorrhizal fungi can contribute to plant–fungal symbioses by fixing atmospheric nitrogen or promoting fungal mycelial growth [10, 75]. Likewise, microorganisms can form unexpected connections with other organisms. For instance, *Metarhizium robertsii*, a common soil fungus, is both an endophyte and an insect pathogen. Its unique position allows it to provide insect-derived nitrogen to its plant host through fungal mycelia in exchange for host carbon [85].

To encapsulate these dimensions of microbiome complexity, a number of bioinformatic approaches have been developed to analyse microbial diversity and interaction networks based on high-throughput DNA sequencing [51, 86]. Nonetheless, most plant microbiome studies have focused on either bacterial or fungal biomes, rather than both. Moreover, most current studies analyse networks connecting coarsely defined (often at the 97% sequence similarity level) operational taxonomic units that can hide meaningful ecological variation, while informatics tools to perform strainor individual-sequence-level analyses are now becoming available [87].

Given the huge number of potential microorganism–microorganism associations, statistical pipelines for analysing network architecture are essential [51, 63, 82]. These can be classified into two types. One is based on the 'snapshot' information of microbiome structure, in which signs of co-occurrence (or non-coexistence) can be statistically examined in each pair of microbial species [29]. For this type of analysis, compositions of microorganisms are evaluated simultaneously for tens or hundreds of plant/soil samples and then network links are inferred in pairs of microorganisms that show strong statistical signs of co-occurrence or segregated distribution [51]. These snapshot analyses are suitable for revealing patterns in microorganism–microorganism co-occurrence, which may indicate interactions or shared environmental preferences (i.e. niche overlap) among microbial species [29, 88]. Although snapshot networks provide less information about functional inter-microorganism interactions than networks inferred with more sophisticated analyses (e.g. time-series analytical methods discussed ahead), they represent overall consequences of direct and indirect interactions, in which host gene expression profiles altered by mycorrhizal fungi, for example, may influence colonization success of some pathogens and/or nitrogen-fixing bacteria [10, 18, 19].

Other emerging analytical methods for time-series community analyses (e.g. empirical dynamic modelling) [89–91], sparse S-map [92] and transfer entropy [93] have the potential to reveal a more dynamic nature of microorganism–microorganism interactions [70]. For example, a time-series study of mouse gut bacteriome networks suggested that both Allobaculum and Ruminococcaceae bacteria positively influenced a Rikenellaceae bacterium when the mice were young, while the Rikenellaceae bacterium had negative impacts on Clostridiales and Ruminococcaceae bacteria in old mice [92]. Those analyses provide ways for inferring not only microorganism–microorganism interaction networks, but also more comprehensive webs of interactions among microbial population dynamics, abiotic environmental fluctuations, and gene expression profiles of respective host and microbial species [70]. They will also help us evaluate how fluctuations in environmental conditions (e.g. temperature, precipitation, and edaphic factors) and/or host properties (e.g. strains and physiological conditions) affect community dynamics [89, 90], as well as how priority effects of early colonizers persist through host plant development. The precision and applicability of these analyses, however, depend on the length and stochastic noise of particular time series as well as properties of the target system's dynamics (e.g. its dimensionality and nonlinearity) [94]. In addition, relative abundance data obtained from high-throughput sequencing should be used with caution as they provide less information than absolute abundance data in inferring microorganism–microorganism and microorganism–host interactions [95]. Methods based on control-DNA calibration of high-throughput sequencing samples [96] provide absolute abundance data and are promising for improving our ability to elucidate how microbial interactions change through time.

7.4 Designing Core Microbiomes

While challenging, informatics of microbial network structure will help us design core microbiomes to be introduced, with top priority, to plants and agroecosystems. As databases and related informatics tools for inferring functional roles of microorganisms are increasing in sophistication and depth [86, 97], microorganisms in network data can be classified, albeit not comprehensively, into functional groups. The basic idea is to use such annotated network data to score microorganisms in terms of their potential positive impact. This can be done in silico by modifying existing network metrics. In Box 7.1, we illustrate how to identify microbial species that are placed, within a microbial network at central positions, interlinking other microorganisms with positive functions for host plants. This approach is also used to score species based on their potential ability to mediate interactions among pathogenic species in a network – species with low scores are prospective in avoiding or slowing the entry of pathogenic species to host plants. Based on the two criteria of functional species recruitment and pathogen/pest blocking, microorganisms in a network can be ranked to nominate candidate core microorganisms.

When each microbial species in a network is scored in terms of its potential for recruiting preferable species and blocking pathogens/pests, the next step is to explore best pairs or sets of core microorganisms. A best pair of core symbionts, for example, would be explored in a microbial network data set by taking into account each microorganism's score for functional species recruitment and pathogen/pest blocking as well as compatibility between two focal microorganisms. In other words, exploring pairs or sets of core microorganisms that potentially interact with each other in facilitative or complementary ways [69, 98], would be a key step for designing core microbiomes. Such strategies of using pairs/sets of microorganisms have been applied in medical studies, wherein inoculated sets of microorganisms had persisted over time and had suppressed pathogens inside the human gut [99]. In plant science, unpredictability in the persistence levels of microorganisms introduced to plants [22, 23] and possible deleterious effects of microbial inoculation to ecosystems [24, 25] are potentially reduced by designing optimal pairs/sets of microorganisms. Pioneering attempts to inoculate multiple microbial species have shown varied effects on plant performance [14, 15] stressing the need for more in planta or in natura studies. The largest challenge in building informatics pipelines for designing core microbiomes is collecting and cataloguing microbial functions. This is a daunting task. The majority of species in databases, including many endophytic bacteria and fungi, have yet to be classified and many known plant-growth-promoting species have been classified as putative saprotrophs [70, 97] potentially causing biases in functional annotations. While an increasing number of studies are estimating microbial functions based on genome structure [86] and experimental inoculation of microbial strains to crop plants [11, 14, 17] we need to make more effort to collect and curate those data systematically. Nonetheless, novel informatics pipelines for maximizing synergistic microbial interactions based on genome data, and integrating multi-omics data (e.g. metagenomic, transcriptomic, and metabolomic data), are becoming available [100, 101]. Moreover, because microbiomes often involve species with the same, or similar, functions (i.e. functional redundancy) [79], we may not need to annotate all microbial species in a network when we explore some of the potential best sets of core microorganisms.

7.4.1 Criterion for Nominating Core Microorganisms

We propose a network theoretical framework for finding optimal core microorganisms based on three-step criteria.

7.4.1.1 Functional Species Recruitment

In terms of potential roles in functional species recruitment, each species (species i) in microorganism–microorganism network data can be scored by a simple index as follows:

$$F_i = \sum_{i \neq k \neq l} \frac{w_k w_l \sigma_{kl}(i)}{\sigma_{kl}},$$

where species k and l are from the set of functional species (MF) in the microbiome (i.e. k, $l \in$ MF), σ_{kl} is the number of shortest paths between species k and l, and $\sigma_{kl}(i)$ is the number of shortest paths between species k and l that pass through the focal species i. Unlike most network theoretical indices [83], this index has user-defined weighting parameters; i.e. w_k and w_l for weighting the functions of species in the microbiome. We incorporated this flexibility into the index because individual farmers may vary in their strategy for prioritizing multiple microbial functions. In the management of agroecosystems suffering from root-knot nematodes, for example, one may want to put higher weighting values on nematophagous endophytes than on mycorrhizal fungi. The weighting parameters may be also used for incorporating results of inoculation experiments comparing plant-growth-promoting abilities among microbial species/strains [11, 68]. Meanwhile, tools for annotating functional roles are increasingly available for both bacteria [86] and fungi [97], although their underlying databases need to be continually enriched.

In practice, we need more ecological information for annotating functional microbial species (MF) and setting species-specific weights (w_k and w_l). In most available microbial networks, a high proportion of species have not yet been ecologically annotated. However, given that species with the same functional roles often coexist in nature (i.e. functional redundancy in microbiomes) [79], the aforementioned network index will help us find core microorganisms even with our current knowledge of microbial functions.

7.4.1.2 Pathogen/Pest Blocking

Core microorganisms can be used not only for recruiting functional symbiont species, but also for blocking pathogens and pests in agroecosystems. By defining a set of plant pathogens/pests (MP) within a microbial network, a species i may be scored as follows:

$$E_i = -\sum_{i \neq m \neq n} \frac{w_m w_n \sigma_{mn}(i)}{\sigma_{mn}},$$

where species m and n are from the pathogen/pest set MP and σ_{mn}, $\sigma_{mn}(i)$, w_m, and w_n are defined as in the aforementioned recruitment index F_i. With user-supplied parameters weighting disease severity among pathogens (w_m and w_n), each microbial species will be evaluated in terms of its potential roles in blocking the entry of the defined pathogens/pests (m, $n \in$ MP) to host plants.

By integrating functional species recruitment and pathogen/pest blocking, the total benefits of a microbial species for its host plant will be scored as follows:

$$B_i = F_i + rE_i,$$

where r is a parameter for balancing two aspects of roles in microbiomes. The index is then standardized so as to vary from 0 to 1 as follows:

$$B_i' = \frac{B_i - B_{\min}}{B_{\max} - B_{\min}},$$

where $B_{\min}$ and $B_{\max}$ are the minimal and maximal scores of B_i within a network, respectively.

7.4.1.3 Core Reinforcement

When each microbial species is evaluated in a network, we can further explore sets of microorganisms that promote the formation of robust microbiomes. In exploring best pairs of core species, we can take into account roles of respective species as well as compatibility between focal two species as follows:

$$R_{ij} = B_i' B_j' C_{ij},$$

where B_i' and B_j' are total benefit scores of species i and j, respectively, and C_{ij} represents co-occurrence patterns observed in microbial network data. Several statistical approaches for evaluating co-occurrence patterns for pairs of microbial species (or strains) (C_{ij}) are now available [29, 51], allowing us to calculate the R_{ij} index and thereby predict reinforcement effects between pairs of microorganisms. However, the index does not represent metabolic interactions involving multiple microbial species. Therefore, the network-based method outlined here should be applied as a complementary, but not an alternative, approach to explore sets of cooperative microorganisms based on metagenomic and metatranscriptomic analyses of metabolic pathways [62].

7.5 Management of Agroecosystems with Core Microbiomes

Emerging research focussing on core microbiomes will offer ways to mediate interactions between plants and their external biotic and abiotic environments. However, diversity and functionality of microorganisms, as well as microbial network structure, depend greatly on crop plant species/varieties and field conditions [3, 21, 33, 52], for example, resident soil biota, particle size, pH, nutrient availability, and tillage cycles. Therefore, we need to accelerate feedback between informatics and field monitoring to understand how crop production can be determined by predictability of function as linked to identity (e.g. similarity in phylum- or order-level compositions of optimal microbiomes among sites or crop plant species/varieties) and regionality in resident microbiome compositions and functions. Moreover, biotic/abiotic conditions can change through time in agroecosystems, potentially having great impact on the success of core microbiome introduction. We, hereafter, discuss how recent technical advances in microfluidics, robotics, and computer sciences are integrated towards next-generation agriculture, through which we can propose core microbiomes optimized for real-time conditions of local farms and plantations.

7.5.1 Logistics of Core Microbiomes

Developing standardized methods for collecting and inoculating core microorganisms onto hosts will require innovative research lines. While many endophytic bacteria and fungi can be easily isolated from the environment [36, 71, 102] there may be core microorganisms that remain unculturable with standard techniques. In such cases, the development of microfluidic chips offers promise because these devices can precisely control density, shape, and size of microbial communities [37, 103]. Physical micro-scale structure prevents the dominance of one species over others, and controlled diffusion of metabolites means that individual bacteria have a greater chance of survival. Such microfluidic devices [104] for culturing microorganisms give hope to the possibility of adding novel microbial species/strains to culture collections. Sophisticated methods for making 'cocktails' of microorganisms for host inoculation are likewise essential as in medical science, faecal microbiome transplantation has been recognized as a promising therapy for otherwise intractable disease [79] (e.g. ulcerative colitis). Analogous approaches for crop plants [7] would involve developing bulk soil transplants or extracted soil microbiome cocktails using centrifuge-based techniques, whose application to soybean has successfully allowed us to enrich symbiont cells from healthy plant tissue [105, 106] (Figure 7.1a). Microbial cocktails could be further characterized and deployed using microdroplet techniques [107], in which pairs or sets of microorganisms are isolated (Figure 7.1b). To realize more complex environmental conditions optimized for bacterial and/or fungal co-cultures, a series of emerging technologies and devices (such as X-ray computed tomography, three-dimensional printing, photothermal etching, and various types of microfluidic chips) will allow us to introduce spatial heterogeneity and chemical gradients at micrometre scales [37, 103].

In embedding core microbiomes in agroecosystems, automated systems for inoculating plants with selected microorganisms are required (Figure 7.1c). Single-cell chemical phenotyping with Raman spectroscopy enables rapid screening of microbial species/strains from mixed microbial cultures [108]. The sorted microorganisms can be automatically inoculated to plants using the Root Chip or layered micro fluidic devices [103, 109, 110] which provide high-throughput platforms for placing sterilized seeds/seedlings on the culture/co-culture media of core microorganisms. Chemicals promoting successful establishment of core microbiomes in host tissue may be simultaneously supplied with such microfluidic devices as inspired by prebiotics, in which galacto-oligosaccharides are used to support Bifidobacterium populations inside host gut [111]. Core-microbiomeinoculated plants might then be introduced to agroecosystems using unmanned planting machines [20].

7.5.2 Portfolios with Multiple Cores

In maximizing contributions of core microbiomes within agroecosystems, biodiversity theory provides key insights [112, 113]. By inoculating differential sets of core microorganisms to individual seeds/seedlings, for example, neighbouring plant

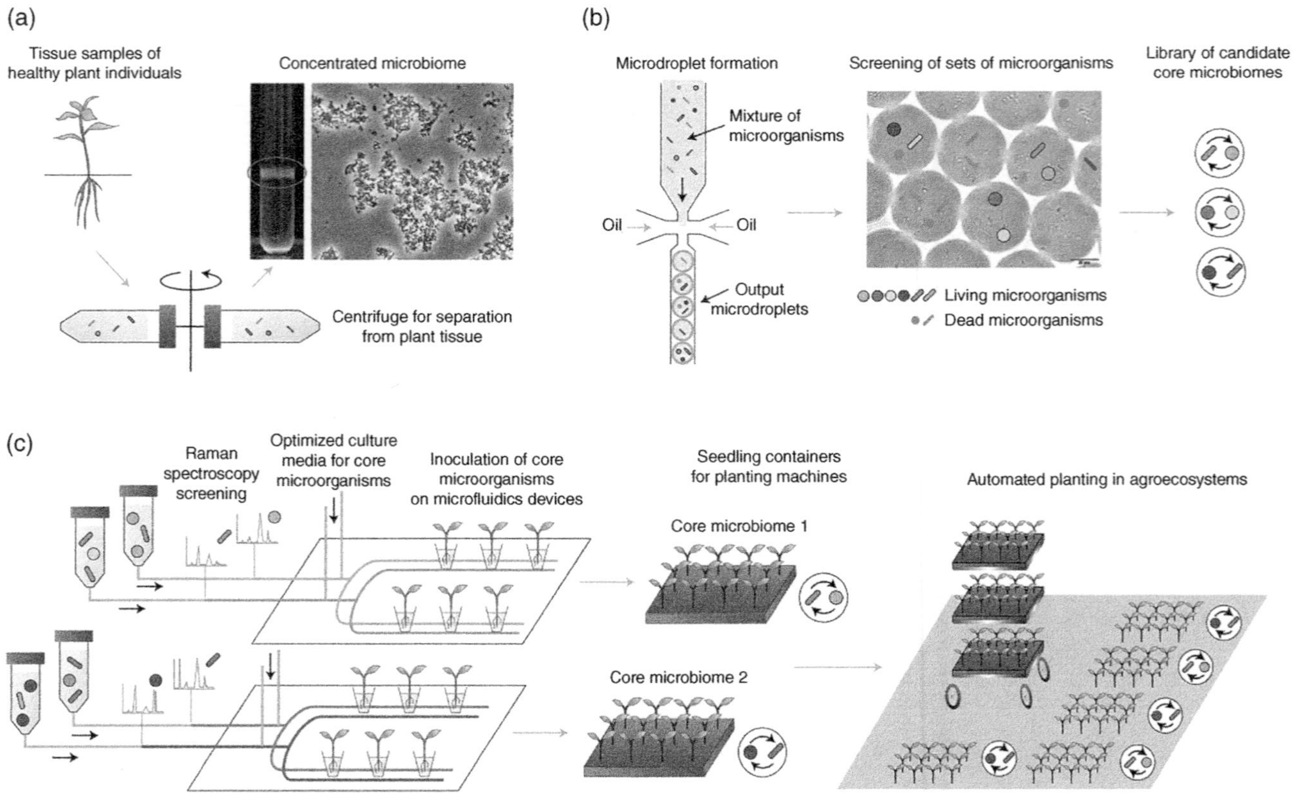

Figure 7.1 Preparing and deploying core microbiomes. (a) Microbiome cocktails. Symbiont cells obtained from healthy plant individuals can be enriched by centrifuge-based methods. (b) Application of microdroplet technologies for isolating and screening core microbiomes. Microdroplet devices that were originally developed for the isolation of single cells will be used for screening pairs/sets of core microorganisms interacting with each other in facilitative ways (i.e. core microbiomes). (c) Deploying core microbiomes. Single-cell phenotyping/screening based on Raman spectroscopy and microfluidics devices can be used for high-throughput inoculation of sets of core microorganisms into seedlings. The use of multiple core microbiomes will increase heterogeneity within agroecosystems, contributing to the recruitment of diverse functional microorganisms and the prevention of pathogen/pest outbreaks [106].

individuals in an agroecosystem may attract more or less different compositions of functional microbial species in the environment (Figure 7.1c). The resultant diversity in established microbiomes may work as insurance for crop production against abrupt weather events and/or the spread of specific pathogen strains at the ecosystem level, as predicted by ecological theories of portfolio effects [113]. Moreover, with high microbial diversity within and around roots, soil phosphorous or nitrogen will be more efficiently supplied to plants as a consequence of functional complementarity among coexisting microorganisms [9, 27]. Accordingly, taking advantage of such functional complementarity and portfolio effects provided by microbiomes, we propose using multiple core microbiomes, which differ in taxonomic and/or functional compositions of constituent microorganisms, within a farmland. When introducing microbial portfolios into real agroecosystems, theoretical studies will help us optimize spatial and temporal configurations of crop plant individuals differing in core microbiomes. Given that spatial distribution of multiple plant varieties within a farmland for example, 'multilines' of wheat, oat, rice, and potatoes [114] can reduce impacts of pathogen outbreaks [38], the best layout of multiple core microbiomes may be explored by simulation studies to prevent the spread of infectious disease. For example, it would be important to examine how spatial autocorrelations in core microbiome structures can determine agroecosystem-level resistance to pathogen outbreaks. Likewise, considering that rotation of crop plant species is a key technique for suppressing pathogen and pest populations and avoiding soil nutrient imbalance [39, 112], studies exploring best microbiome rotation are prospective. Ecosystem scale models of disease and nutrient dynamics will help us design spatiotemporal configurations of multiple core microbiomes, as well as those of multiple crop plant species and genotypes, in agroecosystem management [40].

7.5.3 Smart Farming with AI and Robots

Even if selected and designed core microbiomes are successfully introduced to agroecosystems, they may occasionally be replaced with unfavourable microbiomes after environmental perturbation. Moreover, in continual cropping systems, microbiomes may be easily altered due to outbreaks of native infectious microorganisms adapted to specific host plants. Shifts to disease-related microbiome types ('dysbiosis') have been also reported in medical science [41]. In patients with diarrhoea and inflammatory bowel disease, for example, gut microbiomes are characterized by decreased taxonomic diversity and by a disproportionate increase of virulent bacteria (e.g. *Clostridium difficile*) after antibiotic treatments or other kinds of host physiological perturbation [41]. Innovations in data-driven medicine are expected to provide platforms for predicting and preventing dysbiosis. Artificial intelligence (AI) based on neural networks, which can outperform experienced radiologists in interpreting medical (e.g. X-ray) images [42], has come to integrate genome, epigenome, immunome, metabolome, and microbiome data – also known as 'high-definition medicine', such AI-based methods show ways to prevent and delay disease onset in human individuals [43]. Developing such multi-omics diagnostic platforms for smart farming (or site-adapted agriculture) is a prerequisite for detecting 'early warning signals' [53] of microbiome dysbiosis and related plant physiological disorder in agroecosystems [21]. As Newton's second law was re-discovered by supplying AI with motion-tracking data of various physical systems [54], under lying dynamics of microbiomes may be revealed by machine learning algorithms that integrate time-series observational data of plants in field conditions [55]. Nonetheless, time-series data are much less available in agriculture than in medical science in terms of both quantity and quality; accumulating standardized data with high temporal resolution is a prerequisite for AI-based management of agroecosystem states.

Albeit challenging, emerging technologies for field monitoring of plant physiology and microbiome structure will help us develop platforms for smart farming in the next 5–10 years. Imagery from unmanned aerial vehicles (UAVs), for example, now allows physiological phenotyping and disease monitoring at the individual-plant level, revolutionizing the spatial and temporal resolution of field data [56, 57]. Such UAVs can be equipped with mobile phones and a series of portable analytical devices, providing remote sensing data [58]. In particular, when combined with mobile DNA/RNA sequencers, UAV-based systems might allow us to monitor microbiomes and transcriptomes of plants under field conditions. Expression profiles of phosphate transporter genes, for instance, may be used as indicators of both nutritional and immune conditions of plants [11, 17]. By unifying multiple lines of those monitoring data, AI-based informatics (e.g. several machine learning methods) [55] and transfer learning [59] and novel mathematical methods for tracking biological community stability [91] will allow us to explore best timing for introducing plants with core microbiomes and to forecast disease outbreaks in agroecosystems. Some of those technologies are ready for application in agroecosystems; however, integrating all the different components into a successful interdisciplinary approach is still a fundamental challenge.

References

1 Berendsen, R.L., Pieterse, C.M., and Bakker, P.A. (2012). The rhizosphere microbiome and plant health. *Trends Plant Sci.* 17: 478–486.

2 de Vries, F. and Wallenstein, M. (2017). Below-ground connections underlying above-ground food production: a framework for optimising ecological connections in the rhizosphere. *J. Ecol.* 105: 913–920.

3 Busby, P.E., Soman, C., Wagner, M.R. et al. (2017). Research priorities for harnessing plant microbiomes in sustainable agriculture. *PLoS Biol.* 15: e2001793.

4 Elser, J. and Bennett, E. (2011). Phosphorus cycle: a broken biogeochemical cycle. *Nature* 478: 29–31.

5 Callaway, E. (2016). Devastating wheat fungus appears in Asia for frst time. *Nature* 532: 421–422.

6 Howden, S.M. et al. (2007). Adapting agriculture to climate change. *Proc. Natl. Acad. Sci. U. S. A.* 104: 19691–19696.

7 Dangl, J.L., Horvath, D.M., and Staskawicz, B.J. (2013). Pivoting the plant immune system from dissection to deployment. *Science* 341: 746–751.

8 Robertson, G.P. and Vitousek, P.M. (2009). Nitrogen in agriculture: balancing the cost of an essential resource. *Annu. Rev. Environ. Resour.* 34: 97–125.

9 van der Heijden, M.G. et al. (1998). Mycorrhizal fungal diversity determines plant biodiversity, ecosystem variability and productivity. *Nature* 396: 69–72.

10 Bonfante, P. and Anca, I.-A. (2009). Plants, mycorrhizal fungi, and bacteria: a network of interactions. *Annu. Rev. Microbiol.* 63: 363–383.

11 Hiruma, K. et al. (2016). Root endophyte Colletotrichum tofeldiae confers plant ftness benefts that are phosphate status dependent. *Cell* 165: 464–474.

12 Mendes, R. et al. (2011). Deciphering the rhizosphere microbiome for diseasesuppressive bacteria. *Science* 332: 1097–1100.

13 Zavala-Gonzalez, E.A. et al. (2017). Arabidopsis thaliana root colonization by the nematophagous fungus Pochonia chlamydosporia is modulated by jasmonate signaling and leads to accelerated fowering and improved yield. *New Phytol.* 213: 351–364.

14 Calvo, P., Nelson, L., and Kloepper, J.W. (2014). Agricultural uses of plant biostimulants. *Plant Soil* 383: 3–41.

15 Arnold, A.E. et al. (2003). Fungal endophytes limit pathogen damage in a tropical tree. *Proc. Natl. Acad. Sci. U. S. A.* 100: 15649–15654.

16 Vorholt, J.A., Vogel, C., Carlström, C.I., and Müller, D.B. (2017). Establishing causality: opportunities of synthetic communities for plant microbiome research. *Cell Host Microbe* 22: 142–155.

17 Castrillo, G. et al. (2017). Root microbiota drive direct integration of phosphate stress and immunity. *Nature* 543: 513–518.

18 Jacott, C.N., Murray, J.D., and Ridout, C.J. (2017). Trade-ofs in arbuscular mycorrhizal symbiosis: disease resistance, growth responses and perspectives for crop breeding. *Agronomy* 7: 75.

19 Oldroyd, G.E. (2013). Speak, friend, and enter: signalling systems that promote benefcial symbiotic associations in plants. *Nat. Rev. Microbiol.* 11: 252–263.

20 King, A. (2017). Te future of agriculture. *Nature* 544: 21–23.

21 Schlaeppi, K. and Bulgarelli, D. (2015). The plant microbiome at work. *Mol. Plant Microbe Interact.* 28: 212–217.

22 Girlanda, M. et al. (2001). Impact of biocontrol Pseudomonas fuorescens CHA0 and a genetically modifed derivative on the diversity of culturable fungi in the cucumber rhizosphere. *Appl. Environ. Microbiol.* 67: 1851–1864.

23 Streeter, J.G. (1994). Failure of inoculant rhizobia to overcome the dominance of indigenous strains for nodule formation. *Can. J. Microbiol.* 40: 513–522.

24 Castro-Sowinski, S., Herschkovitz, Y., Okon, Y., and Jurkevitch, E. (2007). Efects of inoculation with plant growth-promoting rhizobacteria on resident rhizosphere microorganisms. *FEMS Microbiol. Lett.* 276: 1–11.

25 Walsh, U. et al. (2003). Residual impact of the biocontrol inoculant Pseudomonas fuorescens F113 on the resident population of rhizobia nodulating a red clover rotation crop. *Microb. Ecol.* 45: 145–155.

26 Paramsothy, S. et al. (2017). Multidonor intensive faecal microbiota transplantation for active ulcerative colitis: a randomised placebo-controlled trial. *Lancet* 389: 1218–1228.

27 Maherali, H. and Klironomos, J.N. (2007). Infuence of phylogeny on fungal community assembly and ecosystem functioning. *Science* 316: 1746–1748.

28 Müller, D.B., Vogel, C., Bai, Y., and Vorholt, J.A. (2016). The plant microbiota: systems-level insights and perspectives. *Annu. Rev. Genet.* 50: 211–234.

29 Toju, H., Yamamoto, S., Tanabe, A.S. et al. (2016). Network modules and hubs in plant-root fungal biomes. *J. R. Soc. Interface* 13: 20151097.

30 Bulgarelli, D. et al. (2015). Structure and function of the bacterial root microbiota in wild and domesticated barley. *Cell Host Microbe* 17: 392–403.

31 Marasco, R., Rolli, E., Fusi, M. et al. (2018). Grapevine rootstocks shape underground bacterial microbiome and networking but not potential functionality. *Microbiome* 6: 3.

32 Edwards, J. et al. (2015). Structure, variation, and assembly of the root-associated microbiomes of rice. *Proc. Natl. Acad. Sci. U. S. A.* 112: 911–920.

33 Hartman, K. et al. (2018). Cropping practices manipulate abundance patterns of root and soil microbiome members paving the way to smart farming. *Microbiome* 6: 14.

34 Arumugam, M. et al. (2011). Enterotypes of the human gut microbiome. *Nature* 473: 174–180.

35 Lloyd-Price, J. et al. (2017). Strains, functions and dynamics in the expanded Human Microbiome Project. *Nature* 550: 61–66.

36 Usuki, F. and Narisawa, K. (2007). A mutualistic symbiosis between a dark septate endophytic fungus, Heteroconium chaetospira, and a nonmycorrhizal plant, Chinese cabbage. *Mycologia* 99: 175–184.

37 Aleklett, K. et al. (2018). Build your own soil: exploring microfuidics to create microbial habitat structures. *ISME J.* 12: 312–319.

38 Suzuki, S.U. and Sasaki, A. (2011). How does the resistance threshold in spatially explicit epidemic dynamics depend on the basic reproductive ratio and spatial correlation of crop genotypes? *J. Theor. Biol.* 276: 117–125.

39 Isbell, F. et al. (2017). Benefts of increasing plant diversity in sustainable agroecosystems. *J. Ecol.* 105: 871–879.

40 Prieto, I. et al. (2015). Complementary efects of species and genetic diversity on productivity and stability of sown grasslands. *Nat. Plants* 1: 15033.

41 Pham, T.A.N. and Lawley, T.D. (2014). Emerging insights on intestinal dysbiosis during bacterial infections. *Curr. Opin. Microbiol.* 17: 67–74.

42 Shen, D., Wu, G., and Suk, H.-I. (2017). Deep learning in medical image analysis. *Annu. Rev. Biomed. Eng.* 19: 221–248.

43 Torkamani, A., Andersen, K.G., Steinhubl, S.R., and Topol, E.J. (2017). Highdefnition medicine. *Cell* 170: 828–843.

44 Kim, S.-W. et al. (2013). Robustness of gut microbiota of healthy adults in response to probiotic intervention revealed by high-throughput pyrosequencing. *DNA Res.* 20: 241–253.

45 Dethlefsen, L. and Relman, D.A. (2011). Incomplete recovery and individualized responses of the human distal gut microbiota to repeated antibiotic perturbation. *Proc. Natl. Acad. Sci. U. S. A.* 108: 4554–4561.

46 Ng, K.M. et al. (2013). Microbiota-liberated host sugars facilitate post-antibiotic expansion of enteric pathogens. *Nature* 502: 96–99.

47 Hacquard, S. et al. (2015). Microbiota and host nutrition across plant and animal kingdoms. *Cell Host Microbe* 17: 603–616.

48 Werner, G.D. and Kiers, E.T. (2015). Order of arrival structures arbuscular mycorrhizal colonization of plants. *New Phytol.* 205: 1515–1524.

49 Braun-Kiewnick, A., Jacobsen, B.J., and Sands, D.C. (2000). Biological control of Pseudomonas syringae pv. syringae, the causal agent of basal kernel blight of barley, by antagonistic Pantoea agglomerans. *Phytopathology* 90: 368–375.

50 Fukami, T. (2015). Historical contingency in community assembly: integrating niches, species pools, and priority efects. *Annu. Rev. Ecol. Evol. Syst.* 46: 1–23.

51 Faust, K. and Raes, J. (2012). Microbial interactions: from networks to models. *Nat. Rev. Microbiol.* 10: 538–550.

52 Peay, K.G. et al. (2017). Convergence and contrast in the community structure of Bacteria, Fungi and Archaea along a tropical elevation–climate gradient. *FEMS Microbiol. Ecol.* 93: 5.

53 Schefer, M., Carpenter, S.R., Dakos, V., and van Nes, E.H. (2015). Generic indicators of ecological resilience: inferring the chance of a critical transition. *Annu. Rev. Ecol. Evol. Syst.* 46: 145–167.

54 Schmidt, M. and Lipson, H. (2009). Distilling free-form natural laws from experimental data. *Science* 324: 81–85.

55 Murphy, K.P. (2012). *Machine Learning: A Probabilistic Perspective.* London: MIT Press.

56 Sugiura, R. et al. (2016). Field phenotyping system for the assessment of potato late blight resistance using RGB imagery from an unmanned aerial vehicle. *Biosyst. Eng.* 148: 1–10.

57 Araus, J.L. and Cairns, J.E. (2014). Field high-throughput phenotyping: the new crop breeding frontier. *Trends Plant Sci.* 19: 52–61.

58 Quesada-González, D. and Merkoçi, A. (2017). Mobile phone-based biosensing: an emerging "diagnostic and communication" technology. *Biosens. Bioelectron.* 92: 549–562.

59 Pan, S.J. and Yang, Q. (2010). A survey on transfer learning. *IEEE Trans. Knowl. Data Eng.* 22: 1345–1359.

60 Brundrett, M.C. (2002). Coevolution of roots and mycorrhizas of land plants. *New Phytol.* 154: 275–304.

61 Foster, K.R., Schluter, J., Coyte, K.Z., and Rakof-Nahoum, S. (2017). The evolution of the host microbiome as an ecosystem on a leash. *Nature* 548: 43–51.

62 Franzosa, E.A. et al. (2014). Relating the metatranscriptome and metagenome of the human gut. *Proc. Natl. Acad. Sci. U. S. A.* 111: 2329–2338.

63 Wei, Z. et al. (2015). Trophic network architecture of root-associated bacterial communities determines pathogen invasion and plant health. *Nat. Commun.* 6: 8413.

64 Weller, D.M. (1988). Biological control of soilborne plant pathogens in the rhizosphere with bacteria. *Annu. Rev. Phytopathol.* 26: 379–407.

65 Morris, C.E. and Monier, J.-M. (2003). The ecological signifcance of bioflm formation by plant-associated bacteria. *Annu. Rev. Phytopathol.* 41: 429–453.

66 Pieterse, C.M. et al. (2014). Induced systemic resistance by benefcial microbes. *Annu. Rev. Phytopathol.* 52: 347–375.

67 Wrather, J.A. and Koenning, S.R. Efects of diseases on soybean yields in the United States 1996 to 2007. *Plant Health Prog.* https://doi.org/10.1094/PHP-2009-0401-01-RS 2009.

68 Masciarelli, O., Llanes, A., and Luna, V. (2014). A new PGPR co-inoculated with Bradyrhizobium japonicum enhances soybean nodulation. *Microbiol. Res.* 169: 609–615.

69 Coyte, K.Z., Schluter, J., and Foster, K.R. (2015). The ecology of the microbiome: networks, competition, and stability. *Science* 350: 663–666.

70 Toju, H. et al. (2017). Species-rich networks and eco-evolutionary synthesis at the metacommunity level. *Nat. Ecol. Evol.* 1: 24.

71 Bai, Y. et al. (2015). Functional overlap of the Arabidopsis leaf and root microbiota. *Nature* 528: 364–369.

72 Tilman, D., Balzer, C., Hill, J., and Befort, B.L. (2011). Global food demand and the sustainable intensifcation of agriculture. *Proc. Natl. Acad. Sci. U. S. A.* 108: 20260–20264.

73 Freilich, S. et al. (2011). Competitive and cooperative metabolic interactions in bacterial communities. *Nat. Commun.* 2: 589.

74 Barea, J.-M., Pozo, M.J., Azcon, R., and Azcon-Aguilar, C. (2005). Microbial co-operation in the rhizosphere. *J. Exp. Bot.* 56: 1761–1778.

75 Bianciotto, V. et al. (1996). An obligately endosymbiotic mycorrhizal fungus itself harbors obligately intracellular bacteria. *Appl. Environ. Microbiol.* 62: 3005–3010.

76 Robertson, G.P. et al. (2017). Cellulosic biofuel contributions to a sustainable energy future: choices and outcomes. *Science* 356: eaal2324.

77 Lundberg, D.S. et al. (2012). Defning the core Arabidopsis thaliana root microbiome. *Nature* 488: 86–90.

78 Beisner, B.E., Haydon, D.T., and Cuddington, K. (2003). Alternative stable states in ecology. *Front. Ecol. Environ.* 1: 376–382.

79 Lozupone, C.A., Stombaugh, J.I., Gordon, J.I. et al. (2012). Diversity, stability and resilience of the human gut microbiota. *Nature* 489: 220–230.

80 Bashan, A. et al. (2016). Universality of human microbial dynamics. *Nature* 534: 259–262.

81 Paredes, S.H. et al. (2018). Design of synthetic bacterial communities for predictable plant phenotypes. *PLoS Biol.* 16: e2003962.

82 Agler, M.T. et al. (2016). Microbial hub taxa link host and abiotic factors to plant microbiome variation. *PLoS Biol.* 14: e1002352.

83 Newman, M.E.J. (2010). *Networks: An Introduction*. New York: Oxford University Press.

84 Hartman, K., van der Heijden, M.G., Roussely-Provent, V. et al. (2017). Deciphering composition and function of the root microbiome of a legume plant. *Microbiome* 5: 2.

85 Behie, S.W. et al. (2017). Carbon translocation from a plant to an insectpathogenic endophytic fungus. *Nat. Commun.* 8: 14245.

86 Langille, M.G. et al. (2013). Predictive functional profling of microbial communities using 16S rRNA marker gene sequences. *Nat. Biotechnol.* 31: 814–821.

87 Callahan, B.J. et al. (2016). DADA2: high-resolution sample inference from Illumina amplicon data. *Nat. Methods* 13: 581–583.

88 Delgado-Baquerizo, M. et al. (2018). A global atlas of the dominant bacteria found in soil. *Science* 359: 320–325.

89 Deyle, E.R., May, R.M., Munch, S.B., and Sugihara, G. (2016). Tracking and forecasting ecosystem interactions in real time. *Proc. R. Soc. Lond. B Biol. Sci.* 283: 20152258.

90 Sugihara, G. et al. (2012). Detecting causality in complex ecosystems. *Science* 338: 496–500.

91 Ushio, M. et al. (2018). Fluctuating interaction network and time-varying stability of a natural fsh community. *Nature* 554: 360–363.

92 Suzuki, K., Yoshida, K., Nakanishi, Y., and Fukuda, S. (2017). An equation-free method reveals the ecological interaction networks within complex microbial ecosystems. *Methods Ecol. Evol.* 8: 1774–1785.

93 Schreiber, T. (2000). Measuring information transfer. *Phys. Rev. Lett.* 85: 461–464.

94 Chang, C.-W., Ushio, M., and Hsieh, C.-H. (2017). Empirical dynamic modeling for beginners. *Ecol. Res.* 32: 785–796.

95 Vandeputte, D. et al. (2017). Quantitative microbiome profling links gut community variation to microbial load. *Nature* 551: 507–511.

96 Smets, W. et al. (2016). A method for simultaneous measurement of soil bacterial abundances and community composition via 16S rRNA gene sequencing. *Soil Biol. Biochem.* 96: 145–151.

97 Nguyen, N.H. et al. (2016). FUNGuild: an open annotation tool for parsing fungal community datasets by ecological guild. *Fungal Ecol.* 20: 241–248.

98 van der Heijden, M.G., de Bruin, S., Luckerhof, L. et al. (2016). A widespread plant-fungal-bacterial symbiosis promotes plant biodiversity, plant nutrition and seedling recruitment. *ISME J.* 10: 389–399.

99 Sheth, R.U., Cabral, V., Chen, S.P., and Wang, H.H. (2016). Manipulating bacterial communities by in situ microbiome engineering. *Trends Genet.* 32: 189–200.

100 Mee, M.T., Collins, J.J., Church, G.M., and Wang, H.H. (2014). Syntrophic exchange in synthetic microbial communities. *Proc. Natl. Acad. Sci. U. S. A.* 111: 2149–2156.

101 Fondi, M. and Liò, P. (2015). Multi-omics and metabolic modelling pipelines: challenges and tools for systems microbiology. *Microbiol. Res.* 171: 52–64.

102 Narisawa, K., Hambleton, S., and Currah, R.S. (2007). Heteroconium chaetospira, a dark septate root endophyte allied to the Herpotrichiellaceae [Chaetothyriales] obtained from some forest soil samples in Canada using bait plants. *Mycoscience* 48: 274–281.

103 Li, R., Lv, X., Zhang, X. et al. (2016). Microfuidics for cell-cell interactions: a review. *Front. Chem. Sci. Eng.* 10: 90–98.

104 Nichols, D. et al. (2010). Use of ichip for high-throughput in situ cultivation of “uncultivable” microbial species. *Appl. Environ. Microbiol.* 76: 2445–2450.

105 Ikeda, S. et al. (2009). Development of a bacterial cell enrichment method and its application to the community analysis in soybean stems. *Microb. Ecol.* 58: 703–714.

106 Toju, H., Peay, K.G., Yamamichi, M. et al. (2018). Core microbiomes for sustainable agroecosystems. *Nat. Plants* 4: 247–257. https://doi.org/10.1038/s41477-018-0139-4.

107 Hosokawa, M. et al. (2015). Droplet-based microfuidics for high-throughput screening of a metagenomic library for isolation of microbial enzymes. *Biosens. Bioelectron.* 67: 379–385.

108 Song, Y., Yin, H., and Huang, W.E. (2016). Raman activated cell sorting. *Curr. Opin. Chem. Biol.* 33: 1–8.

109 Grossmann, G. et al. (2011). The RootChip: an integrated microfuidic chip for plant science. *Plant Cell* 23: 4234–4240.

110 Massalha, H., Korenblum, E., Malitsky, S. et al. (2017). Live imaging of root–bacteria interactions in a microfuidics setup. *Proc. Natl. Acad. Sci. U. S. A.* 114: 4549–4554.
111 Monteagudo-Mera, A. et al. (2016). High purity galacto-oligosaccharides enhance specifc Bifdobacterium species and their metabolic activity in the mouse gut microbiome. *Benefic. Microbes* 7: 247–264.
112 Tilman, D., Cassman, K.G., Matson, P.A. et al. (2002). Agricultural sustainability and intensive production practices. *Nature* 418: 671–677.
113 Yachi, S. and Loreau, M. (1999). Biodiversity and ecosystem productivity in a fuctuating environment: the insurance hypothesis. *Proc. Natl. Acad. Sci. U. S. A.* 96: 1463–1468.
114 Mundt, C. (2002). Use of multiline cultivars and cultivar mixtures for disease management. *Annu. Rev. Phytopathol.* 40: 381–410.

Further Reading

Cassan, F. et al. (2009). Azospirillum brasilense Az39 and Bradyrhizobium japonicum E109, inoculated singly or in combination, promote seed germination and early seedling growth in corn [Zea mays L.] and soybean [Glycine max L.]. *Eur. J. Soil Biol.* 45: 28–35.

Knights, D. et al. (2014). Rethinking "enterotypes". *Cell Host Microbe* 16: 433–437.

Layeghifard, M., Hwang, D.M., and Guttman, D.S. (2017). Disentangling interactions in the microbiome: a network perspective. *Trends Microbiol.* 25: 217–228.

Schefer, M., Carpenter, S., Foley, J.A. et al. (2001). Catastrophic shifs in ecosystems. *Nature* 413: 591–596.

8

Microbiome Mediated

Stress Alleviation in Agroecosystems

CONTENTS

Microbiomics and Sustainable Crop Production, First Edition. Mohammad Yaseen Mir and Saima Hamid.

8.1 Effect of Biotic and Abiotic Stresses on Plants

Plants can be stressed by changes in environmental circumstances. Quality losses in real postharvest product might be directly or indirectly attributed to a mix of stress and stress-induced senescence, since stress is typically defined as any environmental condition potentially adverse to living organisms, save for decay [1]. Stresses pose a major danger to agriculture and are the leading cause of crop failure throughout the world [2]. These stressors cause a slew of morphological, physiological, biochemical, and molecular changes in plants, all of which wreak havoc on their development and production. Plants experience stress when they are subjected to adverse conditions, and they express just a portion of their genetic potential. Stress resistance is a genetic trait that plants use to adapt to adverse environments. Plants can be trained to mobilize defensive responses more quickly or more intensely, resulting in increased stress resistance and stress acquisition. As a result, plant research, particularly in postharvest produce, has turned its attention to the impacts of stress on metabolism and performance.

8.1.1 Biotic and Abiotic Stresses

Other living organisms such as bacteria, fungus, nematodes, protists, insects, viruses, and viroids provide biotic stress on plants. Abiotic stress, on the other hand, is described as the consequences of a variety of environmental conditions, including drought, salt, high temperatures, chemical toxicity, and oxidative stress. Abiotic stressors are linked and have comparable effects. Plants have evolved ways to cope with and adapt to environmental biotic and abiotic stressors.

Similar cell signalling pathways and physiological responses, such as the synthesis of stress proteins, antioxidant upregulation, and the accumulation of suitable solutes, are activated by these different stressors. Multiple signalling pathways that stimulate gene transcription and its downstream machinery therefore govern plant stress responses. Many genes and biochemical molecular processes are involved in a plant's reaction to stress. Stress tolerance molecular control methods are based on the expression of particular stress-related genes and may be achieved through the use of molecular tools (Figure 8.1).

8.1.2 Biotic Stress

Infectious illnesses, such as bacteria, fungus, or yeasts, induce biotic stress, also known as deterioration, in harvested fruit. Plants have a defensive mechanism that helps them cope with biotic stress. The defensive mechanism is divided into two categories: innate and systemic responses. Reactive oxygen species (ROS) are produced after infection, and oxidative bursts restrict pathogen spread [3]. Plants

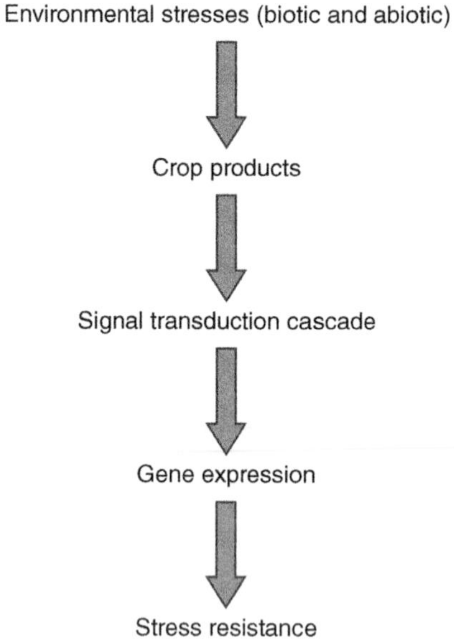

Figure 8.1 Representation of stress responsive mechanisms in horticultural products. *Source:* Madani et al. [2]/with permission of Elsevier.

also enhance cell lignification in response to pathogen attacks. This process prevents parasite invasion and lowers the vulnerability of the host. Morphological and structural barriers, chemical substances, proteins, and enzymes are all examples of biotic stress defences. By preserving goods and providing them strength and stiffness, they give tolerance or resistance to biotic stressors. Specific chemical substances such as – aminobutyric acid (BABA) or benzothiadiazole can be used to increase biotic stress tolerance (BTH).

In biotic stress signalling, plant hormones, salicylic acid (SA), jasmonic acid (JA), and ethylene all play important roles. Multiple transcription factors (TFs) have a role in hormone signalling. Many signal transduction pathways are involved in plant defence against biotic stressors. Abscisic acid (ABA) is the primary hormone involved in abiotic stress sensing [4]. ABA, on the other hand, has a protective effect against biotic stress [5]. Under abiotic and biotic stress, ABA interacts with ethylene in an antagonistic manner, causing the plant to become vulnerable to disease assault. Under abiotic stress, however, ABA levels rise and stomatal closure occurs. As a result, biotic assailants are unable to enter through the stomata. As a result, the plant is shielded against abiotic and biotic stress in such conditions [5]. Plant defence is boosted when kinase protein signals interact with ROS and ABA [5]. PR proteins are important for plant resistance to infections and when plants are damaged; their expression is highly increased. Increased ABA production of certain TFs like as C-repeat binding factors (CBFs) and cup-shaped cotyledon mediated by ABA may result in overexpression of PR genes [4].

8.1.3 Abiotic Stress

Plants can be stressed by changes in the environment. The effects of stress on metabolism and performance have been a significant focus of plant study, as stress is defined as any environmental element that may be harmful to living organisms. When plants are subjected to adverse conditions, they experience stress and only express a portion of their genetic potential. Plants can be sensitized to mobilize defensive responses more quickly or intensely, resulting in increased stress resistance and abiotic stress resistance. Drought, salt, high temperatures, chemical toxicity, and oxidative stress are all major hazards to agriculture, and are the leading cause of horticulture product loss across the world. Abiotic stress causes morphological, physiological, biochemical, and molecular changes in plants that have a negative impact on growth and production. These stressors influence phytochemical accumulation or loss by altering the activity of important enzymes in metabolic pathways. These stressors are frequently intertwined and result in the formation of ROS. Given that ROS have been involved in mechanisms leading to plant stress acclimatization, the involvement of ROS under abiotic stress has become a topic of significant interest. Phytotoxicity can be caused by high ROS levels, although acclamatory signalling can be achieved with very low levels. When it comes to increasing plant production, abiotic stress signalling is crucial. Abiotic stressors cause changes in plants at many levels, from molecular to organ, influencing plant production and altering the pathways involved in secondary metabolite biosynthesis. Multiple signalling pathways that stimulate gene transcription and its downstream machinery therefore govern plant stress responses. Many genes and biochemical-molecular processes play a role in the plant's response to abiotic stress. The expression of particular stress-related genes is being used to further understand the molecular regulatory mechanisms of abiotic stress tolerance, which might lead to the application of molecular tools. Water and ion uptake and transport, such as ion transporters and aquaporins, osmoprotectants, free-radical scavengers, membrane and protein protection, such as heat shock proteins (HSPs) and chaperones, late embryogenesis abundant proteins, and factors involved in signalling cascades and transcriptional control are among the major tolerance mechanisms. HSPs are known to be expressed in plants in response to a variety of abiotic stressors, including water stress, salinity, and osmotic, cold, and oxidative stress. These proteins serve a crucial role in stress resistance and the restoration of cellular homeostasis [6].

8.1.4 Water Stress

One of the most significant abiotic factors affecting horticultural product quality is water stress. It causes drooping, shrivelling, loss of firmness and crispness, colour loss, peel damage, and off-flavouring, among other things. The impact of water loss on horticulture goods influences their shelf life and quality. Many physiological and biochemical systems in plants are harmed when they are subjected to water stress. Even a little amount of water loss from transpiration triggers a cascade of metabolic reactions. Reduced leaf carbon fixation owing to stomatal closure, which can begin at modest plant water shortages, is one of the most significant effects of decreasing water availability. The severity and duration of stress, as well as the plant species and stage of development, all influence how plants respond to water stress at various organizational levels. Water stress prevents plant cell division and growth. Plant cell protection homeostasis systems have been demonstrated to have a positive role in the water stress response, delaying death and enhancing survival under unrealistic stress circumstances. When the water potential outside the plant tissue decreases, plants experience water stress. The plant's water balance is exactly proportional to its water potential. Its gradient is a major factor in water transportation. Reduced water supply due to drought or cold temperatures, as well as increased osmosis due to salt, may induce a reduction in water potential. Plants suffer from severe impacts as a result of water shortages. Loss of cell turgor, lower water potential of plant cells and growth rate, and decreased stomatal conductance, photochemical efficiency, and chlorophyll content are all responses to water stress. When water is removed from the membrane, the usual bilayer structure is disrupted, and the membrane becomes very porous. Displacement of membrane integrity, selectivity, and disruption of cellular compartmentalization may also occur as a result of stress within the lipid bilayer, as well as a loss of activity of enzymes that are largely membrane-based. Furthermore, when dehydrated, cytosolic and organelle proteins may show decreased activity or even total denaturation. Because of the high concentration of cellular electrolytes caused by protoplasm dehydration, cellular metabolism may be disrupted [7]. Under stress, plants can respond and adapt by changing their cellular metabolism and activating numerous defensive systems. Water stress causes the generation and removal of ROS to be activated [7]. Antioxidant enzymes accumulate antioxidant metabolites and organic osmolytes increase, and the concentration of ABA rises as part of the ROS defence system. Cell protection homeostasis systems, such as antioxidant enzymes and osmolytes, have been found to be beneficially engaged in the water stress response, delaying death and enhancing survival under a variety of stress situations.

8.1.5 Transpiration

Water vapour flows from the surface of the plant organ to the surrounding air during horticulture product transpiration. The phenomena are governed by Fick's law, a physical rule. The concentration gradient across the barrier and the surface of the barrier are directly proportional to the flow of any gas in or out of the plant, whereas the barrier's resistance to diffusion is inversely proportionate. The differential of water vapour pressure between plant tissue and the surrounding environment is the driving factor behind transpiration. The physiological degradation of horticultural products is influenced by a number of factors, including transpiration. Water stress in harvested product is common when transpiration isn't balanced by a water supply from the parent plant. It has an impact on how things look, feel, taste, and weigh. The value of degradation varies depending on the commodity. Postharvest losses and poor quality in leaf crops such as lettuce, spinach, and cabbage are thought to be mostly caused by transpiration. Wilting, shrinkage, and sharpness are all caused by evaporation [7].

8.1.6 Water Loss

In horticultural goods, postharvest water loss is caused by respiration and diffusion through the surface. The quality of horticultural goods degrades as a result of water loss during postharvest handling and distribution. It also hastens senescence or ripening symptoms like as softening, membrane degradation, and yellowing. In horticultural goods, the surface area to volume ratio determines sensitivity to water loss. This ratio can range anywhere from 0.2 cm^2 in winter cabbage or turnip to 50 100 cm^2 in leaf crops [8]. A great deal of variation is found across species and cultivars. Maturity, environmental circumstances, and surface features including stomata, lenticels, hair, periderm, and cuticle all play a role in this diversity. Depending on the temperature and relative humidity, the water potential gradient from inside to outside the items fluctuates during storage. The vapour pressure deficit, which is the connection between the difference in agricultural product water activity and the water activity of the atmosphere surrounding it, is the driving factor behind water loss [8]. Horticultural products undergo some metabolic alterations as a result of water loss. The activity of polysaccharide hydrolysing enzymes increases when osmolyte concentration rises in response to water loss. As a result, loss of cell wall integrity and associated increases in soluble sugars has been reported in plant tissues, such as carrots. Polygalacturonase activity is upregulated in response to water stress. Furthermore, water loss causes ethylene production, which causes banana fruit to mature faster and bell pepper fruit to senesce faster [9].

8.1.7 Temperature Stress

8.1.7.1 Chilling Stress

Chilling damage (CI) is a physiological condition that develops when tropical or subtropical fruits are subjected to low but not freezing temperatures. Atypical ripening and damaged regions on the fruit surface, such as pitting, scald, hard lumps in the pulp surrounding the vascular bundles, water soaking of the flesh, and a high vulnerability to decay, are macroscopic signs of CI. In freezing sensitive crops, these alterations are linked to cell wall integrity. The cell membrane is the first cell structure to be impacted by CI. The cell membrane phase transitions from flexible liquid crystalline to a rigid gel form with cooling temperature, increasing cell membrane dysfunction. Increased phospholipase D (PLD) and lipoxygenase (LOX) activities decreased cell membrane integrity and increased CI in cooling damaged plants. Energy is required for CI formation, according to reports. Energy charges were associated with higher pitting occurrence and longer cold storage led to a reduction in adenosine triphosphate (ATP) and adenosine diphosphate (ADP) levels. ATP maintenance at high levels may help to relieve CI. Cucumbers treated with 6-benzylaminopurine, for example, had higher ATP content and were less prone to CI. Chilling stress is also said to enhance ROS production, which affects membrane structure. By activating antioxidant capabilities in the fruit tissues, the destruction of ROS might protect the fruit against CI. Furthermore, the symptoms of CI browning have been connected to enzymatic activity, which is associated with brown pigments. The activity of phenylalanine ammonia lyase (PAL) has been postulated as the limiting factor for low-temperature-induced browning in banana fruit, for example. Plant growth regulators including SA, polyamines (PAs), and nitric oxide (NO) are some of the therapies that might help with CI.

8.1.7.2 Freezing Stress

That deterioration is observed as wilting or softening of plant parts. When a plant freezes, ice formation occurs within the plant tissues, and this ruptures cell membranes, causing loss of cellular integrity and ultimately the death of plant tissue. The primary manifestation of damage from freezing is in the plasma membrane. Thus, membranes play a key role in the ability of a plant cell to withstand or be subject to freezing injury.

Plants survive freezing either by preventing the crystallization of ice within their tissues by supercooling of tissue water (freezing avoidance) or they allow ice to crystallize in the apoplast (freezing tolerance). Freezing avoidance involves supercooling and prevention of ice crystallization in the apoplast. Pure water can be supercooled to a certain point below 0 °C without ice nucleation. Some specialized cell types and organs can use super cooling as a strategy. Many plant species' growth, development, and dispersion are hampered by freezing temperatures, which constitute a key environmental restriction. The difference in sensitivity is caused by the susceptibility of plants to freezing stress. Freezing damage susceptibility varies among horticultural goods. Some people may be able to tolerate cold stress without harm, while others may be permanently harmed. At temperatures below the freezing point of water, freezing injury can occur.

8.1.7.3 Heat Stress

A heat stress might change protein, membrane and RNA structural stability as well as the efficiency of enzyme activities in a cell, causing metabolic imbalance [10] as well as hazardous chemical build-up such as ROS. Heat-stressed unfolded proteins attach molecular chaperones to unfolded proteins, whereas heat stress TF (HSF) subunits trimerize, undergo phosphorylation, and bind to heat stress response promoters in the genome to activate heat stress responses [10]. There's also the possibility that heat stress (which alters cell fluidity) and ROS signalling (mostly respiratory bursts), which are mediated by plasma membrane NADP oxidases in the plasma membrane, might establish an intracellular channel for calcium. In response to heat stress, some data suggests that smRNAs may also be involved. HSPs, pathogen-resistant proteins (PRs), antioxidant enzymes, and polyphenols accumulate as a result of short-term exposure to high-temperature stress [10].

8.1.7.4 Low-Oxygen Atmosphere and High-Carbon-Dioxide Atmosphere

Depending on the quantity of these gases, temperature, exposure time, and commodity, exposing horticultural goods to reduced O_2 and/or higher CO_2 can be helpful or detrimental. Many agricultural goods have been found to benefit from controlled atmosphere (CA) or modified atmosphere (MA) storage, which uses lower O_2 and/or higher CO_2 to retain quality and hence increase shelf life. Delay in ripening, reduced physiological and pathological problems, and the possibility of disinfesting goods are all advantages of CA or MA storage. CA gas composition is monitored, and any deviations from predefined points are rectified. The gas composition of MAs is determined by a balance between plant gas consumption or production and gas diffusion via a permeable membrane. Too far, a lot of study has been done to see how CA and MA storage affects the quality and storability of a variety of horticulture goods and cultivars. Despite the huge economic importance of CA or MA storage, there is a danger that using low O_2 or high CO_2 atmospheres to maintain the quality of fresh horticultural items production during CA or MA storage could harm the produce. To properly analyse storage conditions, a deeper knowledge of fundamental biochemical and physiological reactions to low O_2 or high CO_2 atmospheres is required.

8.1.7.5 Low-Oxygen Atmosphere

By reducing postharvest deterioration, CA and MA can extend the storage life of fruits. CA storage slows the rate of respiration and lowers disease vulnerability. To avoid anaerobic metabolism, at least 1% O_2 must be present in the storage. When oxygen levels are low, mitochondrial respiration slows, and aerobically generated CO_2 drops, resulting in an increase in fermentation, which is linked to low-oxygen repercussions and a rise in total CO_2 production, leading glycolysis to increase, which is known as the Pasteur effect. Fermentation pathways are activated in cells when oxygen levels are low, lowering ATP levels. Nicotinamide adenine dinucleotide (NADH) is used to convert acetaldehyde to ethanol during fermentation. Pyruvate reduction produces lactate as well. As a result, fermentative metabolisms make use of NADH and pyruvate, allowing glycolysis to continue. Under low oxygen levels, most plants make ethanol and lactate. Because of fermentative metabolism, the induction of pyruvate decarboxylase, alcohol dehydrogenase, and lactate dehydrogenase are pathways for anaerobic product build-up [2]. Storage with low oxygen when concentrations of O_2 fall below the Pasteur threshold for lengthy durations of anaerobiosis, physiological problems are often found in apples and pears. In 'Conference' pear and 'Braeburn' apple, low O_2 in storage rooms alters the energy state of the fruit, resulting in negative alterations in membrane phospholipids and an increase in internal diseases. Furthermore, low O_2 levels can promote the growth of anaerobic psychrotrophic bacteria, which can be reduced by utilizing high-oxygen MA packing.

8.1.7.6 High-Carbon-Dioxide Atmosphere

Storage with low oxygen when concentrations of O_2 fall below the Pasteur threshold for lengthy durations of anaerobiosis, physiological problems is often found in apples and pears. In 'Conference' pear and 'Braeburn' apple, low O_2 in storage rooms alters the energy state of the fruit, resulting in negative alterations in membrane phospholipids and an increase in internal diseases. Furthermore, low O_2 levels can promote the growth of anaerobic psychrotrophic bacteria, which can be reduced by utilizing high-oxygen MA packing. Low O_2 concentrations might also exacerbate it. Fruits stored in high-CO_2 environments had greater polyphenol oxidase (PPO) activity than fruits stored in normal CA environments, and PPO activity was found to be positively linked with fruit browning. Core-browning was more common in pears kept at 5% CO_2 compared to 0.7% CO_2, which was linked to lipid peroxidation build-up. Reduced antioxidant enzyme activity might lead to lipid peroxidation and browning [11].

8.1.7.7 Ethylene and Nonethylene Volatiles

Furthermore, CO_2 sensitivity has a hereditary component. Apple cultivars such as 'Gala', 'Golden Delicious', and 'Jonagold' are less vulnerable to CO_2 damage than others. Fruit quality was also harmed by increased CO_2 levels, which caused colour changes and hardness, as well as pectic component solubilization, which could lead to cell wall disintegration. Ethylene is necessary for the synthesis of aromatic volatiles in fruits. During the ripening process, many fruits have been demonstrated to have a relationship between ethylene and volatile production. Various volatile compounds in the apple 'Greensleeves' fruit were reduced or stimulated using ethylene. Ethylene treatment enhanced the synthesis of esters and alcohols, particularly butyl acetate, hexyl acetate, and 2-methylbutyl acetate. TF RIN may play a role in the synthesis of ethylene and aroma volatiles during tomato ripening. The 'Royal Gala' apple fruit produced no detectable ethylene when antisense ACS and 1-aminocyclopropane-1-carboxylic acid oxidase (ACO) were used, resulting in a reduction in aroma volatiles. These findings indicated that during fruit ripening, ethylene modulates enzymes involved in the final phase of ester formation. Furthermore, through enhancing aminotransaminase activity and increasing the relative expression of CmBCAT1 and CmArAT1, ethylene aids in the creation of aromatic volatile chemicals from the catabolism of amino acids such valine, leucine, isoleucine, phenylalanine, and cysteine. Varied storage temperatures have different implications for volatile production. For example, 'Odem' mandarins stored at 2C accumulated 13 volatiles, the majority of which were terpenes, but 'Odem' mandarins stored at 8C accumulated six volatiles, five terpenes, and one terpene. The amount of volatile compounds such as esters and lactones was lowest in peach 'Hujingmilu' fruit preserved at 5C. The levels of 3-methylbutanal, linalool, guaiacol, hexanol, trans-2-hexenal, and trans-3-hexenol in refrigerator-stored tomato fruit changed. The synthesis of seven essential volatile compounds, including 6-methyl-5-hepten-2-one, -ionone, trans-2-hexenal, hexanal, 2-phenylethanol, and 2-methylbutanal, was reduced after freezing exposure. After freezing the tomato, the aroma volatiles of C6 aldehyde and alcohol were reduced, which may be explained by a decrease in the hydroperoxide and alcohol dehydrogenase enzyme activity in the oxylipin pathway. In addition, the effect of low temperature on taste loss could be linked to cultivar. The effects of chilling temperature on the decrease of volatile compounds were more pronounced in tomato fruit from cv Cappricia RZ than in tomato fruit from cv Amoroso RZ. CA and MA storage have an impact on aroma production. Alcohols such as ethanol and acetaldehyde build in MA storage, which are primarily associated with anaerobic or high CO_2 conditions.

8.1.7.8 Light

Appropriate light intensity and quality can influence the quality of horticultural goods after harvest. Low preharvest light intensity reduced total chlorophyll concentration in cucumber skin, resulting in a lower shelf life for greenhouse-grown cucumbers. The shelf life of both intact leaves and fresh-cut product was found to be longer when lettuce was grown under different light intensities. The increased shelf life resulted in increased chlorophyll fluorescence levels. Postharvest illumination of fresh-cut product at low light intensities considerably prolonged the shelf life of fresh-cut product when compared to storage in darkness. Because low-light-treated items contain more sugar, which reduces hunger, they have a longer shelf life. Although light isn't essential for ascorbate synthesis, the amount and intensity of light received over the course of the growing season has an impact on the amount of ascorbate produced. Furthermore, it appeared that exposing the fruit products to light enhanced the synthesis of ascorbate. Ascorbate levels were found to be lower in greenhouse-grown items. Furthermore, continuous irradiation with blue LED light (470 nm) at 50 mol/m^2s intensity increased ascorbate in the juice sacs of three citrus varieties: Satsuma mandarin, Valencia orange, and Lisbon lemon. UV radiation stresses plants and stimulates their defence systems, which leads to the production of phytoalexin compounds. Phytoalexin production may be accompanied by other inducible defences such as cell wall modifications, defence enzymes, and antioxidant activity [2]. The UV portion of the electromagnetic spectrum has wavelengths ranging from 10 to 400 nm. UV light has been used to make items that last a long time (UV-A: 315400 nm). UV radiation dosages of long (UV-A: 315400 nm), medium (UV-B: 280315 nm), and short wave (UV-C: 100280 nm) wavelengths were utilized to produce. The shortest wavelengths in the UV spectrum are the most energetic, making them more effective biocides for surface sterilization of specific food items. As a result of UV radiation, plant development is inhibited, and metabolic activities such as photosynthesis and respiration are altered. Increased ROS production is connected to UV-induced damage, resulting in oxidative stress. Plants, on the other hand, have developed a variety of anti-oxidant defence mechanisms to resist ROS. Plants with higher antioxidant defences are more UV resistant. Furthermore, plants' enzymatic antioxidant system is triggered in parallel with secondary metabolism [12]. UV radiation has improved the quality and shelf life of certain fresh agricultural products after harvest [13]. Low UV exposure results in the production of antifungal chemicals, a delay in ripening, and a drop in CI.

Fruit softening, which is one of the primary variables influencing fruit postharvest life, is delayed by UV-C exposure. UV-C inhibited the activity of enzymes involved in the breakdown of tomato cell walls, delaying fruit softening. UV-C treatment improves the nutritional quality of tomato fruit by increasing ascorbate and total phenolic content. At the genetic level, UV light can influence physiological systems. UV-B increases the expression of genes involved in the flavonoid biosynthesis pathway in parsley, including chalcone synthase and PAL, which are important enzymes in anthocyanin production. Increased ROS production is connected to UV-induced damage, resulting in oxidative stress. Plants, on the other hand, have developed a variety of anti-oxidant defence mechanisms to resist ROS. Plants with higher antioxidant defences are more UV resistant. Furthermore, plants' enzymatic antioxidant system is triggered in parallel with secondary metabolism [14]. UV radiation has improved the quality and shelf life of certain fresh agricultural products after harvest [12]. Low UV exposure results in the production of antifungal chemicals, a delay in ripening, and a drop in CI. By lowering the establishment of ACS in tomato fruit, which occurs when ripening begins, UV-C therapy has disrupted ethylene production. The activation of genes for 1,3-glucanase and PAL was increased after peach fruit was exposed to UV light. Despite the fact that UV-A and UV-B therapy appear to be less harmful than UV-C, they have the impact of increasing antioxidants such as ascorbate, -tocophol, and polyphenols (Yamauchi, 2013). As a physical therapy, hormetic amounts of UV-C radiation have been applied to a variety of fruits and vegetables to extend their postharvest life. According to González-Aguilar et al. [14], hormesis was discovered when low doses of UV-C radiation resulted in beneficial reactions in biological organs. Hormesis is the use of small amounts of potentially harmful chemicals to induce a therapeutic stress response [13]. A number of generated chemical species produce hormetic responses in treated plant tissue. Phytoalexins include scoparone from oranges and resveratrol from grapes. Peach chitinases and glucanases, as well as tomato PAL, are all boosted. Reduced protein synthesis, chloroplast malfunction, and DNA damage have all been demonstrated to be harmful consequences of UV radiation on plant tissues. Low doses of UV, on the other hand, may produce repairable DNA damage, triggering DNA repair processes. Sublethal radiation has the capacity to affect a plant's homeostasis by triggering crucial processes within cells [13]. The effects of UV radiation on the formation of antioxidant molecules and enzymes vary according on hermetic doses, exposure time, and treated horticulture crops [14]. When biological material receives a significant dose of UV irradiation, it can interact with atoms and molecules, usually water, creating ROS in a quick and controlled manner through the univalent reduction of O_2. The principal ROS produced in the cell is O_2, which triggers a series of processes that result in the production of a variety of ROS and the stimulation of anti-oxidative enzymes. H_2O_2, which is produced by the enzyme superoxide dismutase (SOD), is a potent ROS [14]. UV radiation stimulates electrons in photosensitizers, causing ROS like O_2 to develop. After that, SOD dismutates the produced O_2^{2-} to H_2O_2. As a result, H_2O_2 scavenging enzymes such as APX may effectively break down H_2O_2. H_2O_2 may be responsible for increasing the antioxidant status of horticultural crops by boosting gene expression of enzymes involved in antioxidant capacity formation and build-up at lower UV levels rather than directly causing oxidative damage [14].

8.1.7.9 Mechanical Stress

Horticultural items are mechanically stressed during harvest and processing. This form of strain causes bruises as a result of cuts, punctures, and accidents. Cuts can produce transient increases in respiration, wound-induced ethylene production, phenolics synthesis, and plant tissue deterioration [9]. Machine-harvested products are more prone to cut injuries, which occur during the harvesting process.

The severity and extent of bruises is influenced by maturity, water potential, tissue or cellular orientation at the site of the injury, the type of item that causes the bruising force, the intensity and angle of contact, and the temperature of the horticultural product. The most cuts are found in fresh-cut items. Fresh-cut abiotic stress causes discoloration, particularly browning of cut surfaces due to tissue disruption and subsequent oxidative processes, increased respiration and ethylene evolution, flavour and texture loss, weight loss and dehydration, nutrient content decline, development of off-odours, membrane breakdown, and tissue softening. Cut-induced respiration has been linked to increased synthesis of respiratory enzymes as well as a brief increase in aerobic respiration in fresh-cut products. The committed enzyme in phenolic biosynthesis, PAL, is stimulated by wounds, which boosts phenolic levels [9].

8.1.7.10 Oxidative Stress

The development of ROS causes oxidative stress, which kills cell components. The ROS is made up of activated oxygen atoms that aren't usually radicals. ROS are produced by the mitochondrion, which causes fresh products to age prematurely. ROS and free radicals increase oxidative stress by oxidizing cell components. Hydrogen peroxide (H_2O_2), superoxide radical (Od22), hydroxyl radical (dOH), and nitrogen dioxide (NOd) are all important ROS for oxidative stress induction.

8.1.7.11 Mineral Stress

For plants to function at their best, mineral nutrients must be accessible in a balanced and timely manner. Mineral nutrients affect the quality of horticultural products in a variety of ways, but they're especially significant in physiological illnesses. Mineral nutrient imbalances induce some postharvest illnesses in horticultural products. Calcium is the mineral most frequently associated with postharvest issues. Maintaining plant membrane structural and functional integrity, stabilizing cell wall structures, regulating ion transport and selectivity, and managing ion exchange behaviour are all things it does. Mineral imbalances are the cause of salt-stressed plants. These imbalances are caused by salinity's impact on nutrient availability, competitive absorption, transport, or partitioning within the plant, or physiological inactivation of a given nutrient resulting in an increase in the plant's internal need for the required element. Salinity, as well as mineral imbalances, affects plant physiological responses such as osmotic stress and toxic ion stress. As salt levels grow, tomato fruit metabolism and physiology change substantially. It contributes to the sugar content of the fully ripe fruit. Tomatoes that have been cultivated in salty soil produce smaller fruit. The surface-to-volume ratios of smaller fruits are higher, making them more susceptible to water stress [9].

8.2 Molecular and Physiological Responses of Plants Against Stresses

8.2.1 Morpho-Physiological Responses

The mechanisms of drought and heat stress adaptation differ greatly, leading in diverse morphological responses to both stressors. Drought-stressed plants respond by conserving water and increasing uptake. Plants accomplish this by reducing the quantity and area of their leaves while encouraging root development. Long-term heat tolerance strategies include increasing leaf number and area and lowering leaf canopy temperature through increased transpiration. Drought and heat stress have conflicting impacts on various morphological processes. For example, drought stress reduced leaf development, number, and size, whereas heat stress enhanced leaf number and elongation [15–17]. Heat stress reduced the number, length, and diameter of roots, whereas mild drought stress encouraged root formation, which is required for water intake from deeper soil layers [18]. Heat stress led in narrower leaves with a bigger specific leaf area, whereas drought stress resulted in wider leaves with a smaller specific leaf area [19, 20]. Drought, on the other hand, increased the amount of biomass allocated to roots, while heat stress increased the amount of reproductive allocation. Some of the responses to combined stress were similar to drought, whereas others were similar to heat stress. Under combined stress in *Arabidopsis thaliana*, for example, leaf size rose, leaf number decreased, and biomass was allocated mostly to the roots and reproductive regions [21]. Despite the fact that drought and heat stress had different effects on plant vegetative growth, drought, heat, and their combined effects on plant reproductive development were identical. These stressors have been shown to influence *Triticum aestivum*, causing flowering to be delayed as well as a reduction in grain weight and production [17, 22, 23]. When it comes to lowering *Hordeum vulgare* yield, researchers discovered that the combined stress conditions were more detrimental than the individual stressors [24]. Drought and drought–hot combinations reduced the number of spikes, while heat stress had little effect. On the other hand, heat and combined stress increased the number of failed kernels [24]. Drought stress, on the other hand, had no influence on plant endosperm size or nutrient accumulation, whereas heat and drought stressed plants produced larger endosperm with higher starch and protein accumulation [25]. Heat and drought stress impact stomatal characteristics in distinct ways. During heat and drought stress, *Nicotiana tabacum* plants' stomata remained closed, resulting in higher leaf temperatures [26]. Plants that are stressed in a variety of ways have a unique way of reducing leaf temperature. According to Vile et al. [21] *A. thaliana* plants exposed to a combination of heat and drought stress change their leaf orientation by increasing their leaf insertion angle. In *A. thaliana* plants, individual and combined heat and drought stressors increased stomatal density, but heat stress decreased it. However, combined stress reduced stomatal density. This indicates that the impact of combined stress varies based on a plant species' genotypes. The combination of heat and drought stress has been shown to have a greater impact on a variety of physiological processes than either stress alone. Rizhsky et al. [26] discovered that when *N. tabacum* plants were treated to heat and drought stress, photosynthetic activity was inhibited more than when the stressors were administered individually. In *Lotus japonicus*, enhanced lipid peroxidation occurs when heat and drought stress are coupled, and severe changes in the ultra-structure of chloroplasts and mitochondria occur in *T. aestivum* [25, 27, 28]. The combined stress resulted in a greater loss of photosynthetic activity and higher ROS formation in *Populus yunnanensis* [29], as well as a greater reduction in root viability and photochemical efficiency of photosystem II (PS-II) in *Festuca arundinacea* [29, 30]. Individual stresses like heat and dryness reduce photosynthetic activity. Heat stress has less of an effect on photosynthesis, and only exceptionally high temperatures (over 40 °C) are known to be hazardous. The principal mechanisms of heat stress-induced photosynthetic reduction are increased photorespiration [18], decreased RuBisCO activity [31], and decreased PS-II activity [32]. Drought stress and a combination of heat and drought stress lowered photosynthetic activity in tobacco plants by more than 80% [26]. In *Cicer arietinum* leaves, heat stress increased RuBisCO activity, whereas drought stress and combination stress decreased it [33]. Sainz et al. [27] discovered that when *L. japonicus* plants were subjected to a combination of heat and drought stress, PSII function was considerably disturbed. Jiang and Huang [30] looked at how *F. arundinacea* and *Lolium perenne* reacted to combined heat and drought stress and discovered that when the two stresses were combined, the photochemical efficiency of PS-II was reduced more than when the two stresses were applied separately. The alteration of mitochondrial respiration is triggered by drought and heat stress [18]. The rate of nigh time respiration increased with rising temperatures, although plant respiration rates decreased due to drought stress [34]. Drought stress inhibited respiration in *N. tabacum* leaves; however, heat and a combination of drought and heat stress boosted respiration, according to Rizhsky et al. [26].

8.2.2 Molecular Responses

Furthermore, 77, 806, and 332 genes were shown to be frequently altered under drought and heat, drought and combination stress, and heat and combined stress, respectively [26]. In *T. aestivum* (var. Ofanto) plants, the heat stress response was found to be the most prevalent. Heat and drought stress altered a total of 5645 transcripts, with 2037 and 121 transcripts shared by each, and 3150 transcripts unique to combination stress. There were 159 and 779 transcripts that were altered in response to heat and drought stress, respectively, with 90 transcripts being regulated in both circumstances. Rampino et al. [35] studied the gene expression profile of *T. aestivum* plants using cDNA amplified fragment length polymorphism (cDNA-AFLP). According to the research, 380 genes were discovered to be regulated in all three stress scenarios. Out of the 242 increased genes, heat, drought, and combined stress each activated 44, 15, and 90 genes. Individual heat, drought, and combined heat and drought stress all increased the expression of 18 genes, whereas individual heat, drought, and combined heat and drought stress increased the expression of 51 and 24 genes, respectively. As a result, *T. aestivum's* combined stress response differed from that of other species. Transcriptomic analysis of individual and combined stressed *Sorghum bicolor* plants revealed that 1554, 448, and 2043 genes were specifically modulated under heat, drought, and combined stress, respectively, whereas 18, 3021, and 286 genes were found to be common under heat and drought, respectively. Four hundred thirty-eight genes were discovered to be frequently regulated under individual drought, heat, and combination drought and heat stress conditions [36]. The number of heat stress-specific genes clearly outnumbers the number of drought-specific genes in *T. aestivum* and *S. bicolor*. This could be due to the roughness of the heat treatment. Drought-specific molecular responses have been discovered in *A. thaliana*. Moreover, in all three plants, the number of combined stress-specific genes exceeds the number of individual stress-specific genes, showing that the molecular response of these plants to combined stress conditions is largely unique. Plants must achieve a balance between allocating resources and energy to growth and stress adaptation. As a result, when they are exposed to multiple stressors at the same time, they react to the one that is the most detrimental. Gene expression analyses show that the molecular responses in the above-mentioned cases of *T. aestivum* and *S. bicolor* are better suited to heat stress. Under combined drought and heat stress, the activation of heat shock proteins (HSPs), ROS detoxifying enzymes, and photosynthesis and glycolysis enzymes all occurred at the same time [26, 35, 36]. Individual and combined stress increased the expression of genes coding for small HSPs (sHSPs), HSP70, HSP90, and HSP100 in *N. tabacum*, according to Rizhsky et al. [26]. Other genes that contribute to the shared response to individual and combined stress include late embryogenesis 7 (LEA7) genes, dehydrin, photosynthesis-related genes, and genes encoding enzymes involved in the pentose pathway and anthocyanin production [26]. According to functional categorization of genes typically regulated under individual and combined drought and heat stress responses, the most frequently regulated genes in *A. thaliana* are those involved in diverse metabolic functions. Under heat and combined stress, chaperones were the most commonly regulated protein family. Transferases, oxidoreductases, and hydrolases expressing genes were the most commonly regulated genes between drought stress and combined stress response. Despite the fact that a number of commonly regulated genes were found in both the combined and individual stress responses of plants, there were differences in their expression levels in individual and combined stress conditions, indicating that expression was tailored to the combined stress condition. Plants that were combined challenged, for example, demonstrated more activation of HSP coding genes than plants that were drought and heat stressed separately [26]. There were also differences in the types of ROS detoxification genes produced under the three stress scenarios, demonstrating that ROS detoxification pathways are stress-dependent. Heat stress increased the activity of cytosolic ascorbate peroxidase (APX) and thioredoxin peroxidase, for example (TPX). Catalase (CAT) and glutathione peroxidase were upregulated in response to drought stress (GPX). The genes for alternative oxidase (AOX), glutathione reductase (GR), copper–zinc superoxide dismutase (CuZnSOD), and glutathione S transferase (GST) have all been demonstrated to be activated by combined stress. Several different genes were discovered in both individual and combination stress scenarios. Only during heat stress was Sb02g038425 (homologous to *Pseudomonas syringae* pv. maculicola 1 [RPM1] protein) found to be up-regulated. Sb01g021320 (homologous to LEA D34 protein) and Sb05g017950 (*H. vulgare aleurone* 22 [HVA22] related protein) were upregulated in response to drought and combined stress [36]. Similarly, when drought and heat stress were combined, a number of stress-related genes were activated that were not triggered by either stress alone. Among the genes discovered were pathogenesis-related (PR) and phenylalanine ammonia lyase (PAL) proteins. WRKY transcription factors and ethylene response transcriptional co-activator (ERTCA) transcripts were also induced only in response to combined stress [26]. Other genes that are elevated under combined stress include receptor like kinases, protein kinases (MAP3K), small GTP-binding proteins, MYB transcription factors, transporters, aquaporin membrane intrinsic protein (MIP) [26], and genes encoding transcription factor WRKY8, calcium transporter ATPase9, heat shock protein cognate 70, and a disease resistance related protein [35].

8.3 Microbiome Mediated Mitigation of Stress Conditions

Plants are intricately related with their microbiomes, which include bacteria in the rhizosphere, leaf surfaces, endosphere, and other components including pollen and nectar [15, 37, 38]. In response to stressors and environmental stimuli, the structure of the plant microbiome as well as the host services it provides change [39–42]. As a result of millions of years of co-evolution, new research reveals that changes in the microbiome aren't only passive responses of plants, but that plants actively seek collaboration with microbes to resist stress [43–46].

One such technique utilized by plants is the 'cry for assistance' strategy [38, 47], in which plants undergoing abiotic, pest-induced, or pathogen-induced challenges recruit beneficial microbes/traits from the environment utilizing a range of chemical cues to improve their capacity to battle stresses [38, 48]. For nutrient uptake, development, and stress tolerance, plant–microbe interactions are critical [48–52]. During stress, various root exudates (which account for 11–40% of the plant's photosynthesized organic carbon) attract bacteria via mechanisms that are currently poorly understood [53–56]. Changes in the root microbiome are known to improve plant stress tolerance. When exposed to the fungus *Rhizoctonia solani*, sugar beet roots, for example, attract Chitinophaga and Flavobacterium into the endosphere, suppressing the fungus [57]. Leguminous plants produce more flavonoids to attract nitrogen-fixing bacteria when they are subjected to nitrogen (N) limiting conditions [58]. The ability of maize roots to produce benzoxazinoids (tryptophan-derived secondary metabolites) to attract microorganisms that boost plant defence [59, 60] demonstrates that phenolic compounds are essential for plants to attract beneficial bacteria. Crop development and productivity are considerably diminished as a result of biotic and non-biotic stresses. Global food security is jeopardized as a result of this. Plant diseases and pests are expected to cost the global economy $220 billion every year [61]. To counteract some of these stressors, such as soil-borne illnesses, it has been suggested that biological instruments be used [37]. Microbiome engineering has the potential to boost plant productivity in the long run. According to a recent study [62], 1-aminocyclopropane-1-carboxylate (ACC), a stress-related amino acid in plants, can modify the soil microbiome to improve plant tolerance to salt stress. Many other root exudates/signalling cues, on the other hand, could have similar potential but have yet to be discovered. As a result, understanding the relationships between plant microbiomes, signalling molecules, and stressors has become a primary goal. This research uses a combination of metagenomics, metabolomics, and culture-dependent approaches to investigate the role of the microbiome in plant defence. We also propose the 'Defense Biome' idea, which use stress-induced changes in plant microbiomes to help the development of plant pre-, pro-, and post-biotics. Most plant microbiome research relied on assumed functions derived from descriptive genetic data (e.g. metagenomics) until recently [63]. The recent use of metagenomics in combination with metabolomics and a culture-dependent synthetic community (SynComs) strategy has solved this bottleneck in plant–microbe interactions research [64, 65]. Before recreating the SynCom, researchers used a mix of metagenomics and oomycete culture to assess the structure and putative role of plant-associated microbial metagenomes [66]. The SynCom can be introduced into axenic plants and utilized to regulate the plants to study microbial interkingdom interactions and in vitro determination of growth boosting properties in plants. Plant microbiomes from stressed and non-stressed plants can be compared to see if there are any stress-related changes in the microbiome. As a result, it is now possible to establish the biological importance of stress-induced SynComs in plants. It will be feasible to determine how root exudation influences the stress-induced microbiome assembly if root exudates are exposed to metabolomic analysis. Following that, these compounds are isolated and tested in vitro for interactions with plant microbial symbiont and pathogen populations, as well as their impact on plant health. Due to the chemical complexity of varied soil types, major challenges such as root exudation analysis in natural settings still exist. A few plant chemicals influence the microbiota in the rhizosphere. Some substances, such as 2,4-diacetylphloroglucinol [64], peroxidases, oxylipins, benzoxazinoids, flavonoids, coumarins, triterpenes, and aromatic compounds, have been shown to attract helpful bacteria that help plants defend and nourish themselves. The method described above is promising for investigating plant-associated bacteria and metabolites as plant biocontrol agents. In particular, our ability to produce microorganisms for SynComs has vastly increased in recent years. At least 50% of arabidopsis-associated bacteria can be isolated and maintained axenically, according to research [67]. Peroxidases, oxylipins, benzoxazinoids, flavonoids, coumarins, triterpenes, and aromatic chemicals have all been proven to attract beneficial microorganisms that help plants protect and nourish themselves. Plant-associated bacteria and metabolites as plant biocontrol agents could be investigated using the method described above. In recent years, our ability to create microbes for SynComs has greatly improved. According to studies [68], at least 50% of arabidopsis-associated bacteria can be separated and maintained axenically. Genome-wide metagenomic sequencing, when combined with strain-level resolution, may be able to overcome these restrictions and allow the reassembling of entire bacterial genomes [69]. It's also crucial to consider soil habitat-related variables like heterogeneity and morphological characteristics, which might have a big impact on soil microbe activity and composition [70]. The microbe mediated abiotic stress tolerance in plants is shown in Table 8.1.

8.3.1 Improved Understanding of a Microbiome Role in Plant Defence and Immune Systems

Plant immune systems and microbiomes interact in diverse ways, increasing plants' capacity and flexibility to cope with a variety of environments [81]. Pathogens and commensal microbes that survive competition with other soil and plant-associated microbes are then exposed to the plant's innate immune system, specifically the MAMPs triggered immunity (MTI) and effector triggered immunity (ETI), two layers of molecular defence described in the advanced zigzag defence model [82]. Although the interactions between plant commensals and plant immune systems are mostly unknown, they have the potential to have a

Table 8.1 Abiotic stress tolerance in plants mediated by microbes.

Stress	Microbe	Plant	Tolerance strategy	Reference
Drought	*Rhizobium tropici* and *Paenibacillus polymyxa*	*Phaseolus vulgaris*	Upregulation of genes involved in stress tolerance	[71]
Drought	*Bacillus licheniformis*	*Capsicum annuum*	Stress related genes and proteins	[72]
Drought	*Bacillus thuringiensis* AZP2	*Triticum aestivum*	Production of volatile organic compounds	[73]
Drought	*Pseudomonas putida* GAP-P45	Maize	Accumulation of proline improved plantbiomass, relative water content and leaf water potential	[74]
Drought	Consortia containing *Pseudomonas jessenii*, R62, *Pseudomonas synxantha*, R81 and A. nitroguajacolicusstrainYB3, strain YB5	*Oryza sativa*	Accumulation of proline maintained osmotic adjustment and improved plant growth	[75]
Drought	*Azospirillum lipoferum*	Maize	Improved plant growth through accumulation of free amino acids and soluble sugars	[76]
Salt	*Pseudomonas simiae*	*Glycine max*	4-Nitroguaiacol and quinoline promote soybean seed germination	[77]
Salt	Cyanobacteria and cyanobacterial extracts	*Oryza sativa*, *Triticum aestivum*, *Zea mays*, *Gossypium hirsutum*	Phytohormones as elicitor molecule	[78]
Salt	*Pseudomonas koreensis* strain AK-1	*Glycine max* L. Merr	Reduction in Na^+ level and increase in K + level	[79]
Cd, AS, Cu, Pb, and Zn toxicity	*Pseudomonas koreensis* AGB-1	*Miscanthus sinensis*	ACC deaminase, IAA production	[80]

significant impact on plant–stress interactions. MAMPs are also produced by pathogenic microbes, which are identified by plant pattern-recognition receptors (PRRs) in the plasma membrane and trigger intracellular defence systems. Given that both commensals and pathogens can induce MTI and ETI [81], how can a plant distinguish between beneficial and dangerous bacteria's molecular pathways. We propose that commensal microorganisms can reduce plant antagonisms through three key pathways: (i) producing compounds to effectively reduce the hazards, such as enzymes to scavenge ROS production and organic acids to lower the plant's extracellular pH [83]; and (ii) modifying, degrading, or changing MAMP structure and thus reducing MAMP production. The next step is to put these molecular mechanisms to the test in order to better understand how cohesion works [81]. T3SS and T4SS (Type III and IV secretion systems) are uncommon cell surface molecules that activate robust plant defences in pathogens [15, 84]. The next step is to examine these molecular pathways in diverse plant settings to understand how commensal bacteria interact with plant immunity. The plant's innate immune system may be insufficient to keep invaders out of the soil or phyllosphere in rare instances. Microbial communities in the rhizosphere, leaf surfaces, and endosphere protect plants against biotic stresses such as insect herbivory and fungal diseases [85]. In some cases, microbial interaction outperforms host genetics in reducing biotic stresses in plants. This could highlight the role of soil microbiomes in changing plant defence-related metabolic pathways and initiating MAMP-mediated plant defence priming [86]. Due to changes in microbial composition, disease suppressive soils produce higher levels of plant defence gene expression than normal soils [87]. However, in many cases, it's unclear how much microbiomes protect plants from disease as compared to axenic host conditions. Recent study has revealed that the rhizosphere microbiome of crops with higher defence and nutrient absorption has consistent microbial traits [88, 89]. A nitrogen-efficient rice variety's microbiome, for example, has a higher nitrogen metabolic capacity than a rice variety that does not. Tomatoes with high resistance to *Ralstonia solanacearum* (the pathogen that causes tomato wilt) have a higher abundance of Flavobacterium species in the rhizosphere, which limits pathogen growth [44]. In contrast, landrace maize adapted to a nitrogen-depleted soil environment has abundant of diazotrophic bacteria in a carbohydrate-rich mucilage secreted by the plant, which aids nitrogen fixation [87]. According to this notion, plants under stress may balance microbe-mediated nutrition intake and defence. Phosphate starvation response 1 (PHR1) is a direct regulator of a functionally significant group of plant immune system genes and contributes to the formation of root microbiomes in arabidopsis, as recently revealed. Overall, the role of the microbiome in plant defence modulation is still being discovered, but it provides a fantastic chance to learn more about plant microbial survival and growth strategies.

8.3.2 Cry for Help' Strategy for the Applied Plant Stress Probiotics

Plants employ environmental microbiomes to not only prevent pathogens and manage their immune systems, but also to recruit beneficial microbes from the environment to help them cope with specific stresses, a technique known as 'cry for aid' (or 'call to arms') [51]. When plants are subjected to particular conditions, plant–microbe interactions that can be passed down to future generations are feasible. This approach is essential for the survival of both individual plants and their descendants. These bacteria may have evolved to relieve plant stress by activating plant defence signalling pathways [39] and/or decreasing pathogen development and virulence. Microorganisms respond swiftly to changes in root exudates and other root inputs. Drought, light limitation, metal toxicity, and inadequate nutrition [90] are examples of biotic stressors that affect plant root metabolism and the microbial community. In some cases, stress-induced enrichment of specific microbes in root environments (e.g. by dryness) might improve plant stress resistance [91]. Drought increases the abundance of Actinobacteria or some monoderms in the rhizosphere and/or endosphere of plant roots [92, 93], showing that microbes and plants have developed a coadaptive strategy to deal with a particular stress [94]. As phosphate/nitrogen deficit grows, so does plant symbiosis with helpful bacteria for nutrient uptake [95]. Plants adopt the 'cry for help' strategy as an active process, allowing them to profit from microbial interactions during stressful situations, in addition to the typical 'adapt or migrate' strategies. Long-distance signalling mediated by plant hormones is essential for systemic acquired resistance, permitting whole-plant long-term responses to stresses [51]. A number of studies [96, 97] have connected induced resistance in the phyllosphere to microbial changes belowground. The microbial composition of the rhizosphere is altered by leaf injury and hormone delivery, for example [98], with certain bacteria that may help plants defend themselves recruited [38]. The phyllosphere microbiome has been demonstrated to be altered by plant volatile emissions (e.g. volatile organic compounds), while the nature and impact of disease-induced changes in the plant microbiome are still unknown [98]. Overall, more research is needed to better understand the numerous processes that govern interactions between stressed plants and microbiomes, as well as to identify the specific metabolites/cues that drive such interactions.

8.4 Multi-Omics Strategies to Address Stress Alleviation

Ecology of plant–microbe interactions is a complicated and interrelated system. Understanding the fine-tuning and integration of various signals produced by microbial interactions in plants is crucial for crop development. In its surroundings, a plant must contend with a number of biotic and abiotic problems. Plants respond to several stressors with complicated defensive signals, therefore the priority of physiological processes in plants can influence the outcome of plant–microbe interactions [99]. Interactions between microorganisms and plant roots trigger a variety of physiological, biochemical, and molecular responses in proximate and/or distal plant components on a physiological, biochemical, and molecular level. At all levels, such responses are linked to stress; many are similar to stress responses, while others are negative. The tough problem of comprehending changes in plants at the genomic, proteomic, or metabolomic level can be addressed using multi-omics approaches. Multi-omics, a data-driven science that is intertwined with bioinformatics advances, has improved our understanding of microbial community composition and functional behaviour in complex environments like the rhizosphere, where interconnections among microbial communities direct plant responses to stresses. Metagenomics, metatranscriptomics, and metaproteomics are examples of meta-omics technologies that have recently emerged as possible tools for delving deeper into microbial communities and functions in a specific environment [100].

8.4.1 Genomics

Reduced abiotic stress through crop genetics is crucial, but it's a challenging problem that will need a lot of work [101]. Low heritability and environmental factors make breeding programmes much more complex [102]. Cultivar tolerance can be accelerated through strategic marker-assisted breeding. Marker assisted selection requires knowledge of the genetic areas that drive tolerance traits, as well as the availability of molecular markers that are highly linked to them [103]. As a result, a vast amount of genomic data, such as sequenced genomes and expression profiles, promotes stress-relieving breeding [104, 105]. The adoption of genomics-based technology has tremendously aided crop improvement programmes. Molecular markers are increasingly being used in crop development for the accumulation of silicon (Si) in rice to improve the plant's tolerance to abiotic stress. Ma et al. [106] employed PCR-based microsatellite (RM5303) and expressed sequence tag (EST, E60168) markers to locate the Si transporter gene in bulk segregant studies. Plant–microbe interactions, in addition to crop breeding programmes, can greatly reduce abiotic stress in plants. Omics approaches aid in understanding the mechanics of well-established plant–microbe interactions. Tucci et al. [107] investigated the effects of *Trichoderma atroviride* and *Trichoderma harzianum* on tomato plant development and stress reduction by modifying the microbe–plant interaction. *T. atroviride* degrades IAA in the rhizosphere and enhances ACC deaminase activity in the soil to show that it can promote tomato growth and alleviate stress [108]. Gene silencing using the RNAi approach confirmed the presence of an ACC-deaminase putative sequence in Trichoderma [109, 110]. The Lotus japonica genome contains the dicarboxylate transporter LjALMT4, which is responsible for carbohydrate transport, according to Takanashi et al. [111]. Using a gene silencing approach for the Hb1 gene to promote NO production and upregulate the CBF regulon could improve cold tolerance in crops [112]. Kumari et al. [113] found that inoculating soybean with *Pseudomonas sp.* AK-1 and *Bacillus sp.* SJ-5 led in IST to salinity. Salt stress resistance was linked to proline build-up and lipoxygenase activity. According to Koussevitzky et al., [114] the gene Apx1, which codes for cytosolic ascorbate peroxidase 1, is necessary for drought and heat stress tolerance in Arabidopsis. Ectoine, a compatible osmolyte that controls salt tolerance in *Halomonas elongata* OUT30018, is a compatible osmolyte. Three ectoine biosynthetic genes were cloned and introduced to the tobacco plant (*Nicotiana tabacum* L.) cv Bright Yellow 2 (BY2), which boosted hyperosmotic shock tolerance by accumulating ectoine and grew properly under these conditions [115]. By discovering the genes and their regulation, breeders can develop higher stress-tolerance types. Multi-omics approaches generate very efficient and reliable data that can be used to aid in methodical research. Chilling tolerance is a desirable trait in Miscanthus grass varieties; however, it varies widely. Chilling tolerance is classified into distinct kinds based on the expression of genes involved in carbohydrate accumulation. A molecular marker of sensitive genes expressed in sensitive genotypes, such as TF MsCBF3, can predict the decline of tolerance among different varieties [116]. Submergence has put more than 15 million hectares of rainfed lowland rice in Asia at risk [117]. Flooding or anoxia concerns affect 13% of the total land area on the planet [5]. A single big quantitative trait locus (QTL) on chromosome 9 is responsible for rice submergence tolerance [118]. Neeraja et al. [117] used genetic markers for the Sub1 gene in their backcross breeding effort with the recurrent parent Swarna. This Sub1 enables tolerance for sensitive mega kinds. Finally, the involvement of Sub1A in submergence tolerance has been discovered [119]. Through marker assisted backcrossing, this QTL has opened up many opportunities for enhancing submergence tolerance in mega types (MAB). Researchers can obtain access to the mechanism that allows associative interactions to function smoothly through genomic examination of the host and associated microbial communities, particularly phyllosphere-related microbial communities. Numerous studies have looked into the role of several genes from similar bacteria. Plants aid and sustain the biological system at the associative interface by providing crucial molecular equivalents. The interaction process is greatly influenced by plant genotypic diversity. SUNN1 *Medicago truncatula* showed no effect in the presence of associative rhizobia, the roots of SUNN1 [120]. According to studies, associative bacteria's Nod components influence root development and symbiotic relationship [121, 122]. In widely cultivated cereals, there is no way to get nitrogen by nodulation. Mechanical injuries that occur during root development may allow diazotrophic bacteria to colonize root tissues in rare situations [123]. Regardless, little is known about these interactions.

8.4.2 Metagenomics

The culture-independent approach to microbial community analysis, according to Chen and Pachter [124], is a powerful tool for resolving uncultured, unknown microbial diversity in the plant rhizosphere. Users can learn about the distribution of microbial communities in certain settings that include plant growth promoting (PGP), biocontrol, antibiotic producing, and xenobiotic degrading properties by employing metagenomics. Metagenomic libraries were used to identify the whole AcdS operon from uncultivated endophytes, as well as a transcriptional regulator gene called AcdR, which is located upstream of AcdS. In the *Burkholderia* genus, this operon was important [125]. Growing *Escherichia coli* clones at inhibitory NaCl concentrations of 750 mM was used to identify salt tolerance genes in uncultivable bacteria. In two separate clones of bacteria, a putative general stress protein (GspM) with GsiB domain and a probable enoyl-CoA-hydratase (EchM) were revealed to be involved in salt tolerance. Following purification, crotonyl-CoA hydratase activity was identified in EchM [126]. These genes aid recombinant salt-tolerant microorganisms and transgenic plants substantially. A 250-meter-deep acid mine drainage was discovered to possess systems for cold adaptation. Anti-freeze proteins, cold-shock proteins, pathways for creating appropriate solutes, and pH homeostasis were among the genes revealed In the metagenome of acid mine drainage [127]. The discovery of new genes and pathways for cold stress alleviation is aided by metagenomic data. As the number of endophytic microbes that can be effectively cultured is so minimal, we know relatively little about their involvement in root interiors. It is found endophytic bacteria growing on rice roots using a metagenomic technique. Researchers discovered that metabolic pathways associated with endophytic life style and functional characteristics like quorum sensing and ROS detoxification have a role in boosting plant stress tolerance using metagenomic sequences generated from endophytic cell extracts [128]. Following bioinformatics analysis, metaproteomics, like metatranscriptomics, can provide community-wide gene expression patterns, protein abundance, and candidate proteins that can be linked to functions [129]. Diversity profiling and colonization investigations using metagenomics can also show quantitative colonization when stressors are present [129]. When paired with metatranscriptomic analysis [129], this can provide useful information regarding stressor-induced changes in colonizing populations' taxonomic and functional diversity. The combined analyses' findings will be extremely useful in determining how stressors affect colonization.

8.4.3 Transcriptomics

Comparing transcriptome profiles, as per Bräutigam and Gowik [130], can aid in the discovery of various transcripts responsible for differences between two biologically different expressions under different contexts. The use of mRNA sequencing analysis and microarray technologies to provide transcriptome level information is an important technique for analysing plant–microbe interactions [131, 132].

According to a study that used next-generation RNA sequencing, stress response genes are elevated in IAA-producing *Sinorhizobium meliloti* strains [133]. This study investigated the transcription patterns of two *S. meliloti* strains, wild-type strain-1021 and an IAA overproducing derivative RD64. Genes coding for the sigma factor RpoH1 and other stress responses were revealed to produce the IAA overproducing strain of *S. meliloti*. Alavi et al. [134] revealed spermidine as a novel plant growth regulator during abiotic stress by analysing the transcriptome of rapeseed and its parasite *Stenotrophomonas rhizophila*. Different miRNAs in rice, Medicago, Phaseolus, Arabidopsis, and other plants regulate gene expression in response to abiotic circumstances as drought, salinity, and cold [135]. In this study, the transcription patterns of two *S. meliloti* strains were compared: wild-type strain-1021 and an IAA overproducing derivative RD64. Genes coding for the sigma factor RpoH1 and other stress responses were identified to be activated in the IAA overproducing strain of *S. meliloti*. Alavi et al. [134] revealed spermidine as a novel plant growth regulator during abiotic stress using transcriptome analysis of rapeseed and its parasite *S. rhizophila*. Different miRNAs in rice, Medicago, Phaseolus, Arabidopsis, and other plants regulate gene expression in response to abiotic circumstances as drought, salinity, and cold [135]. Bvu-miR13 regulates wDrepeat proteins in cucumbers, which play an essential role in stress tolerance [136]. MiRNAs target stress signalling pathways, which are important for root development, leaf morphogenesis, and stress response, in addition to transcription factors [137]. In silico approaches were used to discover that *Beta vulgaris* plants have thirteen mature miRNAs [138]. MiR398 inhibits the activity of the superoxide dismutases SOD1 and SOD2 in order to minimize ROS and consequent effects of abiotic stress [138]. Among other metabolic activities, miRNAs regulate transcription, auxin homeostasis, and ion transport to alleviate stress [139]. The response of plants to aluminium stress is also influenced by microRNA [140]. Rice subspecies from Japan and India were compared at different degrees of aluminium tolerance. This technique identified 16 different miRNA responses, indicating that the response to aluminium stress is complex. Plants incur irreversible damage as a result of ROS production in response to UVB radiation and flooding (hypoxia) [141]. SOD proteins must be upregulated and miR398 must be down regulated in Arabidopsis to combat oxidative stress [142].

Inducing miR398 expression and suppressing miR395 expression in *Populus tremula* relieved UV-B stress [143]. Sugar beet seedlings and sugar recovery after harvest are both impacted by low temperatures. To construct a transcriptome profile of cold-stressed plants, high-throughput RNA sequencing from leaves and roots was performed [144]. Root tissues were shown to upregulate CBF3 faster than leaf tissues. Cold stress increased the expression of genes associated with jasmonic acid-mediated responses, such as the AP2/ERF family [145].

8.4.4 Proteomics

Proteins play a key role in the expression of plant stress responses because they represent phenotypic trait shaping directly. Proteomic studies in microbes and plants have evolved into powerful tools for studying physiological metabolism and protein–protein interactions. Proteomics has important implications for intra- and inter-microbial species interactions, as well as host–microbe interactions involving host-mediated signalling and associated bacteria's behavioural responses [146]. Various proteins are found as indications of changes in physiological condition due to stress or components responsible for stress reduction in such investigations, resulting in a better understanding of biological system control [124]. Comparing stressed, non-stressed, and microbe-associated plants allows for the identification of protein targets and networks. Plants such as Arabidopsis, wheat (*Triticum aestivum*), durum wheat (*Triticum durum*), barley (*Hordeum vulgare*), maize (*Zea maize*), rice (*Oryza sativa*), soybean (*Soybean max*), common bean (*Phaseolus vulgaris*), pea (*Pisum sativum*), oilseed rape (*Brassica napus*), and potato (*Solanum tuberosum*). These studies looked into dynamic changes in protein functional groups, proteins involved in signalling and regulatory pathways, TFs, protein metabolism, protein–protein interactions at the interface, proteins and enzymes that confer several stress-related compounds, functions of structural proteins associated with the cell wall and cytoskeleton, and bioinformatics-based identification of putative proteins [147]. Chen et al. [148] compared cold tolerant (ZD7) and cold sensitive lines in alfalfa and employed proteome analysis to explore into the processes of cold acclimation (W5). Cassava can modify its metabolism and development when exposed to low temperatures in tropical climes. To better understand the process of cold tolerance and how it may be improved, we undertook a proteomic study. Twenty distinct proteins with similar patterns were found in the apical enlarged leaves of the SC8 and Col1046 cultivars. Variations in proteome profile expression were shown to be strongly linked to photosynthetic activity and peroxiredoxin levels. Electrolyte leakage (EL), chlorophyll content, and MDA accumulation were used to measure the cold tolerance of cassava [147]. A leucine-rich repeat receptor kinase (Srlk) has been discovered to operate as an upstream regulator of salt responsive genes in the *M. truncatula* plant. It was discovered to play a role in both detecting and responding to salt stress [149]. This research uncovered the mechanism for detecting salinity stress. Based on the proteome profile of barley under different water stress scenarios, Ghabooli et al. [150] stated that *Piriformospora indica* alleviates drought stress through photosynthetic stimulation, which releases energy and enhances antioxidant synthesis. A cDNA library of 4M salt-stressed *Dunaliella salina* was used to discover a novel gene, Ds-26-16. *E. coli*, *Haematococcus pluvialis*, and tobacco have all been discovered to be salt tolerant. According to iTRAQ data, Ds-26-16 upregulates TFs for stress responses like ROS alleviation, osmotic balance, and energy metabolism in *E. coli*. Microorganisms are more vulnerable to stress due to their diverse metabolic processes. It's critical in the protocols for deciphering plant-elaborated stress responses. Traditional proteomics, for example, focuses on a single organism, whereas metaproteomics considers several metabolic interactions that occur simultaneously in a system. This could lead to a better understanding of the interconnectedness of different microbial communities and their interactions with the host plant. Procedures for protein extraction from ambient materials are the most significant success milestones in metaproteomics. Protein identification from a range of species has never been easier because to improvements in protein sequencing [99, 151]. The complexity of the metaproteome makes it difficult to resolve and analyse the data. On the other hand, extracting and analysing successful environmental metaproteomes could provide decisive results and develop a better link between omics data and stress response mechanisms among species [152, 153]. The majority of environmental proteome research is now restricted to growing model organisms. They highlight the creatures' remarkable properties, such as their resistance to salinity, sodicity and temperature, a lack of water, harmful metals, and radiation, among other things. We can gain a better understanding of the fine mechanisms used by these species by studying their proteomes. It's also important for confirming the likelihood of them being used to stimulate such metabolism under a variety of circumstances.

In a controlled environment, it's easier to grasp the protein profile, but the expression profile changes with changes in environmental conditions, which isn't the case in laboratory trials. Haloarchea and Halobacteria are gaining a lot of attention these days due to their ability to flourish in high-salt environments. In saline and sodic soils, organisms' ability to generate PGP can be exploited to ease stress. This will assist in increasing yields in areas where there is a lot of stress. Harvesting and utilizing metabolites that provide halotolerance to microorganisms living in high salt stress conditions with other associated stressors could be beneficial to crop development. Inducing the synthesis of beneficial metabolites in these organisms is as simple as cultivating them in a laboratory under in situ stress conditions. These metabolites may confer stress tolerance in plants when administered. In addition, Pseudomonas is the ideal choice for dealing with xenobiotic chemicals due to its unique ability to break down large volumes of carbon sources, particularly xenobiotic ones. The hydrocarbon-degrading ability of Pseudomonas is well known. Pseudomonas proteomic investigations are mostly focused on toxic organic pollutants and recalcitrant xenobiotic compounds [154–156]. Pseudomonas species have been well studied for PGP traits like siderophore production [157, 158], secretion of plant growth stimulating substances [159, 160], and biocontrol of phytopathogenic organisms [161]. Members of this genus have a wide metabolic range and a unique ability to form biofilms, allowing them to thrive in a wide range of conditions, making them the most effective field inoculants [157, 162, 163]. Today, methylotrophic bacteria proteome study is a trendy issue. Methylotrophs account for a significant portion of the phylosphere population, notably on the leaf surface, where the one-carbon substrate methanol is easily available due to transpiration. The most frequent bacteria are pink-pigmented facultative methylotrophic (PPFM), which are being researched for their ability to release plant growth regulators [164–166]. These organisms' PGP potential has been demonstrated in a number of trials in a range of settings [167, 168]. Detailed investigations into these unique phyllosphere community members' proteomic insights led to novel ideas about the role of proteins in organisms' survival mechanisms in relatively harsh environments, such as those found on leaf surfaces, where, in addition to intense radiation, nutrient scarcity is common. Their ability to release plant growth regulators could also be exploited on a large scale. It is required to elucidate deep molecular insights into PGP microbial communities, which are predominantly involved in stress reduction, in order to collect data on processes involved in such activities. The discovery of proteins involved in these processes prompted a spike in molecular-level stress-reduction strategies, in which active molecules were applied directly rather than the entire organism. Protein expression in various plant lines can be studied to improve crops. During cold acclimation, researchers discovered that cold-tolerant sunflower strains showed 14 different proteome expressions [169]. It also discusses potential frost mitigation systems in plants, as well as cross-tolerance mechanisms [170, 171]. As a result, a database of responsive and blocked genotypes for a certain abiotic stress can be used to distinguish stress-sensitive genotypes. Santos et al. [172] used GeLC–MS/MS to profile the proteome of *Araucaria angustifolia* embryogenic cell lines. A total of 106 proteins were identified to be different in protein expression levels between the responsive and blocked type lines under abiotic stress. At an early stage, two proteins were discovered to be linked to cell lines that were only suppressed. These proteins may indicate blocked cell lines in plants at an early stage of development.

8.4.5 Metabolomics

The study of all the metabolites produced by an organism under specific environmental conditions is known as metabolomics. Metabolite content is proportional to the number of cell pathways, which is proportional to the quantity of genetic information. Environmental changes have a profound impact on an organism's metabolome, resulting in acute physiological changes [173]. It's reasonable to infer that species designed to thrive in stressful environments have similar physiological states. Since it's critical to have a full grasp of an organism's metabolome in both normal and stressed physiological states, from which the presence or absence of specific distinctive chemicals of concern can be deduced. As a result, alterations in pathways and stimulation of stress-inducible genes will be easier to detect. Metabolomics is becoming more popular as a research approach for better understanding abiotic stress reactions [174]. Metabolomics research has benefited from high-throughput developments in the realm of molecular detection [175, 176]. Bioactive chemicals are found in the plant metabolome. This finding corresponds to findings that multiple signal molecules are generated by plants to attract and promote important metabolic pathways in invading microbial populations [177, 178]. *Trichoderma* spp. generate auxins, which help plants grow by lowering stress levels [179]. Two secondary Trichoderma metabolites, harzianolide and 6-pentyl-*α*-pyrone, were shown to have auxin-like effects and enhance plant growth in etiolated pea stem [180]. Plant metabolic changes as a result of changing environmental conditions affect the secretion pattern and kind of chemicals produced affecting root colonization [181]. Microbial molecular signalling systems in the rhizosphere are impacted in a similar way, but this has yet to be studied. Plants accumulate many metabolites in response to abiotic stresses, including trehalose, glycine betain, IAA, and others. The accumulation of a single molecule, according to Allen et al. [182], does not imply stress tolerance; rather, tolerance is determined by the modulation of flux to numerous defence and development pathways. Plants' stoichiometry and metabolism have been found to be altered as a means of ensuring optimal fitness [183]. According to *Arabidopsis thaliana* time-series study, metabolic activities respond to abiotic changes more quickly than transcriptional activities [184]. The conditions in the surrounding environment have an impact on the metabolome, which in turn has an impact on the microbial cell's pathways. It is self-evident that this will have a greater impact on their overall performance in the microenvironment and ecosystem, particularly in terms of interactions within and among the individuals who reside there. Microbial metabolic products have been connected to both direct and indirect plant growth stimulation. Many rhizosphere bacteria have been shown to produce plant growth-stimulating proteins including cytokinins and gibberelins, among other things [185, 186]. Microbial metabolites such as IAA, gibberelins, and siderophores are only a few examples [187–189]. In recent years, it has been discovered that *Pseudomonas* sp., *Rhizobium* sp., *Enterobacter* sp., *Pantoea* sp., *Marinobacterium* sp., *Acinetobacter* sp., and *Sinorhizobium* sp. influence wheat germination and seedling growth in saline conditions [162]. Bacillus sp. strains with phosphate solubilizing capacity increased fennel yield and quality in semi-arid saline soil [190]. Phosphate solubilization is generally caused by bacteria's low molecular weight organic acids. Microbial siderophores can also assist plant roots. *Pseudomonas fluorescens* C7, for example, produced siderophores that effectively supplied iron to *Arabidopsis thaliana* [191]. Biogeochemical factors appear to fuel the bacteria's production of siderophores, but they also help to relieve the stress induced by heavy metals [192]. Many microbes rely heavily on their surroundings in order to produce effective siderophores. They discovered that acidic conditions reduced iron intake by sediment and soil magnetotactic bacteria, with Fe^{2+} conversion interfering with the siderophore-mediated iron uptake route as a likely cause [193]. As a result, in situ mechanisms influencing microbial metabolism must be thoroughly investigated. The majority of these metabolites are yet unknown. Cutting-edge metabolomics technology may enable in situ monitoring of these metabolites and environmental interventions in microbial metabolism. In terms of substrate usage and metabolite exchange, many bacteria in the environment are interdependent, which is the foundation of succession. Cometabolism is also important in the biodegradation of refractory and xenobiotic chemicals, where it has been demonstrated that cometabolism is essential. Non-substrate chemicals are oxidized in tandem with the true substrates as bacteria proliferate. Co-metabolizers in this environment are thought to comprise Pseudomonas, Flavobacterium, Bacillus, Microbacterium, Hydrogenobacter, Achromobacter, and Xanthomonas species, as well as members of the Microbacterium, Hydrogenobacter, Achromobacter, and Xanthomonas genera [194]. This feature of cometabolism has substantial implications in the research of biochemical pathways, especially those involved in the metabolism of polycyclic and polyaromatic compounds [195, 196]. They provide details on the enzymes involved in the conversion and the routes they take, making them more likely to be used on a wide scale in locations where native ecosystems have been stressed by the build-up and/or pollution of xenobiotic and refractory substances. Many hydrocarbons, such as *p*-isopropyltoulene, butylcyclohexane, *n*-butylbenzene, ortho and para xylene, are aggressively co-metabolized by them [197, 198]. Individuals from such genera have a great potential for repairing stresses placed on crops by contaminated soils, paving the way for bioremediation. Quantitative metabolomics studies also allow for high precision and accuracy in measuring biological activity [199]. Plant-associated microbes under stress could be assessed using high-throughput mass spectrometric profiling of cellular metabolites to see how much the stressor interferes with general cellular homeostasis. Root exudates and microbially generated signal response molecules are employed to interact in this two-way system [200–202]. Before introducing exogenous microbial metabolites to soil, it's also important to understand how bacteria work. The study includes cellular abundance, biomolecule synthesis under normal and optimal conditions, leakage, plant signals in the cascade, and microbial counter-response. It might be helpful to supplement the rhizosphere with biomolecules that have been down-regulated as a result of the stressor. The same may be said for the possible modulation of stressor-responsive biomolecules, which have an impact on the overall host-microorganism communication process. When a stressor affects plant root exudation, the above-mentioned cascades are not triggered in microbial systems that do not exude. The availability of nucleotide data has been a significant contribution to metabolomics research (NCBI). There is little doubt that research concentrating on systems biology will function more smoothly in the future. Recent studies have shown that microbial-derived biomolecules and/or in vitro generated metabolites can be recruited into the phyllosphere in simulated environments [171]. This is due to the fact that live bacteria fail to express essential genes for PGP function when they are stressed. However, the impact of adequately enriching the rhizosphere with these chemicals must be properly investigated. Understanding the impact of host metabolomics on microbial metabolism after colonization is also a big plus. This could pave the way for the simulation of complex endophytic environments. Furthermore, the Rhizosphere Community is a multifunctional system that involves biogeochemical cycling and nutrient exchange, making it an ideal platform for learning about the microbiology of these systems. Diverse members of the microbial community are operating extensive routes at the same time. Environmental factors, on the other hand, have the greatest impact on the smooth operation of such pathways. Finally, the introduction of a stressor or a change in environmental conditions results in a shift in community structure. Knowing biochemical links within and between ecosystem components is necessary for phenotype-level information in a certain biogeochemical state of occurrence [203].

References

1 Lester, G.E. (2003). Oxidative stress affecting fruit senescence. In: *Postharvest Oxidative Stress in Horticultural Crops* (ed. D.M. Hodges), 113–129. New York: Haworth Press Inc.

2 Madani, B., Mirshekari, A., and Imahori, Y. (2019). Physiological responses to stress. In: *Postharvest Physiology and Biochemistry of Fruits and Vegetables*, 405–423. https://doi.org/10.1016/B978-0-12-813278-4.00020-8.

3 Atkinson, N.J. and Urwin, P.E. (2012). The interaction of plant biotic and abiotic stresses: from genes to the field. *J. Exp. Bot.* 63: 3523–3543.

4 Rejeb, I.B., Pastor, V., and Mauch-Mani, B. (2014). Plant responses to simultaneous biotic and abiotic stress: molecular mechanisms. *Plants* 3: 458–475.
5 Cramer, G.R., Urano, K., Delrot, S. et al. (2011). Effects of abiotic stress on plants: a systems biology perspective. *BMC Plant Biol.* 11: 163–177.
6 Wang, W., Vinocur, B., Shoseyov, O., and Altman, A. (2004). Role of plant heat-shock protein and molecular chaperones in the abiotic stress response. *Trends Plant Sci.* 9: 244–252.
7 Mahajan, S. and Tuteja, N. (2005). Cold, salinity and drought stresses: an overview. *Arch. Biochem. Biophys.* 444: 139–158.
8 Ben-Yehoshua, S. and Rodov, V. (2013). Transpiration and water stress. In: *Postharvest Physiology and Pathology of Vegetables*, 2e (ed. J.A. Bartz and J.K. Brecht), 111–159. New York: Marcel Dekker.
9 Toivonen, P.M.A. and Hodges, D.M. (2011). Abiotic stress in harvested fruits and vegetables. In: *Abiotic Stress in Plants-Mechanisms and Adaptations* (ed. A.K. Shanker and B. Venkateswarlu), 39–58. Europe: InTech.
10 Lurie, S. and Pedreschi, R. (2014). Fundamental aspects of postharvest heat treatments. *Hortic. Res.* 1: 14030.
11 Singh, S.P. (2016). Postharvest oxidative stress in fresh fruits. In: *Advances in Postharvest Fruit and Vegetable Technology* (ed. R.B.H. Wills and J. Golding), 237. CRC Press.
12 Yamauchi, N. (2013). Quality maintenance of postharvest horticultural crops by stress treatments and approach for the elucidation of its mechanism. *J. Jpn. Soc. Hort. Sci.* 82: 1–10.
13 Shama, G. and Alderson, P. (2005). UV homesis in fruits: a concept ripe for commercialization. *Trends Food Sci. Technol.* 16: 128–136.
14 González-Aguilar, G., Villa-Rodriguez, J.A., Ayala-Zavala, J.F., and Yahia, E.M. (2010). Improvement of the antioxidant status of tropical fruits as a secondary response to some postharvest treatment. *Trends Food Sci. Technol.* 21: 475–482.
15 Álvarez-Pérez, S., Lievens, B., and Fukami, T. (2019). Yeast–bacterium interactions: the next frontier in nectar research. *Trends Plant Sci.* 24: 393–401. https://doi.org/10.1016/j.tplants.2019.01.012.
16 Bos, H.J., Tijani-Eniola, H., and Struikl, P.C. (2000). Morphological analysis of leaf growth of maize: responses to temperature and light intensity. *NJAS Wagening. J. Life Sci.* 48: 181–198. https://doi.org/10.1016/S1573-5214[00]80013-5.
17 Prasad, P.V.V., Boote, K.J., and Allen, L.H. (2006). Adverse high temperature effects on pollen viability, seed-set, seed yield and harvest index of grain-sorghum [*Sorghum bicolor* [L.] Moench] are more severe at elevated carbon dioxide due to higher tissue temperatures. *Agric. For. Meteorol.* 139: 237–251. https://doi.org/10.1016/j.agrformet.2006.07.003.
18 Prasad, P.V.V., Staggenborg, S., and Ristic, Z. (2008). Impacts of drought and/or heat stress on physiological, developmental, growth, and yield processes of crop

plants. In: *Response of Crops to Limited Water: Understanding and Modeling Water Stress Effects on Plant Growth Processes: Advances in Agricultural Systems Modeling Series 1* (ed. L.R. Ahuja, V.R. Reddy, S.A. Saseendran and Q. Yu), 301–356. Madison, WI: ASA-CSSA-SSSA.

19 Luomala, E.M., Laitinen, K., Sutinen, S. et al. (2005). Stomatal density, anatomy and nutrient concentrations of Scots pine needles are affected by elevated CO_2 and temperature. *Plant Cell Environ.* 38: 733–749. https://doi.org/10.1111/j.1365-3040.2005.01319.

20 Poorter, H., Niinemets, U., Poorter, L. et al. (2009). Causes and consequences of variation in leaf mass per area [LMA]: a meta-analysis. *New Phytol.* 182: 565–588. https://doi.org/10.1111/j.1469-8137.2009.02830.

21 Vile, D., Pervent, M., Belluau, M. et al. (2012). Arabidopsis growth under prolonged high temperature and water deficit: independent or interactive effects? *Plant Cell Environ.* 35: 702–718. https://doi.org/10.1111/j.1365-3040.2011.02445.

22 Savin, R. and Nicolas, M. (1996). Effects of short periods of drought and high temperature on grain growth and starch accumulation of two malting barley cultivars. *Aust. J. Plant Physiol.* 23: 201. https://doi.org/10.1071/PP9960201.

23 Pradhan, G.P., Prasad, P.V.V., Fritz, A.K. et al. (2012). Effects of drought and high temperature stress on synthetic hexaploid wheat. *Funct. Plant Biol.* 39: 190–198. https://doi.org/10.1071/FP11245.

24 Rollins, J.A., Habte, E., Templer, S.E. et al. (2013). Leaf proteome alterations in the context of physiological and morphological responses to drought and heat stress in barley [*Hordeum vulgare* L.]. *J. Exp. Bot.* 64: 3201–3212. https://doi.org/10.1093/jxb/ert158.

25 Szucs, A., Jäger, K., Jurca, M.E. et al. (2010). Histological and microarray analysis of the direct effect of water shortage alone or combined with heat on early grain development in wheat [*Triticum aestivum*]. *Physiol. Plant.* 140: 174–188. https://doi.org/10.1111/j.1399-3054.2010.01394.

26 Rizhsky, L., Liang, H., and Mittler, R. (2002). The combined effect of drought stress and heat shock on gene expression in tobacco. *Plant Physiol.* 130: 1143–1151. https://doi.org/10.1104/pp.006858.

27 Sainz, M., Díaz, P., Monza, J., and Borsani, O. (2010). Heat stress results in loss of chloroplast Cu/Zn superoxide dismutase and increased damage to photosystem II in combined drought-heat stressed *Lotus japonicus*. *Physiol. Plant.* 140: 46–56. https://doi.org/10.1111/j.1399-3054.2010.01383.

28 Grigorova, B., Vassileva, V., Klimchuk, D. et al. (2012). Drought, high temperature, and their combination affect ultrastructure of chloroplasts and mitochondria in wheat [*Triticum aestivum* L.] leaves. *J. Plant Interact.* 7: 204–213. https://doi.org/10.1080/17429145.2011.654134.

29 Li, X., Yang, Y., Sun, X. et al. (2014). Comparative physiological and proteomic analyses of poplar [*Populus yunnanensis*] plantlets exposed to high temperature and drought. *PLoS One* 9: e107605. https://doi.org/10.1371/journal.pone.0107605.

30 Jiang, Y. and Huang, B. (2001). Physiological responses to heat stress alone or in combination with drought: a comparison between tall fescue and perennial ryegrass. *HortScience* 36: 682–686.

31 Salvucci, M.E. and Crafts-Brandner, S.J. (2004). Inhibition of photosynthesis by heat stress: the activation state of Rubisco as a limiting factor in photosynthesis. *Physiol. Plant.* 120: 179–186.

32 Yang, X., Wen, X., Gong, H. et al. (2007). Genetic engineering of the biosynthesis of glycinebetaine enhances thermotolerance of photosystem II in tobacco plants. *Planta* 225: 719–733. https://doi.org/10.1007/s00425-006-0380-3.

33 Awasthi, R., Kaushal, N., Vadez, V. et al. (2014). Individual and combined effects of transient drought and heat stress on carbon assimilation and seed filling in chickpea. *Funct. Plant Biol.* 41: 1148–1167. https://doi.org/10.1071/FP13340.

34 Bryla, D.R., Bouma, T.J., Hartmond, U., and Eissenstat, D.M. (2001). Influence of temperature and soil drying on respiration of individual roots in citrus: integrating greenhouse observations into a predictive model for the field. *Plant Cell Environ.* 24: 781–790. https://doi.org/10.1046/j.1365-3040.2001.00723.x.

35 Rampino, P., Mita, G., Fasano, P. et al. (2012). Plant physiology and biochemistry novel durum wheat genes up-regulated in response to a combination of heat and drought stress. *Plant Physiol. Biochem.* 56: 72–78. https://doi.org/10.1016/j.plaphy.2012.04.006.

36 Johnson, S.M., Lim, F.-L., Finkler, A. et al. (2014). Transcriptomic analysis of *Sorghum bicolor* responding to combined heat and drought stress. *BMC Genomics* 15: 456. https://doi.org/10.1186/1471-2164-15-456.

37 Liu, H., Carvalhais Lilia, C., Mark, C. et al. (2017). Inner plant values: diversity, colonization and benefits from endophytic bacteria. *Front. Microbiol.* 8: 2552.

38 Liu, H. and Brettell, L.E. (2019). Plant defense by VOC-induced microbial priming. *Trends Plant Sci.* 24: 187–189.

39 Timm, C.M., Carter, K.R., Carrell, A.A. et al. (2018). Abiotic stresses shift belowground Populus-associated bacteria toward a core stress microbiome. *mSystems* 3: e00070-17.

40 Santos-Medellín, C., Edwards, J., Liechty, Z. et al. (2017). Drought stress results in a compartment-specific restructuring of the rice root-associated microbiomes. *MBio* 8: e00764-17. https://doi.org/10.1128/mBio.00764-17.

41 Ferrando, L. and Fernández Scavino, A. (2015). Strong shift in the diazotrophic endophytic bacterial community inhabiting rice [Oryza sativa] plants after flooding. *FEMS Microbiol. Ecol.* 91: fiv104.

42 McSpadden Gardener, B.B. and Weller, D.M. (2001). Changes in populations of rhizosphere bacteria associated with take-all disease of wheat. *Appl. Environ. Microbiol.* 67: 4414–4425.

43 Durán, P., Thiergart, T., Garrido-Oter, R. et al. (2018). Microbial interkingdom interactions in roots promote Arabidopsis survival. *Cell* 175: 973–983. https://doi.org/10.1016/j.cell.2018.10.020.

44 Kwak, M.-J., Kong, H.G., Choi, K. et al. (2018). Rhizosphere microbiome structure alters to enable wilt resistance in tomato. *Nat. Biotechnol.* 36: 1100–1109. https://doi.org/10.1038/nbt.4232.

45 Lau, J.A. and Lennon, J.T. (2012). Rapid responses of soil microorganisms improve plant fitness in novel environments. *Proc. Natl. Acad. Sci. U. S. A.* 109: 14058–14062.

46 Lee, B., Lee, S., and Ryu, C.M. (2012). Foliar aphid feeding recruits rhizosphere bacteria and primes plant immunity against pathogenic and non-pathogenic bacteria in pepper. *Ann. Bot.* 110: 281–290. https://doi.org/10.1093/aob/mcs055.

47 Chen, Y., Wang, J., Yang, N. et al. (2018). Wheat microbiome bacteria can reduce virulence of a plant pathogenic fungus by altering histone acetylation. *Nat. Commun.* 9: 3429. https://doi.org/10.1038/s41467-018-05683-7.

48 Bakker, P.A., CMJ, P., de Jonge, R., and Berendsen, R.L. (2018). The soil-borne legacy. *Cell* 172: 1178–1180. https://doi.org/10.1016/j.cell.2018.02.024.

49 Carrión, V.J., Perez-Jaramillo, J., Cordovez, V. et al. (2019). Pathogen-induced activation of diseasesuppressive functions in the endophytic root microbiome. *Science* 366: 606–612. https://doi.org/10.1126/science.aaw9285.

50 Berendsen, R.L., Pieterse, C.M., Bakker, P.A. et al. (2012). The rhizosphere microbiome and plant health. *Trends Plant Sci.* 17: 478–486. https://doi.org/10.1016/j.tplants.2012.04.001.

51 Lu, T., Ke, M., Lavoie, M. .Y.J. et al. (2018). Rhizosphere microorganisms can influence the timing of plant flowering. *Microbiome* 6: 231. https://doi.org/10.1186/s40168-018-0615-0.

52 Chialva, M., Salvioli di Fossalunga, A., Daghino, S. et al. (2018). Native soils with their microbiotas elicit a state of alert in tomato plants. *New Phytol.* 220: 1296–1308. https://doi.org/10.1111/nph.15014.

53 Hubbard, C.J., Li, B., McMinn, R. et al. (2018). The effect of rhizosphere microbes outweighs host plant genetics in reducing insect herbivory. *Mol. Ecol.* 28: 1801–1811. https://doi.org/10.1111/mec.14989.

54 Neal, A.L., Ahmad, S., Gordon-Weeks, R., and Ton, J. (2012). Benzoxazinoids in root exudates of maize attract Pseudomonas putida to the rhizosphere. *PLoS One* 7: e35498. https://doi.org/10.1371/journal.pone.0035498.

55 Stringlis, I.A., Yu, K., Feussner, K. et al. (2018). MYB72-dependent coumarin exudation shapes root microbiome assembly to promote plant health. *Proc. Natl. Acad. Sci. U. S. A.* 115: E5213–E5222. https://doi.org/10.1073/pnas.1722335115.

56 Carvalhais, L.C., Dennis, P.G., Badri, D.V. et al. (2015). Linking jasmonic acid signaling, root exudates, and rhizosphere microbiomes. *Mol. Plant-Microbe Interact.* 28: 1049–1058. https://doi.org/10.1094/MPMI-01-15-0016-R.

57 Gu, Y., Wang, X., Friman, V.-P. et al. (2016). Pathogen invasion indirectly changes the composition of soil microbiome via shifts in root exudation profile. *Biol. Fertil. Soils* 52: 997–1005.

58 Hassan, S. and Mathesius, U. (2012). The role of flavonoids in root–rhizosphere signalling: opportunities and challenges for improving plant–microbe interactions. *J. Exp. Bot.* 63: 3429–3444.

59 Lombardi, N., Stefania, V., David, T. et al. (2018). Root exudates of stressed plants stimulate and attract Trichoderma soil fungi. *Mol. Plant-Microbe Interact.* 31: 982–994.

60 Kudjordjie, E.N., Sapkota, R., Steffensen, S.K. et al. (2019). Maize synthesized benzoxazinoids affect the host associated microbiome. *Microbiome* 7: 59. https://doi.org/10.1186/s40168-019-0677-7.

61 Hu, L., CAM, R., Cadot, S. et al. (2018). Root exudate metabolites drive plant-soil feedbacks on growth and defense by shaping the rhizosphere microbiota. *Nat. Commun.* 9: 2738. https://doi.org/10.1038/s41467-018-05122-7.

62 Chakraborty, S. and Newton, A.C. (2011). Climate change, plant diseases and food security: an overview. *Plant Pathol.* 60: 2–14.

63 Liu, H., Khan, M.Y., Carvalhais, L.C. .M.D.B. et al. (2019). Soil amendments with ethylene precursor alleviate negative impacts of salinity on soil microbial properties and productivity. *Sci. Rep.* 9: 6892. https://doi.org/10.1038/s41598-019-43305-4.

64 Berendsen, R.L., Vismans, G., Yu, K. et al. (2018). Disease-induced assemblage of a plant-beneficial bacterial consortium. *ISME J.* 12: 1496–1507. https://doi.org/10.1038/s41396-018-0093-1.

65 Castrillo, G., Teixeira, P., Paredes, S. et al. (2017). Root microbiota drive direct integrationof phosphate stress and immunity. *Nature* 543: 513–518. https://doi.org/10.1038/nature21417.

66 Lanoue, A., Vincent, B., Gunnar, J.H. et al. (2010). De novo biosynthesis of defense root exudates in response to Fusarium attack in barley. *New Phytol.* 185: 577–588.

67 Huang, A.C., Jiang, T., Liu, Y.X. et al. (2019). A specialized metabolic network selectively modulates Arabidopsis root microbiota. *Science* 364: eaau6389. doi: 10.1126/science.aau6389.

68 Schulz-Bohm, K., Gerards, S., Hundscheid, M. et al. (2018). Calling from distance: attraction of soil bacteria by plant root volatiles. *ISME J.* 12: 1252–1262. https://doi.org/10.1038/s41396-017-0035-3.

69 Bai, Y., Müller, D., Srinivas, G. et al. (2015). Functional overlap of the Arabidopsis leaf and root microbiota. *Nature* 528: 364–369. https://doi.org/10.1038/nature16192.

70 Aleklett, K., Kiers, E., Ohlsson, P.P.O. et al. (2018). Build your own soil: exploring microfluidics to create microbial habitat structures. *ISME J.* 12: 312–319. https://doi.org/10.1038/ismej.2017.184.

71 Figueiredo, M.V.B., Burity, H.A., Martinez, C.R., and Chanway, C.P. (2008). Alleviation of drought stress in common bean (Phaseolus vulgaris L.) by coinoculation with Paenibacillus polymyxa and Rhizobium tropici. *Appl. Soil Ecol.* 40: 182–188. https://doi.org/10.1016/j.apsoil.2008.04.005.

72 Lim, J.H. and Kim, S.D. (2013). Induction of drought stress resistance by multifunctional PGPR Bacillus licheniformis K11 in pepper. *Plant Pathol. J.* 29: 201–208. https://doi.org/10.5423/PPJ.SI.02.2013.0021.

73 Timmusk, S., El-Daim, I.A.A., Copolovici, L. et al. (2014). Drought-tolerance of wheat improved by rhizosphere bacteria from harsh environments: enhanced biomass production and reduced emissions of stress volatiles. *PLoS One* 9: e96086. https://doi.org/10.1371/journal.pone.0096086.

74 Sandhya, V., Ali, S.K.Z., Grover, M. et al. (2010). Effect of plant growth promoting Pseudomonas spp. on compatible solutes antioxidantstatus and plant growth of maize under drought stress. *Plant Growth Regul.* 62: 21–30.

75 Gusain, Y.S., Singh, U.S., and Sharma, A.K. (2015). Bacterial mediated amelioration ofdrought stress in drought tolerant and susceptible cultivars of rice (Oryzasativa L.). *Afr. J. Biotechnol.* 14: 764–773.

76 Bano, Q., Ilyas, N., Bano, A. et al. (2013). Effect of Azospirillum inoculation on maize (zea mays l.) under drought stress. *Pak. J. Bot.* 45: 13–20.

77 Vaishnav, A., Kumari, S., Jain, S. et al. (2016). PGPR-mediated expression of salt tolerance gene in soybean through volatiles under sodium nitroprusside. *J. Basic Microbiol.* 56: 1274–1288. https://doi.org/10.1002/jobm.201600188.

78 Singh, S. (2014). A review on possible elicitor molecules of cyanobacteria: their role in improving plant growth and providing tolerance against biotic or abiotic stress. *J. Appl. Microbiol.* 117: 1221–1244. https://doi.org/10.1111/jam.12612.

79 Kasotia, A., Varma, A., and Choudhary, D.K. (2015). Pseudomonas-mediated mitigation of salt stress and growth promotion in Glycine max. *Agric. Res.* 4: 31–41. https://doi.org/10.1007/s40003-014-0139-1.

80 Babu, A.G., Shea, P.J., Sudhakar, D. et al. (2015). Potential use of Pseudomonas koreensis AGB-1 in association with Miscanthus sinensis to remediate heavy metal(loid)-contaminated mining site soil. *J. Environ. Manag.* 151: 160–166. https://doi.org/10.1016/j.jenvman.2014.12.045.

81 Bollet, C., Davin-Regli, A., and De Micco, P. (1995). A simple method for selective isolation of Stenotrophomonas maltophilia from environmental samples. *Appl. Environ. Microbiol.* 61: 1653–1654. https://doi.org/10.1128/aem.61.4.1653-1654.1995.

82 Ranjan, R., Rani, A., Metwally, A. et al. (2016). Analysis of the microbiome: advantages of whole genome shotgun versus 16S amplicon sequencing. *Biochem. Biophys. Res. Commun.* 469: 967–977. https://doi.org/10.1016/j.bbrc.2015.12.083.

83 Young, I.M. and Crawford, J.W. (2004). Interactions and selforganization in the soil-microbe complex. *Science* 304: 1634–1637.

84 Teixeira, P.J.P.L., Colaianni, N.R., Fitzpatrick, C.R., and Dangl, J.L. (2019). Beyond pathogens: microbiota interactions with the plant immune system. *Curr. Opin. Microbiol.* 49: 7–17. https://doi.org/10.1016/j.mib.2019.08.003.

85 (a) Jones, J.D.G. and Dangl, J.L. (2006). The plant immune system. *Nature* 444: 323–329; (b) Yu, K. et al. (2019). Rhizosphere-associated Pseudomonas suppress local root immune responses by gluconic acid-mediated lowering of environmental pH. *Curr. Biol.* 29: 3913–3920.

86 Levy, A., Salas Gonzalez, I., Mittelviefhaus, M. et al. (2018). Genomic features of bacterial adaptation to plants. *Nat. Genet.* 50: 138–150. https://doi.org/10.1038/s41588-017-0012-9.

87 Elhady, A., Shimaa, A., Johannes, H. et al. (2018). Rhizosphere microbiomes modulated by pre-crops assisted plants in defense against plant-parasitic nematodes. *Front. Microbiol.* 9: 1133.

88 Ritpitakphong, U., Falquet, L., Vimoltust, A. et al. (2016). The microbiome of the leaf surface of Arabidopsis protects against a fungal pathogen. *New Phytol.* 210: 1033–1043. https://doi.org/10.1111/nph.13808.

89 Zhang, J., Liu, Y.X., Zhang, N. et al. (2019). NRT1.1B is associated with root microbiota composition and nitrogen use in field-grown rice. *Nat. Biotechnol.* 37: 676–684. https://doi.org/10.1038/s41587-019-0104-4.

90 Dudenhöffer, J.-H., Stefan, S., Alexandre, J. et al. (2016). Systemic enrichment of antifungal traits in the rhizosphere microbiome after pathogen attack. *J. Ecol.* 104: 1566–1575.

91 Fitzpatrick, C.R., Copeland, J., Wang, P.W. et al. (2018). Assembly and ecological function of the root microbiome across angiosperm plant species. *Proc. Natl. Acad. Sci. U. S. A.* 115: E1157–E1165. https://doi.org/10.1073/pnas.1717617115.

92 Ham, B.-K., Chen, J., Yan, Y., and Lucas, W.J. (2018). Insights into plant phosphate sensing and signaling. *Curr. Opin. Biotechnol.* 49: 1–9. https://doi.org/10.1016/j.copbio.2017.07.005.

93 Xu, L., Dan, N., Zhaobin, D. et al. (2018). Drought delays development of the sorghum root microbiome and enriches for monoderm bacteria. *Proc. Natl. Acad. Sci. U. S. A.* 115: E4284–E4293.

94 Terhorst, C.P., Lennon, J.T., and Lau, J.A. (2014). The relative importance of rapid evolution for plant-microbe interactions depends on ecological context. *Proc. Natl. Acad. Sci. U. S. A.* 281: 20140028. http://dx.doi.org/10.1098/rspb.2014.0028.

95 Pieterse, C.M.J., Leon-Reyes, A., der Ent, V. et al. (2009). Networking by small-molecule hormones in plant immunity. *Nat. Chem. Biol.* 5: 308–316. https://doi.org/10.1038/nchembio.164.

96 Pineda, A., Kaplan, I., and Bezemer, T.M. (2017). Steering soil microbiomes to suppress aboveground insect pests. *Trends Plant Sci.* 22: 770–778. https://doi.org/10.1016/j.tplants.2017.07.002.

97 Carvalhais, L.C., Dennis, P.G., Badri, D.V. et al. (2013). Activation of the jasmonic acid plant defence pathway alters the composition of rhizosphere bacterial communities. *PLoS One* 8: e56457.

98 Farré-Armengol, G., Filella, I., Llusia, J., and Peñuelas, J. (2016). Bidirectional interaction between phyllospheric microbiotas and plant volatile emissions. *Trends Plant Sci.* 21: 854–860. https://doi.org/10.1016/j.tplants.2016.06.005.

99 Schenk, P.M., Carvalhais, L.C., and Kazan, K. (2012). Unraveling plant–microbe interactions: can multi-species transcriptomics help? *Trends Biotechnol.* 30: 177–184. https://doi.org/10.1016/j.tibtech.2011.11.002.

100 de Castro, A.P., Sartori, A., Silva, M.R. et al. (2013). Combining "omics" strategies to analyze the biotechnological potential of complex microbial environments. *Curr. Protein Pept. Sci.* 14: 447–458.

101 Grainger, C.M. and Rajcan, I. (2013). Characterization of the genetic changes in a multi-generational pedigree of an elite Canadian soybean cultivar. *Theor. Appl. Genet.* 127: 211–229. https://doi.org/10.1007/s00122-013-2211-9.

102 Manavalan, L.P., Guttikonda, S.K., Tran, L.S.P., and Nguyen, H.T. (2009). Physiological and molecular approaches to improve drought resistance in soybean. *Plant Cell Physiol.* 50: 1260–1276. https://doi.org/10.1093/pcp/pcp082.

103 Xu, Y., Lu, Y., Xie, C. et al. (2012). Whole-genome strategies for marker-assisted plant breeding. *Mol. Breed.* 29: 833–854. https://doi.org/10.1007/s11032-012-9724-.

104 Sonah, H., Deshmukh, R.K., Sharma, A. et al. (2011). Genome-wide distribution and organization of microsatellites in plants: an insight into marker development in *Brachypodium*. *PLoS One* 6: 21298. https://doi.org/10.1371/journal.pone.0021298.

105 Tomar, R.S.S., Deshmukh, R.K., Naik, K., and Tomar, S.M.S. (2014). Development of chloroplast-specific microsatellite markers for molecular characterization of alloplasmic lines and phylogenetic analysis in wheat. *Plant Breed.* 133: 12–18. https://doi.org/10.1111/pbr.12116.

106 Ma, J.F., Mitani, N., Nagao, S. et al. (2004). Characterization of the Silicon uptake system and molecular mapping of the Silicon transporter gene in rice. *Plant Physiol.* 136: 3284–3289. https://doi.org/10.1104/pp.104.047365.

107 Tucci, M., Ruocco, M., De Masi, L. et al. (2011). The beneficial effect of *Trichoderma* spp. on tomato is modulated by the plant genotype. *Mol. Plant Pathol.* 12: 341–354. https://doi.org/10.1111/j.1364-3703.2010.00674.x.

108 Gravel, V., Antoun, H., and Tweddell, R.J. (2007). Growth stimulation and fruit yield improvement of greenhouse tomato plants by inoculation with *Pseudomonas putida* or *Trichoderma atroviride*: possible role of indole acetic acid [IAA]. *Soil Biol. Biochem.* 39: 1968–1977. https://doi.org/10.1016/j.soilbio.2007.02.015.

109 Kubicek, C.P., Herrera-Estrella, A., Seidl-Seiboth, V. et al. (2011). Comparative genome sequence analysis underscores mycoparasitism as the ancestral life style of *Trichoderma*. *Genome Biol.* 12: R40. https://doi.org/10.1186/gb-2011-12-4-r40.

110 Viterbo, A., Landau, U., Kim, S. et al. (2010). Characterization of ACC deaminase from the biocontrol and plant growth-promoting agent *Trichoderma asperellum* T203. *FEMS Microbiol. Lett.* 305: 42–48. https://doi.org/10.1111/j.1574-6968.2010.01910.x.

111 Takanashi, K., Sasaki, T., Kan, T. et al. (2016). A dicarboxylate transporter, LjALMT4, mainly expressed in nodules of Lotus japonicus. *Mol. Plant-Microbe Interact.* 29: 584–592. https://doi.org/10.1094/MPMI-04-16-0071-R.

112 Sehrawat, A., Gupta, R., and Deswal, R. (2013). Nitric oxide-cold stress signalling cross-talk, evolution of a novel regulatory mechanism. *Proteomics* 13: 1816–1835. https://doi.org/10.1002/pmic.201200445.

113 Kumari, S., Vaishnav, A., Jain, S. et al. (2015). Bacterial-mediated induction of systemic tolerance to salinity with expression of stress alleviating enzymes in soybean [*Glycine max* L. Merrill]. *J. Plant Growth Regul.* 34: 558–573. https://doi.org/10.1007/s00344-015-9490-0.

114 Koussevitzky, S., Suzuki, N., Huntington, S. et al. (2008). Ascorbate peroxidase 1 plays a key role in the response of *Arabidopsis thaliana* to stress combination. *J. Biol. Chem.* 283: 34197–34203. https://doi.org/10.1074/jbc.M806337200.

115 Nakayama, H., Yoshida, K., and Ono, H. (2000). Ectoine, the compatible solute of *Halomonas elongata*, confers hyperosmotic tolerance in cultured tobacco cells. *Plant Physiol.* 122: 1239–1247. https://doi.org/10.1104/pp.122.4.1239.

116 Purdy, S.J., Maddison, A.L., Jones, L.E. et al. (2013). Characterization of chilling-shock responses in four genotypes of *Miscanthus* reveals the superior tolerance of *M. x giganteus* compared with *M. sinensis* and *M. sacchariflorus*. *Ann. Bot.* 111: 999–1013. https://doi.org/10.1093/aob/mct059.

117 Neeraja, C.N., Maghirang-Rodriguez, R., Pamplona, A. et al. (2007). A marker-assisted backcross approach for developing submergence-tolerant rice cultivars. *Theor. Appl. Genet.* 115: 767–776. https://doi.org/10.1007/s00122-007-0607-0.

118 Toojinda, T., Siangliw, M., Tragroonrung, S., and Vanavichit, A. (2003). Molecular genetics of submergence tolerance in rice: QTL analysis of key traits. *Ann. Bot.* 91: 243–253. https://doi.org/10.1093/aob/mcf072.

119 Swarbrick, P.J., Schulze-Lefert, P., and Scholes, J.D. (2006). Metabolic consequences of susceptibility and resistance in barley leaves challenged with powdery mildew. *Plant Cell Environ.* 29: 1061–1076. https://doi.org/10.1111/j.1365-3040.2005.01472.x.

120 Jin, J., Watt, M., and Mathesius, U. (2012). The auto regulation gene SUNN mediates changes in root organ formation in response to nitrogen through alteration of shoot-to-root auxin transport. *Plant Physiol.* 159: 489–500. https://doi.org/10.1104/pp.112.194993.

121 Olah, B., Briere, C., Becard, G. et al. (2005). Nod factors and a diffusible factor from arbuscular mycorrhizal fungi stimulate lateral root formation in *Medicago truncatula* via the DMI1/DMI2 signalling pathway. *Plant J.* 44: 195–207. https://doi.org/10.1111/j.1365-313X.2005.02522.x.

122 Oldroyd, G.E.D. (2013). Speak, friend, and enter: signalling systems that promote beneficial symbiotic associations in plants. *Nat. Rev. Microbiol.* 11: 252–263. https://doi.org/10.1038/nrmicro2990.

123 Gaiero, J.R., McCall, C.A., Thompson, K.A. et al. (2013). Inside the root microbiome: bacterial root endophytes and plant growth promotion. *Am. J. Bot.* 100: 1738–1750. https://doi.org/10.3732/ajb.1200572.

124 Chen, K. and Pachter, L. (2005). Bioinformatics for whole genome shotgun sequencing of microbial communities. *PLoS Comput. Biol.* 1: 106–112. https://doi.org/10.1371/journal.pcbi.0010024.

125 Nikolic, B., Schwab, H., and Sessitsch, A. (2011). Metagenomic analysis of the 1-aminocyclopropane-1-carboxylate deaminase gene [acdS] operon of an uncultured bacterial endophyte colonizing *Solanum tuberosum* L. *Arch. Microbiol.* 193: 665–676. https://doi.org/10.1007/s00203-011-0703-z.

126 Kapardar, R.K., Ranjan, R., Grover, A. et al. (2010). Identification and characterization of genes conferring salt tolerance to *Escherichia coli* from pond water metagenome. *Bioresour. Technol.* 101: 3917–3924. https://doi.org/10.1016/j.biortech.2010.01.017.

127 Liljeqvist, M., Ossandon, F.J., Gonzalez, C. et al. (2015). Metagenomic analysis reveals adaptations to a cold-adapted lifestyle in a low-temperature acid mine drainage stream. *FEMS Microbiol. Ecol.* 91: fiv011. https://doi.org/10.1093/femsec/fiv011.

128 Sessitsch, A., Hardoim, P., Döring, J. et al. (2012). Functional characteristics of an endophyte community colonizing roots as revealed by metagenomic analysis. *Mol. Plant-Microbe Interact.* 25: 28–36. https://doi.org/10.1094/MPMI-08-11-0204.

129 Turner, T.R., Ramakrishnan, K., Walshaw, J. et al. (2013). Comparative metatranscriptomics reveals kingdom level changes in the rhizosphere microbiome of plants. *ISME J.* 7: 2248–2258. https://doi.org/10.1038/ismej.2013.119.

130 Bräutigam, A. and Gowik, U. (2010). What can next generation sequencing do for you? Next generation sequencing as a valuable tool in plant research. *Plant Biol.* 12: 831–841. https://doi.org/10.1111/j.1438-8677.2010.00373.x.

131 Akpinar, B.A., Kantar, M., and Budak, H. (2015). Root precursors of microRNAs in wild emmer and modern wheats show major differences in response to drought stress. *Funct. Integr. Genomics* 15: 587–598. https://doi.org/10.1007/s10142-015-0453-0.

132 Wang, Y., Hu, B., Du, S. et al. (2016). Proteomic analyses reveal the mechanism of *Dunaliella salina* Ds-26-16 gene enhancing salt tolerance in *Escherichia coli*. *PLoS One* 11: e0153640. https://doi.org/10.1371/journal.pone.0153640.

133 Defez, R., Esposito, R., Angelini, C., and Bianco, C. (2016). Overproduction of indole-3-acetic acid in free-living rhizobia induces transcriptional changes resembling those occurring in nodule bacteroids. *Mol. Plant-Microbe Interact.* 29: 484–495. https://doi.org/10.1094/MPMI-01-16-0010-R.

134 Alavi, P., Starcher, M.R., Zachow, C. et al. (2013). Root-microbe systems: the effect and mode of interaction of stress protecting agent [SPA] *Stenotrophomonas* rhizophila DSM14405T. *Front. Plant Sci.* 4: 141. https://doi.org/10.3389/fpls.2013.00141.

135 Trindade, I., Capitao, C., Dalmay, T. et al. (2010). miR398 and miR408 are up-regulated in response to water deficit in *Medicago truncatula*. *Planta* 231: 705–716. https://doi.org/10.1007/s00425-009-1078-0.

136 Curaba, J., Singh, M.B., and Bhalla, P.L. (2014). miRNAs in the crosstalk between phytohormone signalling pathways. *J. Exp. Bot.* 65: 1425–1438. https://doi.org/10.1093/jxb/eru002.

137 Li, J.L., Cui, J., and Cheng, D.Y. (2015). Computational identification and characterization of conserved miRNAs and their target genes in beet [*Beta vulgaris*]. *Genet. Mol. Res.* 14: 9103–9108. https://doi.org/10.4238/2015.August.7.19.

138 Kantar, M., Lucas, S.J., and Budak, H. (2011). miRNA expression patterns of Triticum dicoccoides in response to shock drought stress. *Planta* 233: 471–484. https://doi.org/10.1007/s00425-010-1309-4.

139 Li, Y.F., Zheng, Y., Addo-Quaye, C. et al. (2010). Transcriptome-wide identification of microRNA targets in rice. *Plant J.* 62: 742–759. https://doi.org/10.1111/j.1365-313X.2010.04187.x.

140 Lima, J.C., Arenhart, R.A., Margis-Pinheiro, M., and Margis, R. (2011). Aluminum triggers broad changes in microRNA expression in rice roots. *Genet. Mol. Res.* 10: 2817–2832. https://doi.org/10.4238/2011.November.10.4.

141 Blokhina, O. and Fagerstedt, K.V. (2010). Reactive oxygen species and nitric oxide in plant mitochondria: origin and redundant regulatory systems. *Physiol. Plant.* 138: 447–462. https://doi.org/10.1111/j.1399-3054.2009.01340.x.

142 Jia, X., Sun, C., Zuo, Y. et al. (2016). Integrating transcriptomics and metabolomics to characterise the response of *Astragalus membranaceus* Bge. var. mongolicus [Bge.] to progressive drought stress. *BMC Genomics* 17: 188. https://doi.org/10.1186/s12864-016-2554-0.

143 Moliterni, V.M.C., Paris, R., Onofri, C. et al. (2015). Early transcriptional changes in *Beta vulgaris* in response to low temperature. *Planta* 242: 187–201. https://doi.org/10.1007/s00425-015-2299-z.

144 Licausi, F., Ohme-Takagi, M., and Perata, P. (2013). APETALA2/Ethylene Responsive Factor [AP2/ERF] transcription factors: mediators of stress responses and developmental programs. *New Phytol.* 199: 639–649. https://doi.org/10.1111/nph.12291.

145 Kosova, K., Vitamvas, P., Urban, M.O. et al. (2015). Biological networks underlying abiotic stress tolerance in temperate crops – a proteomic perspective. *Int. J. Mol. Sci.* 16: 20913–20942. https://doi.org/10.3390/ijms160920913.

146 Silva-Sanchez, C., Li, H., and Chen, S. (2015). Recent advances and challenges in plant phosphoproteomics. *Proteomics* 15: 1127–1141. https://doi.org/10.1002/pmic.201400410.

147 An, F., Li, G., Li, Q.X. et al. (2016). The comparatively proteomic analysis in response to cold stress in Cassava Plantlets. *Plant Mol. Biol. Report.* 34: 1095–1110. https://doi.org/10.1007/s11105-016-0987-x.

148 Chen, J., Han, G., Shang, C. et al. (2015). Proteomic analyses reveal differences in cold acclimation mechanisms in freezing-tolerant and freezing-sensitive cultivars of alfalfa. *Front. Plant Sci.* 6: 105. https://doi.org/10.3389/fpls.2015.00105.

149 de Lorenzo, L., Merchan, F., Laporte, P. et al. (2009). A novel plant leucine-rich repeat receptor kinase regulates the response of *Medicago truncatula* roots to salt stress. *Plant Cell* 21: 668–680. https://doi.org/10.1105/tpc.108.059576.

150 Ghabooli, M., Khatabi, B., Ahmadi, F.S. et al. (2013). Proteomics study reveals the molecular mechanisms underlying water stress tolerance induced by *Piriformospora indica* in barley. *J. Proteome* 94: 289–301. https://doi.org/10.1016/j.jprot.2013.09.017.

151 Cordwell, S.J., Wilkins, M.R., Cerpa-poljak, A. et al. (1995). Cross species identification of proteins separated by 2-dimensional gel-electrophoresis using matrix-assisted laser-desorption ionization time-of-flight mass-spectrometry and aminoacid-composition. *Electrophoresis* 16: 438–443. https://doi.org/10.1002/elps.1150160171.

152 Schulze, W.X. (2005). Protein analysis in dissolved organic matter: what proteins from organic debris, soil leachate and surface water can tell us – a perspective. *Biogeosciences* 2: 75–86. https://doi.org/10.5194/bg-2-75-2005.

153 Schweder, T., Markert, S., and Hecker, M. (2008). Proteomics of marine bacteria. *Electrophoresis* 29: 2603–2616. https://doi.org/10.1002/elps.200800009.

154 Lupi, C.G., Colangelo, T., and Mason, C.A. (1995). Two-dimensional gel electrophoresis analysis of the response of *Pseudomonas putida* KT2442 to 2-Chlorophenol. *Appl. Environ. Microbiol.* 61: 2863–2872.

155 Reardon, K.F. and Kim, K.H. (2002). Two-dimensional electrophoresis analysis of protein production during growth of *Pseudomonas putida* F1 on toluene, phenol, and their mixture. *Electrophoresis* 23: 2233–2241. https://doi.org/10.1002/1522-2683[200207]23:14<2233::AID-ELPS2233<3.0.CO;2-B.

156 Kim, S.I., Choi, J.S., and Kahng, H.Y. (2007). A proteomics strategy for the analysis of bacterial biodegradation pathways. *OMICS* 11: 280–294. https://doi.org/10.1089/omi.2007.0019.

157 Acebo-Guerrero, Y., Hernández-Rodríguez, A., Vandeputte, O. et al. (2015). Characterization of *Pseudomonas chlororaphis* from *Theobroma cacao* L. rhizosphere with antagonistic activity against *Phytophthora palmivora* [Butler]. *J. Appl. Microbiol.* 119: 1112–1126. https://doi.org/10.1111/jam.12910.

158 Ferret, C., Sterckeman, T., Cornu, J.Y. et al. (2014). Siderophore-promoted dissolution of smectite by fluorescent *Pseudomonas. Environ. Microbiol. Rep.* 6: 459–467. https://doi.org/10.1111/175.

159 Cunrath, O., Gasser, V., Hoegy, F. et al. (2015). A cell biological view of the siderophore pyochelin iron uptake pathway in *Pseudomonas aeruginosa. Environ. Microbiol.* 17: 171–185. https://doi.org/10.1111/1462-2920.125448-2229.12146.

160 Pereira, S.I. and Castro, P.M. (2014). Diversity and characterization of culturable bacterial endophytes from *Zea mays* and their potential as plant growth-promoting agents in metal-degraded soils. *Environ. Sci. Pollut. Res. Int.* 24: 14110–14123. https://doi.org/10.1007/s11356-014-3309-6.

161 Balcazar, W., Rondón, J., Rengifo, M. et al. (2015). Bioprospecting glacial ice for plant growth promoting bacteria. *Microbiol. Res.* 177: 1–7. https://doi.org/10.1016/j.micres.2015.05.001.

162 Sorty, A.M., Meena, K.K., Choudhary, K. et al. (2016). Effect of plant growth promoting bacteria associated with halophytic weed [*Psoralea corylifolia* L.] on germination and seedling growth of wheat under saline conditions. *Appl. Biochem. Biotechnol.* 180: 872–882. https://doi.org/10.1007/s12010-016-2139-z.

163 Arevalo-Ferro, C., Reil, G., Gorg, A. et al. (2005). Biofilm formation of *Pseudomonas putida* IsoF: the role of quorum sensing as assessed by proteomics. *Syst. Appl. Microbiol.* 28: 87–114. https://doi.org/10.1016/j.syapm.2004.10.005.

164 Meena, K.K., Kumar, M., Kalyuzhnaya, M.G. et al. (2012). Epiphytic pink-pigmented methylotrophic bacteria enhance germination and seedling growth of wheat [*Triticum aestivum*] by producing phytohormone. *Antonie Van Leeuwenhoek* 101: 777–786. https://doi.org/10.1007/s10482-011-9692-9.

165 Araujo, W.L., Santos, D.S., Dini-Andreote, F. et al. (2015). Genes related to antioxidant metabolism are involved in *Methylobacterium mesophilicum*-soybean interaction. *Antonie Van Leeuwenhoek* 108: 951–963. https://doi.org/10.1007/s10482-015-0548-6.

166 Dourado, M.N., Camargo Neves, A.A., Santos, D.S., and Araújo, W.L. (2015). Biotechnological and agronomic potential of endophytic pink-pigmented methylotrophic *Methylobacterium* spp. *Biomed. Res. Int.* 2015: 909016. https://doi.org/10.1155/2015/909016.

167 Tani, A., Takai, Y., Suzukawa, I. et al. (2012). Practical application of methanol-mediated mutualistic symbiosis between *Methylobacterium* species and a roof greening moss, *Racomitrium japonicum*. *PLoS One* 7: e33800. https://doi.org/10.1371/journal.pone.0033800.

168 Yim, W., Seshadri, S., Kim, K. et al. (2013). Ethylene emission and PR protein synthesis in ACC deaminase producing *Methylobacterium* spp. inoculated tomato plants [*Lycopersicon esculentum* Mill.] challenged with *Ralstonia solanacearum* under greenhouse conditions. *Plant Physiol. Biochem.* 67: 95–104. https://doi.org/10.1016/j.plaphy.2013.03.002.

169 Balbuena, T.S., Salas, J.J., Martínez-Force, E. et al. (2011). Proteome analysis of cold acclimation in sunflower. *J. Proteome Res.* 10: 2330–2346. https://doi.org/10.1021/pr101137q.

170 Yuan, H., Zhang, T., Liu, X. et al. (2015). Sumoylation of CCAAT/enhancer-binding protein a is implicated in hematopoietic stem/progenitor cell development through regulating runx1 in zebrafish. *Sci. Rep.* 5: 9011. https://doi.org/10.1038/srep09011.

171 Meng, C., Zeleznik, O.A., Thallinger, G.G. et al. (2016). Dimension reduction techniques for the integrative analysis of multi-omics data. *Brief. Bioinform.* 17: 628–641. https://doi.org/10.1093/bib/bbv108.

172 Santos, R.D., Patel, M., Cuadros, J., and Martins, Z. (2016). Influence of mineralogy on the preservation of amino acids under simulated Mars conditions. *Icarus* 277: 342–353. https://doi.org/10.1016/j.icarus.2016.05.029.

173 Bundy, J.G., Willey, T.L., Castell, R.S. et al. (2005). Discrimination of pathogenic clinical isolates and laboratory strains of *Bacillus cereus* by NMR-based metabolomic profiling. *FEMS Microbiol. Lett.* 242: 127–136. https://doi.org/10.1016/j.femsle.2004.10.048.

174 Jorge, T.F., Rodrigues, J.A., Caldana, C. et al. (2015). Mass-spectrometry-based plant metabolomics: metabolite responses to abiotic stress. *Mass Spectrom. Rev.* 35: 620–649. https://doi.org/10.1002/mas.21449.

175 Hollywood, K., Brison, R.D., and Goodacre, R. (2006). Metabolomics: current technologies and future trends. *Proteomics* 6: 4716–4723. https://doi.org/10.1002/pmic.200600106.

176 Morrow, K.J. Jr. (2010). Mass spec central to metabolomics. *Gen. Eng. Biotechnol. News* 30: 1–3.

177 Zhang, X.S. and Cheng, H.P. (2006). Identification of Sinorhizobium meliloti early symbiotic genes by use of a positive functional screen. *Appl. Environ. Microbiol.* 72: 2738–2748. https://doi.org/10.1128/AEM.72.4.2738-2748.2006.

178 Micallef, S.A., Channer, S., Shiaris, M.P., and Colón-Carmona, A. (2009). Plant age and genotype impact the progression of bacterial community succession in the *Arabidopsis* rhizosphere. *Plant Signal. Behav.* 4: 777–780. https://doi.org/10.1093/jxb/erp053.

179 Contreras-Cornejo, H.A., Macías-Rodríguez, L., Cortés-Penagos, C., and López-Bucio, J. (2009). *Trichoderma virens*, a plant beneficial fungus, enhances biomass production and promotes lateral root growth through an auxin-dependent mechanism in *Arabidopsis*. *Plant Physiol.* 149: 1579–1592. https://doi.org/10.1104/pp.108.130369.

180 Vinale, F., Sivasithamparam, K., Ghisalberti, E.L. et al. (2008). A novel role for *Trichoderma* secondary metabolites in the interactions with plants. *Physiol. Mol. Plant Pathol.* 72: 80–86. https://doi.org/10.1016/j.pmpp.2008.05.005.

181 Martínez-Cortés, T., Pomar, F., Merino, F., and Novo-Uzal, E. (2014). A proteomic approach to *Physcomitrella patens* rhizoid exudates. *J. Plant Physiol.* 171: 1671–1678. https://doi.org/10.1016/j.jplph.2014.08.004.

182 Allen, D.K., Libourel, I.G., and Shachar-Hill, Y. (2009). Metabolic flux analysis in plants: coping with complexity. *Plant Cell Environ.* 32: 1241–1257. https://doi.org/10.1111/j.1365-3040.2009.01992.x.

183 Rivas-Ubach, A., Sardans, J., Perez-Trujillo, M. et al. (2012). Strong relationship between elemental stoichiometry and metabolome in plants. *Proc. Natl. Acad. Sci. U. S. A.* 109: 4181–4186. https://doi.org/10.1073/pnas.1116092109.

184 Caldana, C., Degenkolbe, T., Cuadros-Inostroza, A. et al. (2011). High-density kinetic analysis of the metabolomic and transcriptomic response of *Arabidopsis* to eight environmental conditions. *Plant J.* 67: 869–884. https://doi.org/10.1111/j.1365-313X.2011.04640.x.

185 Williams, P.M. and de Mallorca, M.S. (1982). Abscisic acid and gibberellin-like substances in roots and root nodules of *Glycine max*. *Plant Soil* 65: 19–26. https://doi.org/10.1007/BF02376799.

186 Robin, A., Mougel, C., Siblot, S. et al. (2006). Effect of ferritin overexpression in tobacco on the structure of bacterial and pseudomonad communities associated with the roots. *FEMS Microbiol. Ecol.* 58: 492–502. https://doi.org/10.1111/j.1574-6941.2006.00174.x.

187 Tringe, S.G., von Mering, C., Kobayashi, A. et al. (2005). Comparative metagenomics of microbial communities. *Science* 308: 554–557. https://doi.org/10.1126/science.1107851.

188 Raes, J. and Bork, P. (2008). Molecular eco-systems biology: towards an understanding of community function. *Nat. Rev. Microbiol.* 6: 693–699. https://doi.org/10.1038/nrmicro1935.

189 Jiao, N., Herndl, G.J., Hansell, D.A. et al. (2010). Microbial production of recalcitrant dissolved organic matter: long-term carbon storage in the global ocean. *Nat. Rev. Microbiol.* 8: 593–599. https://doi.org/10.1038/nrmicro2386.

190 Mishra, B.K., Meena, K.K., Dubey, P.N. et al. (2016). Influence on yield and quality of fennel [*Foeniculum vulgare* Mill.] grown under semi-arid saline soil, due to application of native phosphate solubilizing rhizobacterial isolates. *Ecol. Eng.* 97: 327–333. https://doi.org/10.1016/j.ecoleng.2016.10.034.

191 Vansuyt, G., Robin, A., Briat, J.F. et al. (2007). Iron acquisition from Fe-pyoverdine by *Arabidopsis thaliana*. *Mol. Plant-Microbe Interact.* 20: 441–447. https://doi.org/10.1094/MPMI-20-4-0441.

192 Diels, L., van der Lelie, N., and Bastiaens, L. (2002). New developments in treatment of heavy metal contaminated soils. *Rev. Environ. Sci. Biotechnol.* 1: 75–82. https://doi.org/10.1023/A:1015188708612.

193 Sorty, A.M. and Shaikh, N.R. (2015). Novel co-enrichment method for isolation of magnetotactic bacteria. *J. Basic Microbiol.* 55: 520–526. https://doi.org/10.1002/jobm.201400457.

194 Beam, H.W. and Perry, J.J. (1973). Co-Metabolism as a factor in microbial degradation of cycloparaffinic hydrocarbons. *Arch. Microbiol.* 91: 87–90.

195 Horvath, R.S. (1972). Microbial co-metabolism and the degradation of organic compounds in nature. *Bacteriol. Rev.* 36: 146–155.

196 Chauhan, A., Faziurrahman, Oakeshott, J.G., and Jain, R.K. (2008). Bacterial metabolism of polycyclic aromatic hydrocarbons: strategies for bioremediation. *J. Ind. Microbiol.* 48: 95–113. https://doi.org/10.1007/s12088-008-0010-9.

197 Davis, J.B. and Raymond, R.L. (1961). Oxidation of alkyl-substituted cyclic hydrocarbons by a nocardia during growth on n-alkanes. *Appl. Microbiol.* 9: 383–388.

198 Raymond, R.L., Jamison, V.W., and Hudson, J.O. (1967). Microbial hydrocarbon co-oxidation. I. oxidation of mono- and dicyclic hydrocarbons by soil isolates of the genus *Nocardia*. *Appl. Microbiol.* 15: 857–865.

199 Noack, S. and Wiechert, W. (2014). Quantitative metabolomics: a phantom? *Trends Biotechnol.* 5: 238–244. https://doi.org/10.1016/j.tibtech.2014.03.006.

200 Oldroyd, G.E.D. and Downie, J.A. (2008). Coordinating nodule morphogenesis with rhizobial infection in legumes. *Annu. Rev. Plant Biol.* 59: 519–546. https://doi.org/10.1146/annurev.arplant.59.032607.092839.

201 Inceoglu, O.L., Al-Soud, W.A., Salles, J.F. et al. (2011). Comparative analysis of bacterial communities in a potato field as determined by pyrosequencing. *PLoS One* 6: e23321. https://doi.org/10.1371/journal.pone.0023321.

202 Peiffer, J.A., Spor, A., Koren, O. et al. (2013). Diversity and heritability of the maize rhizosphere microbiome under field conditions. *Proc. Natl. Acad. Sci. U. S. A.* 110: 6548–6553. https://doi.org/10.1073/pnas.1302837110.

203 Breitling, R., Vitkup, D., and Barrett, M.P. (2008). New surveyor tools for charting microbial metabolic maps. *Nat. Rev. Microbiol.* 6: 156–161. https://doi.org/10.1038/nrmicro1797.

Index

Microbiomics and Sustainable Crop Production, First Edition. Mohammad Yaseen Mir and Saima Hamid.